INTRODUCTION TO
SPACE DYNAMICS

INTRODUCTION TO
SPACE DYNAMICS

by William Tyrrell Thomson
Professor Emeritus of Engineering
University of California, Santa Barbara

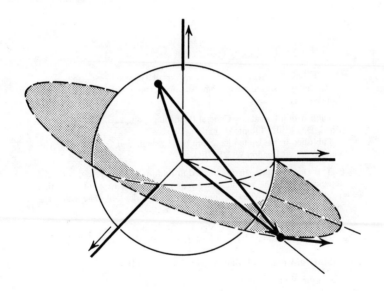

DOVER PUBLICATIONS, INC., NEW YORK

Published in Canada by General Publishing Company, Ltd., 30 Lesmill
Road, Don Mills, Toronto, Ontario.
Published in the United Kingdom by Constable and Company, Ltd., 10
Orange Street, London WC2H 7EG.

This Dover edition, first published in 1986, is an unabridged, corrected
republication of the second (corrected) printing, 1963, of the work first
published by John Wiley & Sons, Inc., New York, 1961. The author has
written a new Preface to the Dover Edition.

Manufactured in the United States of America
Dover Publications, Inc., 31 East 2nd Street, Mineola, N.Y. 11501

Library of Congress Cataloging-in-Publication Data

Thomson, William Tyrrell.
    Introduction to space dynamics.

    Reprint. Originally published: New York : Wiley, c1961. With new pref.
    Bibliography: p.
    Includes index.
    1. Astrodynamics.   I. Title.
TL1050.T47   1986   629.4′11   85-31147
ISBN 0-486-65113-4

to Patricia

Preface to the Dover Edition

Sputnik, the world's first artificial satellite, was successfully launched into orbit around the earth on October 4, 1957, thus initiating the Space Age. *Introduction to Space Dynamics* was first published four years later, in 1961. Although it is already a quarter century later, the basic nature of the text makes it still useful and appropriate for instruction today.

Fundamental principles of dynamics are invariable, and only the problems to which they are applied change with time. The development of the high-speed electronic computer has radically changed the way we solve these problems today. Computers have opened a new means of tackling a host of more complex problems and have solved these problems with the speed and accuracy demanded by modern technology. They have not, however, eliminated the necessity for the analytical study of basic principles. It is for this reason that the author has called upon Dover Publications to make available once again this textbook for the many students, engineers, and scientists who wish to study in this field.

1985   WILLIAM T. THOMSON

Preface to the First Edition

Developments in the exploration of space have brought to the forefront many new problems in science and technology. The new environment and the requirements of high speeds, large energy inputs, accurate guidance, reliability, and a host of other considerations confront the engineer with formidable and challenging problems. Motion in outer space poses unusual dynamic problems, the solution of which requires a thorough knowledge and understanding of the pertinent dynamical principles and techniques of analysis. Three-dimensional attitude problems involving gyroscopic phenomena play an important part in the behavior of guidance instrumentation and in establishing the motion of satellites, missiles, and space vehicles. The analysis of variable mass systems under rotation, effects of vibration due to flexibility, and optimization of flight conditions are all dynamical problems requiring the use of advanced analytical techniques. For most problems, the high speeds encountered in space motion are not sufficiently great (in a relativistic sense) to invalidate Newton's laws of motion, and a knowledge of classical mechanics serves as adequate foundation for a description of the phenomena encountered.

As any experienced student of mechanics will know, the difficulties of dynamics lie not so much in the comprehension of the formal theories of mechanics as in their application to the solution of actual problems. The formulation of the problem and the analytical details required for its solution are, in general, difficult and often beyond the present state of knowledge, as in the case of nonlinear problems. The problems of space dynamics soon become too complicated for adequate discussion in an introductory textbook. However, the underlying principles involved and the techniques of analysis which can be employed are capable of thorough discussion and exposition in terms of simplified versions of these problems. It is from this viewpoint that this textbook was written.

In acquiring a thorough understanding of dynamics, it is necessary to associate with each mathematical concept or operation a logical intuitive feeling which must serve as guide for the mode of analysis. Confidence in approaching new problems is generally acquired through experience in the mastering of a variety of dynamical situations. Problem solving by students is therefore essential. An effort has been made to supply problems of sufficient variety and difficulty, and some of these serve to extend the discussions of the text.

The first two chapters cover the essentials of vector algebra and kinematics, including the general case of space motion. In the third chapter the subject of transformation of coordinates is presented. Central force problems and orbit theory are discussed in Chapter 4 under particle dynamics. They lead to many interesting problems of satellite and space vehicle motions. No attempt is made here to discuss precision orbit calculations or perturbing effects other than those due to impulsive thrust and the oblateness of earth. Chapter 5 on gyrodynamics forms the basis for rigid body dynamics. Gyroscopic phenomena are examined and physically interpreted. The subject of Chapter 5 leads naturally to the theories of gyroscopic instruments and their oscillatory behavior. The instrument gyro, gyrocompass, inertial platform, and inertial navigation are some of the subjects considered in Chapter 6. Chapter 7 on space vehicle motion deals primarily with the attitude of spinning rockets, missiles, and satellites. The effect of thrust misalignment, unbalance in inertia, variable mass, and changing configuration are analyzed. Special techniques for the transformation of motion from body coordinates to inertial coordinates are covered in this chapter. Optimization with respect to performance of rockets is a subject which can be treated at great length. Chapter 8 is a brief introduction to this subject, first in terms of multistage rockets in vertical flight and, later, in terms of single-stage rockets moving along curved trajectories. Aerodynamic forces are omitted in the discussion, not because they are unimportant but because their inclusion renders the problem unmanageable from the analytical point of view. Finally, the last chapter presents the generalized theories of Hamilton and Lagrange which unite the field of mechanics in an over-all formulation. The presentation is from the variational approach which has the advantage of providing greater generality and clarity.

This book can be used at the intermediate or graduate level of instruction. Although matrices, dyadics, Laplace transformations, and the calculus of variations are occasionally encountered, it is my belief that they are introduced in a manner which is understandable to the beginner and will serve as an introduction to the use of these mathematical techniques.

Finally I wish to acknowledge my indebtedness to many persons with whom I have enjoyed working and learning—students, colleagues, and associates at Space Technology Laboratories.

September, 1961 WILLIAM T. THOMSON

Contents

INTRODUCTION TO
SPACE DYNAMICS

Introduction

CHAPTER I

1.1 Basic Concepts

The basic concepts of mechanics are space, time, and mass or force, which are more or less understood intuitively by most beginning students. To be useful for quantitative analysis, however, these notions must be viewed as mathematical concepts related to each other by fundamental laws. Such a formulation was introduced by Sir Isaac Newton (1642–1727) in his three laws of motion which have become the foundation for the classical or Newtonian mechanics.

Newton's laws were formulated for a single particle. Our notion of a particle is that of a material body of infinitely small dimensions. In mechanics we can expand this notion to include any material body whose dimensions are small in comparison with distances or lengths involved in defining its position or motion. Thus planets can be considered to be particles when their position in the solar system is under consideration.

In defining the position of a particle in space, a frame of reference is necessary. For this purpose, three mutually perpendicular lines intersecting at a common point called the origin are adequate. The position of the particle can then be defined in terms of distances along these lines.

To describe the motion of a particle, the concept of time is required. By noting the position of the particle as a function of time, its motion—described by its displacement, velocity, and acceleration—is completely defined.

Force is known to us intuitively as a push or pull. It represents the action of one body on another, exerted by contact or through a distance as in the case of gravitational or magnetic force. To describe a force, it is necessary to know its magnitude, its direction, and its point of application.

Mass is a property possessed by all material bodies. It is a property which describes the effort necessary in giving the body a change in motion. Its precise definition is embodied in Newton's second law, which may be stated as follows. "A particle acted upon by a force $\mathbf{F}$ will move with an acceleration $\mathbf{a}$ proportional to and in the direction of the force; the ratio $\mathbf{F}/\mathbf{a}$ being constant for any particle." Thus for a given particle we can write,

$$\frac{\mathbf{F_1}}{\mathbf{a_1}} = \frac{\mathbf{F_2}}{\mathbf{a_2}} = \frac{\mathbf{F_3}}{\mathbf{a_3}} = \text{constant}$$

where $\mathbf{F}$ and $\mathbf{a}$ are in consistent set of units. The ratio $\mathbf{F}/\mathbf{a}$ which is found to be a constant for any given particle, is a property of the particle, which is designated as mass. We can therefore write Newton's second law as,

$$\mathbf{F} = m\mathbf{a}$$

Newton's first law is a special case of the second law when $\mathbf{F} = 0$. It states that if no force acts on the particle, it will remain at rest or continue to move in a straight line with constant velocity. The equilibrium concept in statics is based on the first law, and the converse of the above statement requires that the resultant of all the forces acting on a particle in equilibrium must be zero.

The extension of Newton's laws to a group of particles necessarily involves the action between particles. Actual bodies can be viewed as a group of particles, and to deduce the behavior of such bodies, Newton introduced his third law which states that, "For every action, there is an equal and opposite reaction." Thus if particle 1 exerts a force $\mathbf{f_{12}}$ on particle 2, particle 2 must exert a force $\mathbf{f_{21}}$ on particle 1, where $\mathbf{f_{21}} = -\mathbf{f_{12}}$.

From his interest in astronomy, Newton formulated the law of gravitation between two particles. The law states that any two particles attract one another with a force of magnitude,

$$F = G\,\frac{m_1 m_2}{r^2}$$

where m_1 and m_2 are the masses of the particles, r is the distance between them, and G is the universal constant of gravitation. Application of this law to a particle on the earth's surface gives us an understanding of the relationship existing between mass and weight. Letting M and m be the mass of the earth and that of another body at the earth's surface, a distance equal to the radius $r = R$ from the center, the attraction of the earth on the body, which is called weight, is given by the above equation,

$$W = m\,\frac{GM}{R^2}$$

If this force is not opposed by a support of some kind, the body will fall toward the center of the earth with an acceleration g. Thus from Newton's second law, we can write,

$$W = mg$$

and by comparison with the previous equation, we arrive at the result,

$$g = \frac{GM}{R^2}$$

We find then that the acceleration of gravity g will vary inversely with R^2. Since m is a property of the body which is fixed, its weight will then vary with g or R, and so we find a given body weighing somewhat different amounts at different places on the earth's surface. At the earth's surface, g is very nearly 32.2 ft/sec^2.

So far we have avoided one very important question regarding the frame of reference used in the measurement of the motion. Newton assumed that there was a frame of reference whose absolute motion was zero. He considered such an *inertial frame* fixed relative to the stars to be one of absolute zero motion, and his laws of motion to be valid when referred to such a reference. Controversies regarding the existence of such a reference frame of absolute zero motion led to the formulation of the theories of relativity for which Newtonian mechanics is a special case.

In arriving at the concept of weight, it was necessary to measure g relative to the surface of the earth which is not at rest. Thus the acceleration of the earth's surface due to its rotation must be accounted for in a more exact analysis. For many problems this error is insignificant, in which case the earth's surface will be found to be an adequate reference. There are other problems, however, such as navigation for space flight, where the earth's surface cannot be considered stationary. In general, problems in space dynamics are involved with rotating and accelerating coordinates, and the subject of relative motion and transformation of coordinates plays an important role.

1.2 Scalar and Vector Quantities

In our discussion so far we have encountered two types of physical quantities. The first type can be adequately expressed by a single number denoting so many units; i.e., temperature, density, mass, time, and energy. Such quantities are known as *scalar* quantities.

The second type cannot be fully represented by a number only, and further information is required. For instance, a displacement in space requires in addition to its numerical value a statement as to its starting point and its direction. Such additional information is necessary to describe completely many physical quantities such as force, velocity, and acceleration. Physical quantities possessing magnitude and direction and satisfying certain necessary requirements are called *vector* quantities.

Vectors are further subdivided into *free* and *bound* vectors. Free vectors are those which can be shifted about as long as their magnitude and direction are not altered. Vectors in rigid body mechanics are in general free

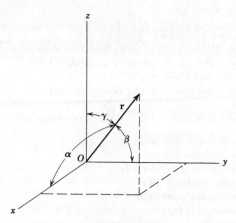

Fig. I.2–I. Right-handed coordinate system.

vectors. Bound vectors are those which cannot be moved without altering the results sought. For instance, in determining the stress distribution in a deformable solid, the applied force cannot be moved without altering the results.

Vector quantity will be denoted by a bold face roman letter, whereas a scalar quantity will be indicated by a light-face italic letter. Frequently we are not concerned with the direction of the vector quantity, in which case the italic letter will be used to indicate its magnitude. This situation may occur when all the vectors are collinear or when we are concerned with components of a vector along specified directions. To distinguish vector quantities at the blackboard or in hand-written material, an overhead bar or arrow can be used.

Graphically, a vector is represented by a straight line with an arrow referenced to some coordinate system. When rectangular axes x, y, z are used, the right-handed coordinate system of Fig. 1.2–1 will be adopted. A left-handed coordinate system would have the Oz axis in the opposite sense, or the x and y axes interchanged.

1.3 Properties of a Vector

Addition

Not every quantity possessing magnitude and direction can be treated as a vector. To qualify as a vector, such quantities must obey the law of composition which states that the sum of two vectors is represented by the diagonal of a parallelogram formed by the two vectors as sides. Subtraction can be viewed as an addition of the vector directed in the opposite sense. Moreover, vector addition must be commutative (independent of the order of summation).

As an example of a directed quantity that will not obey the law of composition, we can mention the finite angular rotation of a rigid body. It has magnitude equal to the angle of rotation, which can be directed along the axis of rotation. However, two such rotations along different axes are not commutative and will not add up to the diagonal of a parallelogram, as we can readily demonstrate by rotating a book 90° about the x and y axes and repeating the procedure in the reverse order about y and x.

Aside from being commutative, vector addition is associative (may be grouped in any order), and these two properties enable the parallelogram law to be successively applied to any pair of vectors for the addition of several vectors.

Resolution

Resolution of a vector is the reverse of composition. Since different components could have the same resultant, a vector can be resolved in an infinite number of different ways. Resolution of a vector into rectangular components often leads to a simpler formulation.

Unit vector

A vector $\mathbf{r}$ multiplied by a scalar n is equal to $n\mathbf{r}$. Its direction is unchanged and its magnitude is n times the original magnitude. The unit vector $\mathbf{1}$, or $\mathbf{i}$, $\mathbf{j}$, $\mathbf{k}$, is constantly used to define the orientation of a vector quantity. A vector $\mathbf{r}$, expressed in terms of its rectangular components, becomes

$$\mathbf{r} = r_x\mathbf{i} + r_y\mathbf{j} + r_z\mathbf{k}$$
$$= r(\mathbf{i} \cos \alpha + \mathbf{j} \cos \beta + \mathbf{k} \cos \gamma)$$
$$= r\mathbf{1}$$

The unit vector in the direction **r** is thus identified as

$$1 = \frac{r}{r} = (i \cos \alpha + j \cos \beta + k \cos \gamma)$$

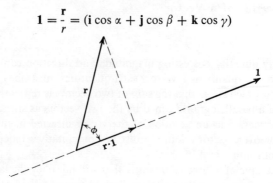

Fig. 1.3-1. Component of **r** along **1** by dot product.

Scalar "dot" product

The *dot* product of two vectors **a** and **b** with angle ϕ between them is a scalar quantity defined by the equation,

$$\mathbf{a} \cdot \mathbf{b} = ab \cos \phi$$

The result is not dependent on the order of multiplication and, hence, the dot product is commutative.

$$\mathbf{a} \cdot \mathbf{b} = \mathbf{b} \cdot \mathbf{a}$$

The equation suggests a convenient procedure for determining the component of a vector **r** along any chosen direction **1**, the result being,

$$\mathbf{r} \cdot \mathbf{1} = r \cos \phi$$

as shown in Fig. 1.3–1.

Vector "cross" product

The *cross product* of two vectors **a** and **b** is a vector defined by the equation,

$$\mathbf{a} \times \mathbf{b} = (ab \sin \phi)\mathbf{1}$$

where **1** is a unit vector in a direction perpendicular to both **a** and **b**. From Fig. 1.3–2 it is seen that the magnitude is equal to the product of the length of one of the two vectors and the projection of the other on a line perpendicular to the first vector, which is equal to the area of the parallelogram formed with **a** and **b** as sides.

To establish the direction of the cross-product vector, the three-finger rule of the right hand, as illustrated in Fig. 1.3–2, is helpful. The first vector **a** is represented by the thumb; the second vector **b** (or its component $b \sin \phi$ perpendicular to **a**) by the index finger; and the product vector perpendicular to the previous two by the third finger. We note here that the cross product is noncommutative, and the following rule holds:

$$\mathbf{a} \times \mathbf{b} = -\mathbf{b} \times \mathbf{a}$$

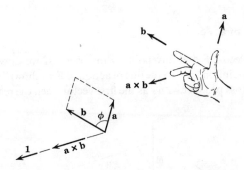

Fig. I.3–2. Cross product of two vectors is normal to the plane of the two vectors.

Products in rectangular components

When resolved into rectangular components, the dot and cross products become,

$$\mathbf{a} \cdot \mathbf{b} = (a_x\mathbf{i} + a_y\mathbf{j} + a_z\mathbf{k}) \cdot (b_x\mathbf{i} + b_y\mathbf{j} + b_z\mathbf{k})$$

$$= a_xb_x + a_yb_y + a_zb_z$$

$$\mathbf{a} \times \mathbf{b} = (a_x\mathbf{i} + a_y\mathbf{j} + a_z\mathbf{k}) \times (b_x\mathbf{i} + b_y\mathbf{j} + b_z\mathbf{k})$$

$$= (a_xb_y - a_yb_x)(\mathbf{i} \times \mathbf{j}) + (a_xb_z - a_zb_x)(\mathbf{i} \times \mathbf{k}) + (a_yb_z - a_zb_y)(\mathbf{j} \times \mathbf{k})$$

$$= (a_xb_y - a_yb_x)\mathbf{k} + (a_zb_x - a_xb_z)\mathbf{j} + (a_yb_z - a_zb_y)\mathbf{i}$$

The result of the cross product can be conveniently expressed by the following determinant.

$$\mathbf{a} \times \mathbf{b} = \begin{vmatrix} \mathbf{i} & \mathbf{j} & \mathbf{k} \\ a_x & a_y & a_z \\ b_x & b_y & b_z \end{vmatrix}$$

Multiple products

Certain multiple products of vectors are occasionally encountered and we list two of the most common ones in the following:

$$\mathbf{a} \cdot (\mathbf{b} \times \mathbf{c}) = \mathbf{b} \cdot (\mathbf{c} \times \mathbf{a}) = \mathbf{c} \cdot (\mathbf{a} \times \mathbf{b})$$

This product can be interpreted geometrically as being equal to the volume of a parallelopiped of sides a, b, and c.

The second multiple product is,

$$\mathbf{a} \times (\mathbf{b} \times \mathbf{c}) = (\mathbf{a} \cdot \mathbf{c})\mathbf{b} - (\mathbf{a} \cdot \mathbf{b})\mathbf{c}$$

1.4 Moment of a Vector

Consider a vector $\mathbf{F}$ and any point O in space. If we draw a vector $\mathbf{r}$ from O to any point on $\mathbf{F}$ or on the line of action of $\mathbf{F}$, as shown in Fig. 1.4–1, and form the cross product with $\mathbf{r}$ as the first vector, then the result will be a

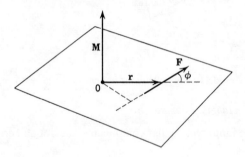

Fig. 1.4–1.   Moment vector $\mathbf{M}$ normal to plane of $\mathbf{r}$ and $\mathbf{F}$ is $\mathbf{r} \times \mathbf{F}$.

moment $\mathbf{M}$ about an axis through O in a direction perpendicular to the plane containing $\mathbf{r}$ and $\mathbf{F}$, the direction of which is indicated by the unit vector $\mathbf{1}$.

$$\mathbf{M} = \mathbf{r} \times \mathbf{F} = (Fr \sin \phi)\mathbf{1}$$

The moment $\mathbf{M}$ will be independent of where $\mathbf{r}$ terminates on $\mathbf{F}$ or on its line of action, as can easily be shown.

1.5 Angular Velocity Vectors

It was pointed out previously that finite angles of rotation, although representable by a vector, are not commutative and, hence, will not obey the rules of vector addition. Infinitesimal rotations however can be shown to be commutative and to possess all the properties of a vector. To show this, consider the displacement of a point p due to two infinitesimal rotations $\omega_1\, dt$ and $\omega_2\, dt$ about any two axes, where ω_1 and ω_2 are their respective rotational speeds. Let the direction of each of the axes of rotation be indicated by unit vectors 1_1 and 1_2, as shown in Fig. 1.5–1, and we

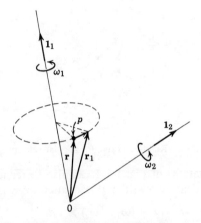

Fig. 1.5–1. Angular velocity represented by vector.

will perform the rotations in the order 1 and 2, then repeat in the reverse order to examine the final result.

Because of the infinitesimal rotation $(\omega_1\, dt)1_1$, the end of the displacement vector $\mathbf{r}$ defining the position p will be displaced by an amount,

$$(\omega_1\, dt)1_1 \times \mathbf{r} \tag{1.5–1}$$

and the new position is defined by the vector,

$$\mathbf{r}_1 = \mathbf{r} + (\omega_1\, dt)1_1 \times \mathbf{r} \tag{1.5–2}$$

Next allow the second infinitesimal rotation $(\omega_2\, dt)1_2$, in which case the final position of p is defined by the vector $\mathbf{r}_2$.

$$\mathbf{r}_2 = \mathbf{r}_1 + (\omega_2\, dt)1_2 \times \mathbf{r}_1$$
$$= \mathbf{r} + (\omega_1\, dt)1_1 \times \mathbf{r} + (\omega_2\, dt)1_2 \times \mathbf{r} + (\omega_2\, dt)1_2 \times (\omega_1\, dt)1_1 \times \mathbf{r} \tag{1.5–3}$$

Neglecting the second-order term $(\omega_2\omega_1\, dt\, dt)$, we arrive at the result,

$$\mathbf{r}_2 = \mathbf{r} + (\omega_1 1_1 + \omega_2 1_2)\, dt \times \mathbf{r} \tag{1.5–4}$$

If we repeat the operation in the reverse order, we will find the equation for $\mathbf{r}_2$ to be identical to the previous case, indicating that infinitesimal rotations are commutative. In effect, we have represented an angular velocity by a vector $\boldsymbol{\omega} = \omega\mathbf{1}$ according to the right-hand screw convention, as shown in Fig. 1.5–2. The fingers in this case indicate the rotation sense, and the positive direction of the vector $\boldsymbol{\omega}$ is represented by the thumb.

Since angular velocities obey all the rules of vectors, they can be compounded by the parallelogram rule to a single resultant vector. Thus the

Fig. 1.5–2.   Right-hand rule for angular velocity.

two infinitesimal rotations of Fig. 1.5–1 can also be reduced to a single rotation which is evident by rewriting the equation for $\mathbf{r}_2$ as,

$$\mathbf{r}_2 = \mathbf{r} + (\boldsymbol{\omega}_1 + \boldsymbol{\omega}_2) \times \mathbf{r}\, dt \tag{1.5–5}$$

1.6   Derivative of a Vector

In differentiating a vector, the usual rules of the limiting process apply.

$$\frac{d\mathbf{r}}{dt} = \lim_{\Delta t \to 0} \frac{(\mathbf{r} + \Delta\mathbf{r}) - \mathbf{r}}{\Delta t} = \lim_{\Delta t \to 0} \frac{\Delta\mathbf{r}}{\Delta t} \tag{1.6–1}$$

If the vector $\mathbf{r}$ is referenced to a fixed coordinate system, the $\Delta\mathbf{r}$ is the vector change relative to the coordinates which is also the total change, and Eq. 1.6–1 is the total derivative of $\mathbf{r}$.

If the vector $\mathbf{r}$ is referenced to a rotating coordinate system such as the one shown in Fig. 1.6–1, the vector $\mathbf{r}$ remaining stationary relative to the rotating axes will undergo a change

$$\Delta\theta r \sin\phi$$

along the tangent to the dotted circle, and its rate of change is established by the limit,

$$\lim_{\Delta t \to 0} \left(\frac{\Delta\theta}{\Delta t}\right) r \sin\phi = (\omega r \sin\phi)\mathbf{1}$$

where $\mathbf{1}$ is a unit vector along the tangent.

Since this expression is equal to the cross product of $\boldsymbol{\omega}$ and $\mathbf{r}$, we conclude that due to rotation $\boldsymbol{\omega}$ of the coordinate system the vector $\mathbf{r}$ undergoes a rate of change of

$$\boldsymbol{\omega} \times \mathbf{r} \qquad (1.6\text{--}2)$$

This term occurs in addition to the vector change relative to the coordinate system, so that the total derivative relative to inertial axes is

$$\left(\frac{d\mathbf{r}}{dt}\right)_{\text{inertial}} = \left[\frac{d\mathbf{r}}{dt}\right]_{\substack{\text{rel. to} \\ \text{coord.}}} + \boldsymbol{\omega} \times \mathbf{r} \qquad (1.6\text{--}3)^{*}$$

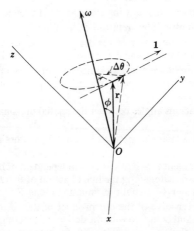

Fig. 1.6–1. Differentiation of a vector $\mathbf{r}$ referenced to rotating coordinates.

Equation 1.6–3 applies to any vector quantity and is of fundamental importance to dynamics where body-fixed axes are often used.

We mention finally that the derivatives of the dot and cross products are treated as in products of scalar quantities, except that the order of the cross product must be maintained. These rules are illustrated by the following equations.

$$\frac{d}{dt}(\mathbf{a} \cdot \mathbf{b}) = \mathbf{a} \cdot \frac{d\mathbf{b}}{dt} + \frac{d\mathbf{a}}{dt} \cdot \mathbf{b} \qquad (1.6\text{--}4)$$

$$\frac{d}{dt}(\mathbf{a} \times \mathbf{b}) = \mathbf{a} \times \frac{d\mathbf{b}}{dt} + \frac{d\mathbf{a}}{dt} \times \mathbf{b} \qquad (1.6\text{--}5)$$

* A convenient notation to distinguish between differentiation in the inertial and rotating coordinates is to use brackets, or parenthesis around the latter, i.e.,

$$\frac{d\mathbf{r}}{dt} = \left[\frac{d\mathbf{r}}{dt}\right] + \boldsymbol{\omega} \times \mathbf{r}$$

This notation will be used in this book.

PROBLEMS

1. Determine the unit vector along $\mathbf{r} = 3\mathbf{i} - 2\mathbf{j} + 2\mathbf{k}$.
2. Determine the angles α, β, and γ between $\mathbf{r}$ of Prob. 1 and the x, y, and z axes.
3. Determine the dot product of the vectors $\mathbf{r}_1 = 4\mathbf{i} - 3\mathbf{j} + \mathbf{k}$; $\mathbf{r}_2 = -2\mathbf{i} + 2\mathbf{j} + 3\mathbf{k}$.
4. Determine the angle between $\mathbf{r}_1$ and $\mathbf{r}_2$ of Prob. 3.
5. Determine the cross product $\mathbf{r}_1 \times \mathbf{r}_2$ of Prob. 3. Show by means of dot products between the cross product vector and $\mathbf{r}_1$ and $\mathbf{r}_2$ that the cross product is perpendicular to each of vectors $\mathbf{r}_1$ and $\mathbf{r}_2$.
6. Show by vector means that any inscribed angle in a semicircle is a right angle.
7. What is the geometric interpretation of $(\mathbf{a} + \mathbf{b})^2$.
8. If a and b are absolute values of vectors $\mathbf{a}$ and $\mathbf{b}$, express the scalar quantity a^2b^2 by using only the dot and cross products of $\mathbf{a}$ and $\mathbf{b}$.
9. Write the equation of a plane, a distance c from a given point O, in terms of the radius vector $\mathbf{r}$ from O to any point in the plane, and the unit vector $\mathbf{1}$ normal to the plane.
10. Determine the various angles between the spatial diagonals of a unit cube of sides $\mathbf{i}$, $\mathbf{j}$, and $\mathbf{k}$.
11. Find the area of a triangle specified by the two vectors $\mathbf{r}_1 = 3\mathbf{i} + 4\mathbf{j}$; $\mathbf{r}_2 = -5\mathbf{i} + 7\mathbf{j} + 2\mathbf{k}$.
12. Show that the moment $\mathbf{r} \times \mathbf{F}$ about O is independent of how $\mathbf{r}$ is drawn from O as long as it terminates along the line of action of $\mathbf{F}$. (*Hint:* Consider two vectors $\mathbf{r}_1$ and $\mathbf{r}_2 = \mathbf{r}_1 + \mathbf{r}_3$ both terminating along $\mathbf{F}$.)
13. Determine the derivative of the dot product of the vectors $\mathbf{r}_1 = 2t\mathbf{i} - 3t^2\mathbf{j}$; $\mathbf{r}_2 = -4\mathbf{i} + 2t\mathbf{j}$, and compare with the derivative determined from Eq. 1.6–4.
14. Repeat Prob. 13 for the cross product, using Eq. 1.6–5.
15. A vector $\mathbf{r} = 3t\mathbf{i} - 4\mathbf{j} + \mathbf{k}$ is referenced to a coordinate system which is rotating at a speed $\boldsymbol{\omega} = 2t\mathbf{k}$. Determine its derivative.
16. Prove the equations given on p. 8 for the multiple products.
17. Complete the derivation for Eq. 1.5–3 when the order of rotation is 2 and 1. Under what conditions are the two equations equal?
18. Determine the relative magnitudes of the terms of Eq. 1.5–3 when $\mathbf{r} = 4\mathbf{i} + 3\mathbf{j}$; $\omega_1 = 2$; $\omega_2 = -1$; $\mathbf{1}_1 = \mathbf{i}$ and $\mathbf{1}_2 = \mathbf{j}$.
19. In Prob. 18, determine the value of dt necessary to reduce the term $(\omega_2 \, dt)\mathbf{1}_2 \times (\omega_1 \, dt)\mathbf{1}_1 \times \mathbf{r}$ to 1% of the magnitude of the smallest vector component.

Kinematics

CHAPTER 2

2.1 Velocity and Acceleration

The subject of kinematics is the study of motion. It is concerned with space and time, and with the time rate of change of vector quantities relating to the geometry of motion.

We consider first the motion of a point in a fixed coordinate system xyz. The position of a point p which is in continuous motion along a curve such as s in Fig. 2.1–1 is specified by its position vector $\mathbf{r}$, the magnitude and direction of which are functions of time. In time Δt, $\mathbf{r}$ changes to $\mathbf{r} + \Delta\mathbf{r}$, and its velocity $\mathbf{v}$ is given by the time derivative,

$$\mathbf{v} = \lim_{\Delta t \to 0} \frac{(\mathbf{r} + \Delta\mathbf{r}) - \mathbf{r}}{\Delta t} = \lim_{\Delta t \to 0} \frac{\Delta\mathbf{r}}{\Delta t} = \frac{d\mathbf{r}}{dt} \qquad (2.1\text{–}1)$$

The direction of $\mathbf{v}$ can be shown to coincide with the limiting direction of $\Delta\mathbf{r}$ as it approaches zero, or the tangent to the curve s at p. By rewriting $\mathbf{v}$ in the form,

$$\mathbf{v} = \lim_{\Delta t \to 0} \frac{\Delta\mathbf{r}}{\Delta s} \frac{\Delta s}{\Delta t}$$

the limiting value of $\Delta\mathbf{r}/\Delta s$ is a unit vector along the tangent to the curve, so that the velocity can also be written as,

$$\mathbf{v} = \frac{ds}{dt} \mathbf{1}_t \qquad (2.1\text{–}2)$$

If $\mathbf{r}$ is represented in terms of its rectangular components, we obtain

$$\mathbf{r} = r_x\mathbf{i} + r_y\mathbf{j} + r_z\mathbf{k} \qquad (2.1\text{--}3)$$

where r_x, r_y, and r_z are components of $\mathbf{r}$ along the fixed xyz coordinates, and $\mathbf{i}$, $\mathbf{j}$, and $\mathbf{k}$ are their corresponding unit vectors. Differentiating, we obtain

$$\frac{d\mathbf{r}}{dt} = \dot{r}_x\mathbf{i} + \dot{r}_y\mathbf{j} + \dot{r}_z\mathbf{k} \qquad (2.1\text{--}4)$$

where $\mathbf{i}$, $\mathbf{j}$, and $\mathbf{k}$, are treated as constants.

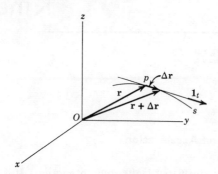

Fig. 2.1–1. Time rate of change of a vector $\mathbf{r}$.

Acceleration is the time rate of change of velocity $\mathbf{v}$, and by observing the vector change from $\mathbf{v}$ to $\mathbf{v} + \Delta\mathbf{v}$ in time Δt, we obtain

$$\mathbf{a} = \lim_{\Delta t \to 0} \frac{(\mathbf{v} + \Delta\mathbf{v}) - \mathbf{v}}{\Delta t} = \lim_{\Delta t \to 0} \frac{\Delta\mathbf{v}}{\Delta t} = \frac{d\mathbf{v}}{dt} = \dot{v}_x\mathbf{i} + \dot{v}_y\mathbf{j} + \dot{v}_z\mathbf{k} \quad (2.1\text{--}5)$$

2.2 Plane Motion (Radial and Transverse Components)

Consider a particle p moving along a curve s fixed in a stationary Oxy plane, as shown in Fig. 2.2–1. The position of p is established by the position vector

$$\mathbf{r} = r\mathbf{1}_r \qquad (2.2\text{--}1)$$

where $\mathbf{1}_r$ is a unit vector which is always oriented along $\mathbf{r}$.

To determine the velocity of p, we differentiate $\mathbf{r}$, recognizing that $\mathbf{1}_r$ changes in direction,

$$\frac{d\mathbf{r}}{dt} = \dot{r}\mathbf{1}_r + r\frac{d\mathbf{1}_r}{dt} \qquad (2.2\text{--}2)$$

The unit vector 1_r is one of fixed magnitude which is rotating with angular velocity $\dot{\theta}$ about an axis through the origin perpendicular to the xy plane. Its derivative or its velocity is the cross product of the vectors $\dot{\theta}$ and 1_r, which is a vector perpendicular to r, or

$$\frac{d1_r}{dt} = \dot{\theta} \times 1_r = \dot{\theta}1_\theta \qquad (2.2\text{--}3)$$

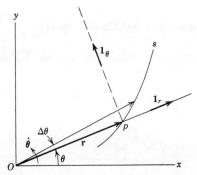

Fig. 2.2-1. Unit vectors 1_r and 1_θ moving with r in the x, y plane.

where 1_θ is a unit vector in the direction perpendicular to r.

Equation 2.2–2 may then be written as

$$\frac{d\mathbf{r}}{dt} = \mathbf{v} = \dot{r}1_r + \dot{\theta} \times \mathbf{r}$$
$$= \dot{r}1_r + r\dot{\theta}1_\theta \qquad (2.2\text{--}4)$$

which expresses the velocity in terms of the radial and transverse components.

We can view this problem as that of a point p moving along a set of rotating axes with direction 1_r and 1_θ. The point p always moves along the 1_r axis, and its relative velocity along it is $\dot{r}$ which corresponds to the first term of Eq. 2.2–4. The second term $\dot{\theta} \times \mathbf{r}$ is the velocity of the coincident point p due to rotation $\dot{\theta}$.

The acceleration of p can be determined by differentiating Eq. 2.2–4.

$$\mathbf{a} = \ddot{\mathbf{r}} = \ddot{r}1_r + \dot{r}\frac{d1_r}{dt} + (r\ddot{\theta} + \dot{r}\dot{\theta})1_\theta + r\dot{\theta}\frac{d1_\theta}{dt} \qquad (2.2\text{--}5)$$

As before, the derivative of a unit vector rotates it 90° and multiplies it with its angular rate $\dot{\theta}$, so that

$$\frac{d1_\theta}{dt} = \dot{\theta} \times 1_\theta = -\dot{\theta}1_r \qquad (2.2\text{--}6)$$

Thus Eq. 2.2–5 reduces to

$$\mathbf{a} = (\ddot{r} - r\dot{\theta}^2)\mathbf{1}_r + (r\ddot{\theta} + 2\dot{r}\dot{\theta})\mathbf{1}_\theta \qquad (2.2\text{--}7)$$

which expresses the acceleration in terms of the radial and transverse components. We note here the term $2\dot{r}\dot{\theta}$, which is known as the Coriolis acceleration, and which will be referred to again in the more general case.

2.3 Tangential and Normal Components

To resolve the acceleration into tangential and normal components, we start with Eq. 2.1–2.

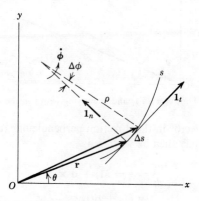

Fig. 2.3–I. Unit vectors $\mathbf{1}_n$ and $\mathbf{1}_t$ moving along curve s, in x, y plane.

$$\mathbf{v} = \dot{s}\mathbf{1}_t \qquad (2.3\text{--}1)$$

Differentiating and noting that

$$\dot{\mathbf{1}}_t = \dot{\phi}\mathbf{1}_n \qquad (2.3\text{--}2)$$

where $\mathbf{1}_n$ is a unit vector along the radius of curvature which is normal to the tangent to the curve s at p, and $\dot{\phi}$ is the angular rate of the radius of curvature as shown in Fig. 2.3–1, we obtain

$$\mathbf{a} = \ddot{s}\mathbf{1}_t + \dot{s}\dot{\phi}\mathbf{1}_n \qquad (2.3\text{--}3)$$

Since the length Δs is related to the radius of curvature ρ and to the angle $\Delta\phi$ swept out by ρ,

$$\Delta s = \rho\, \Delta\phi$$
$$\dot{s} = \rho\dot{\phi} \qquad (2.3\text{--}4)$$

the acceleration, Eq. 2.3-3, can be expressed in the following alternate forms.

$$\mathbf{a} = \ddot{s}\mathbf{1}_t + \frac{\dot{s}^2}{\rho}\mathbf{1}_n \tag{2.3-5}$$

$$= \ddot{s}\mathbf{1}_t + \rho\dot{\phi}^2\mathbf{1}_n \tag{2.3-6}$$

$$= \ddot{s}\mathbf{1}_t + \dot{\boldsymbol{\phi}} \times \dot{\mathbf{s}} \tag{2.3-7}$$

PROBLEMS

1. The position of a particle in a plane is given by the equations

$$x = 6t \qquad y = 4t^2$$

Determine the rectangular components of its velocity and acceleration as a function of time.

2. A projectile is fired with speed v_0 at an angle θ_0 above the horizon. Neglecting air friction and the rotation of the earth, the acceleration components are $a_y = -g$ and $a_x = 0$. Determine the equation for its trajectory, the range R, and the maximum height H.

3. A rock is thrown at an angle of 45° with the horizontal, and it just clears a wall 24 ft high and 40 ft away. Determine the initial speed of the rock.

4. Find the greatest distance that a stone can be thrown inside a horizontal tunnel 15 ft high with an initial speed of 80 ft/sec. Find also the corresponding time of flight.

5. For a gun with a given muzzle speed v_0, show that a point (x, y) can be hit by two different trajectories with initial elevation θ_1 and θ_2.

6. Derive the equation for the envelope of a series of trajectories of a projectile fired with constant speed v_0 and varying angles θ_0.

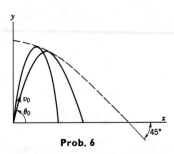

Prob. 6

7. A point moves so that $r = 20 + 10t$ and $\theta = 0.20t^2$. Determine the radial and transverse components of the velocity and acceleration at $t = 2$ sec.

8. A spiral fixed in a plane is given by the equation $r = 10e^{0.2\theta}$. If a particle moves along the spiral according to the equation $\theta = 0.5t^2$, determine the radial and transverse components of the velocity and acceleration at $t = 2.0$ sec.

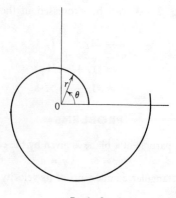

Prob. 8

9. In Prob. 8, determine the tangential and normal components of the velocity and acceleration, and compare the resultant for the two problems.

10. A particle moves along the circumference of a circle of radius R at constant speed v_0. Determine its radial and transverse velocity with respect to an origin on the circumference. Find the radial and transverse components of the acceleration and show that their resultant passes through the center of the circle.

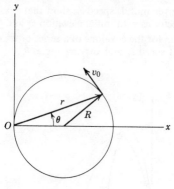

Prob. 10

2.4 Plane Motion Along a Curve Rotating About 0 (Relative Motion)

A point p moves along a curve s which is rotating with angular velocity $\boldsymbol{\omega}$ and angular acceleration $\dot{\boldsymbol{\omega}}$. We will attach axes 1 and 2 to curve s so that 1, 2, and s rotate together with $\boldsymbol{\omega}$ and $\dot{\boldsymbol{\omega}}$, as shown in Fig. 2.4–1.

The velocity of p may be determined as follows. If the point p remained fixed on s, its velocity due to the angular velocity $\boldsymbol{\omega}$ of the curve and attached coordinate axes is $\boldsymbol{\omega} \times \mathbf{r}$ which is perpendicular to $\mathbf{r}$. The motion of p along s results in an additional velocity $d\mathbf{r}/dt = \dot{s}\mathbf{1}_t$ directed along the tangent to s. This latter component is the velocity relative to the curve s

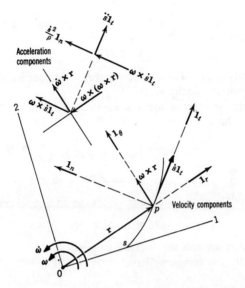

Fig. 2.4–1. Motion along a rotating curve s in a plane.

and can be determined as if the curve s were held stationary. Thus the sum of the above two components represents the velocity of point p

$$\mathbf{v} = \left[\frac{d\mathbf{r}}{dt}\right] + \boldsymbol{\omega} \times \mathbf{r}$$

$$= \dot{s}\mathbf{1}_t + \boldsymbol{\omega} \times \mathbf{r} \qquad (2.4\text{–}1)$$

and the above equation conforms to Eq. 1.6–3 for the derivative of a vector referenced to a rotating coordinate system.*

The acceleration of point p is determined by the derivative of $\mathbf{v}$ or of its two components in Eq. 2.4–1. Since the coordinates are rotating with angular speed $\boldsymbol{\omega}$, the acceleration is given by the equation

$$\mathbf{a} = \left[\frac{d\mathbf{v}}{dt}\right] + \boldsymbol{\omega} \times \mathbf{v} \qquad (2.4\text{–}2)$$

which can be interpreted in terms of the two components of $\mathbf{v}$.

* Differentiation relative to the moving coordinates is indicated by [ ].

Considering first the component $\ddot{s}\mathbf{1}_t$, if the curve is held stationary and p is allowed to move along it, the acceleration relative to the curve s would be

$$\ddot{s}\mathbf{1}_t + \frac{\dot{s}^2}{\rho}\mathbf{1}_n \qquad (2.4\text{--}3)$$

as in Eq. 2.3–5. The rotation of the coordinate system at a rate $\boldsymbol{\omega}$ introduces an additional term

$$\boldsymbol{\omega} \times \dot{s}\mathbf{1}_t \qquad (2.4\text{--}4)$$

which is normal to $\dot{s}\mathbf{1}_t$. Thus we obtain three components of acceleration from the tangential velocity vector.

The remaining velocity vector $\boldsymbol{\omega} \times \mathbf{r}$ is treated similarly. Relative to the coordinates we have

$$\left[\frac{d}{dt}(\boldsymbol{\omega} \times \mathbf{r})\right] = \boldsymbol{\omega} \times \dot{\mathbf{r}} + \dot{\boldsymbol{\omega}} \times \mathbf{r}$$

$$= \boldsymbol{\omega} \times \dot{s}\mathbf{1}_t + \dot{\boldsymbol{\omega}} \times \mathbf{r} \qquad (2.4\text{--}5)$$

The rotation of the coordinates introduces the additional component

$$\boldsymbol{\omega} \times (\boldsymbol{\omega} \times \mathbf{r}) \qquad (2.4\text{--}6)$$

which is directed towards the negative $\mathbf{r}$ direction.

Summing all these terms, we have

$$\mathbf{a} = \left(\ddot{s}\mathbf{1}_t + \frac{\dot{s}^2}{\rho}\mathbf{1}_n\right) + (\dot{\boldsymbol{\omega}} \times \mathbf{r} + \boldsymbol{\omega} \times \boldsymbol{\omega} \times \mathbf{r}) + 2\boldsymbol{\omega} \times \dot{s}\mathbf{1}_t$$

$$= \left(\ddot{s}\mathbf{1}_t + \frac{\dot{s}^2}{\rho}\mathbf{1}_n\right) + (\dot{\omega}r\mathbf{1}_\theta - r\omega^2\mathbf{1}_r) + 2\omega\dot{s}\mathbf{1}_n \qquad (2.4\text{--}7)$$

In this equation the first two terms represent the acceleration of p relative to the curve s. The next two terms represent the acceleration of the coincident point due to angular velocity and angular acceleration of the coordinate system and the curve s fixed to it. The last term is known as the Coriolis acceleration, and it is perpendicular to the relative velocity $\dot{s}\mathbf{1}_t$. The various components of Eq. 2.4–7 are shown in Fig. 2.4–1.

2.5 General Case of Space Motion

We now consider the general case of the motion of a particle p moving with respect to a rigid body which is itself in motion with respect to a fixed coordinate system, as shown in Fig. 2.5–1.

We will designate the fixed coordinate system by capital letters X, Y, Z, and attach a set of axes x, y, z on the moving body, calling them the *body axes*. Thus the motion of the rigid body is established by the motion of the origin of the body axes x, y, z, and a rotation $\boldsymbol{\omega}$ with respect to XYZ. The vector $\boldsymbol{\omega}$ will, in general, vary in magnitude and direction, both of which can be referenced with respect to the fixed X, Y, Z axes. Thus the absolute motion of the particle p, referred to the X, Y, Z axes, will

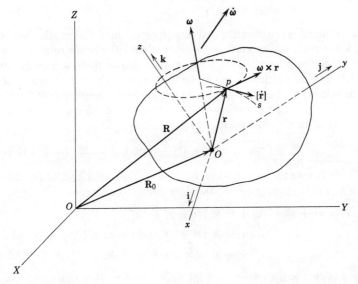

Fig. 2.5–1. General case of space motion in terms of body axes x, y, z, and inertial axes X, Y, Z.

be equal to the motion of the particle relative to the body axes x, y, z (or the body itself) plus the motion of the coincident point (a point on the body momentarily coinciding with p), which is further equal to the motion of the moving origin and the additional motion of the coincident point due to rotations $\boldsymbol{\omega}$ and $\dot{\boldsymbol{\omega}}$.

To visualize the motion, let the motion of the particle p with respect to the rigid body be indicated along a curve s fixed in the body. Thus, this curve is also fixed with respect to the body axes x, y, z. An observer sitting on the body would see only the motion of p along the curve s.

Let the position of p relative to the x, y, z axes be represented by the vector

$$\mathbf{r} = x\mathbf{i} + y\mathbf{j} + z\mathbf{k} \qquad (2.5\text{–}1)$$

where $\mathbf{i}$, $\mathbf{j}$, and $\mathbf{k}$ are unit vectors along x, y, z, and, hence, must be treated as variables due to their changing direction. If we differentiate **r**, we will obtain

$$\dot{\mathbf{r}} = [\dot{x}\mathbf{i} + \dot{y}\mathbf{j} + \dot{z}\mathbf{k}] + \left(x\frac{d\mathbf{i}}{dt} + y\frac{d\mathbf{j}}{dt} + z\frac{d\mathbf{k}}{dt}\right) \qquad (2.5\text{-}2)$$

Since $d\mathbf{i}/dt = \boldsymbol{\omega} \times \mathbf{i}$, $d\mathbf{j}/dt = \boldsymbol{\omega} \times \mathbf{j}$, $d\mathbf{k}/dt = \boldsymbol{\omega} \times \mathbf{k}$, this expression can be written as

$$\dot{\mathbf{r}} = [\dot{x}\mathbf{i} + \dot{y}\mathbf{j} + \dot{z}\mathbf{k}] + \boldsymbol{\omega} \times (x\mathbf{i} + y\mathbf{j} + z\mathbf{k})$$
$$= [\dot{\mathbf{r}}] + \boldsymbol{\omega} \times \mathbf{r} \qquad (2.5\text{-}3)$$

Interpreting this equation, the first term $[\dot{\mathbf{r}}]$ represents differentiation holding **i**, **j**, **k** constant and, hence, is the velocity of p relative to the rotating x, y, z axes, or the velocity which an observer stationed on the x, y, z axes is able to detect as the particle moves along the curve s. The second term $\boldsymbol{\omega} \times \mathbf{r}$ is the velocity of the coincident point relative to the origin, due to rotation $\boldsymbol{\omega}$. Finally, we add to the above the velocity $\mathbf{v}_o$ of the moving origin, in which case the absolute velocity of p becomes

$$\mathbf{v} = \mathbf{v}_o + [\dot{\mathbf{r}}] + \boldsymbol{\omega} \times \mathbf{r} \qquad (2.5\text{-}4)$$

To determine the acceleration, we start with the velocity $\dot{\mathbf{r}}$, of Eq. 2.5-3, relative to the moving origin, and differentiate once more

$$\ddot{\mathbf{r}} = [\ddot{x}\mathbf{i} + \ddot{y}\mathbf{j} + \ddot{z}\mathbf{k}] + \boldsymbol{\omega} \times [\dot{x}\mathbf{i} + \dot{y}\mathbf{j} + \dot{z}\mathbf{k}]$$
$$+ \dot{\boldsymbol{\omega}} \times (x\mathbf{i} + y\mathbf{j} + z\mathbf{k}) + \boldsymbol{\omega} \times [\dot{x}\mathbf{i} + \dot{y}\mathbf{j} + \dot{z}\mathbf{k}]$$
$$+ \boldsymbol{\omega} \times \boldsymbol{\omega} \times (x\mathbf{i} + y\mathbf{j} + z\mathbf{k}) \qquad (2.5\text{-}5)$$

The first two terms result from the differentiation of the first term $[\dot{x}\mathbf{i} + \dot{y}\mathbf{j} + \dot{z}\mathbf{k}]$, whereas the differentiation of the second term $\boldsymbol{\omega} \times (x\mathbf{i} + y\mathbf{j} + z\mathbf{k})$ results in the remaining three terms. We can now group these terms together as

$$\ddot{\mathbf{r}} = [\mathbf{a}] + \boldsymbol{\omega} \times \boldsymbol{\omega} \times \mathbf{r} + \dot{\boldsymbol{\omega}} \times \mathbf{r} + 2\boldsymbol{\omega} \times [\mathbf{v}] \qquad (2.5\text{-}6)$$

where $[\mathbf{v}] = [\dot{\mathbf{r}}] = [\dot{x}\mathbf{i} + \dot{y}\mathbf{j} + \dot{z}\mathbf{k}]$ is the velocity of p relative to the body axes

$[\mathbf{a}] = [\ddot{x}\mathbf{i} + \ddot{y}\mathbf{j} + \ddot{z}\mathbf{k}]$ is the acceleration of p relative to the body axes

We now add the acceleration $\mathbf{a}_0$ of the origin, to $\ddot{\mathbf{r}}$ to obtain the total acceleration

$$\mathbf{a} = \mathbf{a}_0 + [\mathbf{a}] + \boldsymbol{\omega} \times \boldsymbol{\omega} \times \mathbf{r} + \dot{\boldsymbol{\omega}} \times \mathbf{r} + 2\boldsymbol{\omega} \times [\mathbf{v}] \qquad (2.5\text{-}7)$$

The terms $\boldsymbol{\omega} \times \boldsymbol{\omega} \times \mathbf{r}$ and $\dot{\boldsymbol{\omega}} \times \mathbf{r}$ are the acceleration of the coincident point and $2\boldsymbol{\omega} \times [\mathbf{v}]$ is the Coriolis acceleration which is directed normal to the plane containing the vectors $\boldsymbol{\omega}$ and relative velocity $[\mathbf{v}]$, as given by the three-finger rule.

The vector Eqs. 2.5–4 and 2.5–7 are in the most compact form for defining the general case of space motion. All special cases can be deduced directly from these equations.

PROBLEMS

1. A wheel having a 2-ft radius rolls on a belt which is moving to the right with a speed of 1 ft/sec. To an observer standing on the ground, the center of the wheel appears to move to the left with a speed of 3 ft/sec. Determine the velocity of point p.

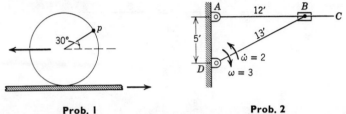

Prob. I   Prob. 2

2. At a given instant, bar AC is horizontal and bar DB has an angular velocity of 3 rad/sec clockwise and angular acceleration of 2 rad/sec² counterclockwise. Determine the velocity of the slider B relative to bar AC and the angular velocity and angular acceleration of AC.

3. Pin p which slides on arm OC is made to move along the slot AB in a disk. If the disk is held stationary and OC has an angular velocity of 3 rad/sec clockwise, when $\theta = 30°$, determine the velocity of p relative to OC and also its absolute velocity.

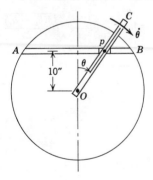

Prob. 3

4. Determine the absolute acceleration of p of Prob. 3 if, in addition to the data given, the arm OC has an angular acceleration of 4 rad/sec² in the counterclockwise direction. Specify all its components.

5. If in Probs. 3 and 4 the data given are relative to the disk which is rotating with angular velocity and acceleration of 2 rad/sec and 4 rad/sec² both clockwise, determine the absolute velocity and acceleration of *p*.

6. A particle moves with velocity *kr* outward along a spoke of a wheel rotating with angular speed $\dot\theta$ and angular acceleration $\ddot\theta$. Determine its absolute velocity and acceleration, identifying each component by a diagram.

7. An airplane travels overhead at constant altitude *h* and constant horizontal speed *v*. Determine the angular velocity $\dot\theta$ and angular acceleration $\ddot\theta$ of the line of sight of a tracking device on the ground, where θ is the angle measured from zenith. How fast is the distance to the plane increasing at θ.

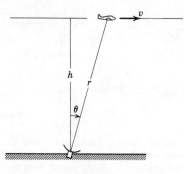

Prob. 7

8. A bomber flying at constant speed v_b and constant altitude h_b sees an enemy plane flying in the same direction with velocity v_e and at a lower altitude h_e. Assuming that $v_b > v_e$, show that the angle θ of the line of sight at the instant at which the bomb should be released must be

$$\theta = \tan^{-1} \frac{1}{(v_b - v_e)} \sqrt{\frac{g(h_b - h_e)}{2}}$$

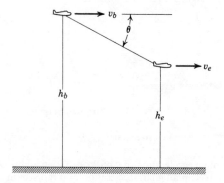

Prob. 8

9. An airplane traveling at constant altitude h and constant speed v is observed by a radar station a distance s from the vertical plane of travel of the airplane. Determine the angular velocity of the radar dish about the horizontal and vertical axes from the instant when the plane is closest to the station.

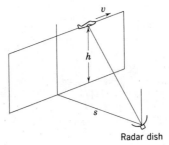

Prob. 9

10. In Prob. 9 determine the angular rates if $v = 450$ mph, $h = 12,000$ ft, $s = \frac{1}{2}$ mile, and $t = 15$ sec.

11. Fuel flows out along the impeller blade of a turbo pump at a speed of 100 ft/sec and acceleration of 120 ft/sec² relative to the blade. If the turbo wheel is running clockwise at 2400 rpm, determine the absolute velocity and acceleration of the fuel just before it leaves the impeller.

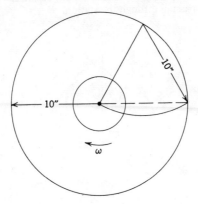

Prob. 11

12. The large wheel of the quick-return mechanism rotates in the counterclockwise direction at a constant speed of 120 rpm. Determine the velocity of point p as a function of the angle θ and the relative velocity of the slider s.

13. Determine the angular acceleration of the arm Op of Prob. 12 when the crank arm makes an angle 30° with the horizontal in the first quadrant.

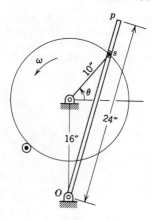

Prob. 12

14. A bomber traveling with constant speed v at constant altitude h spots an enemy plane at a lower altitude h_e and traveling along a line perpendicular to his. Assuming h_e and v_e to be constant, determine how far ahead of the intersecting vertical and perpendicular planes the bomb should be released.

15. A satellite in a S-to-N polar orbit of 120 min is observed to travel from horizon to horizon in 25 min and pass directly over an observation station at latitude 35° N. Determine the direction of the path of the satellite relative to the meridian plane passing through the station.

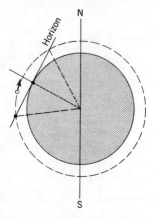

Prob. 15

16. A satellite traveling in a circular polar orbit of altitude h around the earth has a period of 120 min. An observer on the earth's surface tracks the satellite and finds that it is moving at a rate of 15°/min when directly overhead. Determine its altitude h.

2.6 Motion Relative to the Rotating Earth

The center of the earth as it moves around the sun is accelerating toward the sun. We, however, neglect this acceleration and place a set of non-rotating axes with origin at the earth's center as an inertial system.

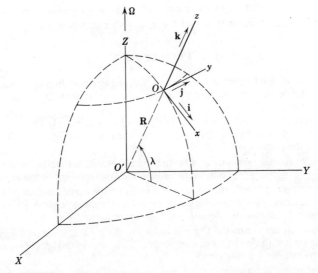

Fig. 2.6–1. Motion relative to earth using axes x, y, z fixed to earth's surface.

We are often interested in the motion of a body relative to the earth's surface. Placing a set of axes at a given point O on the earth's surface, we orient the z axis along the plumb line which, for simplicity, we assume to be equal to the geocentric line. The x and the y axes then lie in the horizontal plane, and we orient x along the meridian plane pointing south and y along the latitude line pointing east as shown in Fig. 2.6–1.

The acceleration of O is $\mathbf{\Omega} \times \mathbf{\Omega} \times \mathbf{R} = -(\Omega^2 R \sin \lambda \cos \lambda)\mathbf{i} - (\Omega^2 R \cos^2 \lambda)\mathbf{k}$ where the angular velocity of the earth-fixed coordinate system x, y, z is $\mathbf{\Omega}$, the earth's rotational velocity with components:

$$\mathbf{\Omega} = -(\Omega \cos \lambda)\mathbf{i} + (\Omega \sin \lambda)\mathbf{k} \qquad (2.6\text{–}1)$$

The acceleration relative to the inertial system is

$$\mathbf{a} = \frac{\mathbf{F}}{m} = \mathbf{\Omega} \times \mathbf{\Omega} \times \mathbf{R} + [\mathbf{a}] + \mathbf{\Omega} \times \mathbf{\Omega} \times \mathbf{r} + 2\mathbf{\Omega} \times [\mathbf{v}] \quad (2.6\text{–}2)$$

If the only force acting on the body is its weight, then $\mathbf{F}/m = -g\mathbf{k}$, and the acceleration [a] relative to the rotating earth is

$$[\mathbf{a}] = -g\mathbf{k} - \mathbf{\Omega} \times \mathbf{\Omega} \times \mathbf{R} - \mathbf{\Omega} \times \mathbf{\Omega} \times \mathbf{r} - 2\mathbf{\Omega} \times [\mathbf{v}] \quad (2.6\text{--}3)$$

Since Ω^2 is $(0.729 \times 10^{-4} \text{ rad/sec})^2$, it can be neglected, leaving

$$[\mathbf{a}] = -g\mathbf{k} - 2\mathbf{\Omega} \times [\mathbf{v}] \qquad\qquad (2.6\text{--}4)$$

Displacements relative to the earth can be found by two integrations of Eq. 2.6–4. expressed in terms of the x, y, and z components.

PROBLEMS

1. Determine, due to the earth's rotation, the angular deviation of a plumb line from the geocentric line at latitude $\lambda°$ N. At what latitude is this deviation a maximum?

2. A particle is dropped from rest at a height of 1 mile at latitude 32° N. Neglecting air friction and using the x, y, z coordinate system of Fig. 2.6–1, determine where the particle will land.

3. A bullet is fired vertically at a latitude of 50° N with a muzzle speed of 2000 ft/sec. Neglecting air friction, determine the landing point of the bullet.

4. A rocket is fired vertically upward at a point on the earth's surface of latitude $\lambda°$ N. Determine the Coriolis deviation at its maximum height h. What is the numerical value of this deviation if $h = 150$ miles and $\lambda = 35°$ N.

5. In Prob. 4, determine the Coriolis deviation during the downward flight and compare it with that at maximum height. (*Caution:* Initial lateral velocity for downward flight is not zero.)

6. An airplane is traveling with speed v due south with constant altitude at latitude $\lambda°$ N. Determine the Coriolis acceleration relative to earth.

7. At latitude $\lambda°$ S, a projectile is fired with speed v_0 in the east-west vertical plane at an elevation θ_0 in the easterly direction. Determine the latitude deviation.

Transformation
of Coordinates

CHAPTER 3

3.1 Transformation of Displacements

Consider a case where the position of a particle p in space is defined in terms of the displacement vector $\mathbf{r}$ relative to the moving coordinates xyz, shown in Fig. 3.1–1. If the displacement of the origin of the moving coordinate system is $\mathbf{R_0}$, the displacement of p relative to the fixed coordinates X, Y, Z is,

$$\mathbf{R} = \mathbf{R_0} + \mathbf{r} \tag{3.1–1}$$

Letting unit vectors along the fixed and moving axes be designated by $\mathbf{i}, \mathbf{j}, \mathbf{k}$ and $\mathbf{i}', \mathbf{j}', \mathbf{k}'$ respectively, the above equation can be written as,

$$(X\mathbf{i} + Y\mathbf{j} + Z\mathbf{k}) = (X_0\mathbf{i} + Y_0\mathbf{j} + Z_0\mathbf{k}) + (x\mathbf{i}' + y\mathbf{j}' + z\mathbf{k}') \tag{3.1–2}$$

We can determine the component of the above vector in any direction by forming the dot product of the above equation with a unit vector in the desired direction. For example, the X component is obtained by the dot product of the above equation with $\mathbf{i}$, etc. The three rectangular components along the fixed coordinates are, then,

$$
\begin{aligned}
X &= X_0 + x\mathbf{i} \cdot \mathbf{i}' + y\mathbf{i} \cdot \mathbf{j}' + z\mathbf{i} \cdot \mathbf{k}' \\
Y &= Y_0 + x\mathbf{j} \cdot \mathbf{i}' + y\mathbf{j} \cdot \mathbf{j}' + z\mathbf{j} \cdot \mathbf{k}' \\
Z &= Z_0 + x\mathbf{k} \cdot \mathbf{i}' + y\mathbf{k} \cdot \mathbf{j}' + z\mathbf{k} \cdot \mathbf{k}'
\end{aligned}
\tag{3.1–3}
$$

29

where the dot product of the various unit vectors represents the direction cosine between the coordinates.

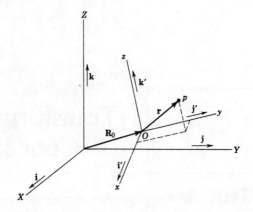

Fig. 3.1–1. Transformation between coordinates x, y, z and X, Y, Z.

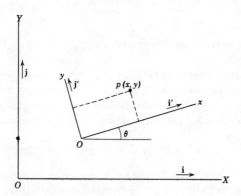

Fig. 3.1–2. Coordinate transformation in a plane.

For plane motion with $Z = z = 0$ shown in Fig. 3.1–2, the direction cosines involved are,

$$\mathbf{i} \cdot \mathbf{i}' = \cos \theta = \mathbf{j} \cdot \mathbf{j}'$$

$$\mathbf{i} \cdot \mathbf{j}' = -\sin \theta = -\mathbf{j} \cdot \mathbf{i}'$$

so that Eq. 3.1–3 reduces to,

$$X - X_0 = x \cos \theta - y \sin \theta$$
$$Y - Y_0 = x \sin \theta + y \cos \theta$$

$$(3.1–4)$$

A convenient way to express the above equations is by the following matrix equation,*

$$\begin{bmatrix} X - X_0 \\ Y - Y_0 \end{bmatrix} = \begin{bmatrix} \cos\theta & -\sin\theta \\ \sin\theta & \cos\theta \end{bmatrix} \begin{bmatrix} x \\ y \end{bmatrix} \tag{3.1-5}$$

The square matrix with the direction cosines for elements is called the transfer matrix which, in this case, transforms the body coordinates to the fixed inertial coordinates.

To obtain the inverse transformation from the fixed coordinate system to the moving coordinate system, we can start with Eq. 3.1-2 arranged as follows:

$$(x\mathbf{i}' + y\mathbf{j}') = (X - X_0)\mathbf{i} + (Y - Y_0)\mathbf{j} \tag{3.1-6}$$

and form the dot product with $\mathbf{i}'$ and $\mathbf{j}'$

$$\begin{aligned} x &= (X - X_0)\mathbf{i} \cdot \mathbf{i}' + (Y - Y_0)\mathbf{j} \cdot \mathbf{i}' \\ y &= (X - X_0)\mathbf{i} \cdot \mathbf{j}' + (Y - Y_0)\mathbf{j} \cdot \mathbf{j}' \end{aligned} \tag{3.1-7}$$

The above equations in matrix notation become,

$$\begin{aligned} \begin{bmatrix} x \\ y \end{bmatrix} &= \begin{bmatrix} \mathbf{i} \cdot \mathbf{i}' & \mathbf{j} \cdot \mathbf{i}' \\ \mathbf{i} \cdot \mathbf{j}' & \mathbf{j} \cdot \mathbf{j}' \end{bmatrix} \begin{bmatrix} X - X_0 \\ Y - Y_0 \end{bmatrix} \\ &= \begin{bmatrix} \cos\theta & \sin\theta \\ -\sin\theta & \cos\theta \end{bmatrix} \begin{bmatrix} X - X_0 \\ Y - Y_0 \end{bmatrix} \end{aligned} \tag{3.1-8}$$

which is the inverse of Eq. 3.1-5. The transfer matrix is here the inverse of the transfer matrix of Eq. 3.1-5.

$$\begin{bmatrix} \cos\theta & \sin\theta \\ -\sin\theta & \cos\theta \end{bmatrix} = \begin{bmatrix} \cos\theta & -\sin\theta \\ \sin\theta & \cos\theta \end{bmatrix}^{-1}$$

3.2 Transformation of Velocities

The velocity of any arbitrary *fixed point* on the moving coordinate system x, y, z with respect to the fixed-axis system is,

$$\dot{\mathbf{R}} = \dot{\mathbf{R}}_0 + \boldsymbol{\omega} \times \mathbf{r} \tag{3.2-1}$$

This equation indicates that we can start with the displacement equation in terms of the rectangular components and differentiate, holding x, y, z as

* See Appendix A.

constants. For instance, if we differentiate Eq. 3.1–4, with x and y held constant, we obtain the velocity equation,

$$\dot{X} - \dot{X}_0 = -(x \sin\theta + y \cos\theta)\dot{\theta}$$
$$\dot{Y} - \dot{Y}_0 = (x \cos\theta - y \sin\theta)\dot{\theta}$$

$$(3.2–2)$$

Comparing with Eq. 3.1–4, these equations can also be written as,

$$\dot{X} - \dot{X}_0 = -(Y - Y_0)\dot{\theta}$$
$$\dot{Y} - \dot{Y}_0 = (X - X_0)\dot{\theta}$$

$$(3.2–3)$$

3.3 Instantaneous Center

If we define an instantaneous center as a point of zero velocity, we can find its coordinates in the XY plane by letting $\dot{X} = \dot{Y} = 0$ in Eq. 3.2–3.

$$X_{ic} = X_0 - \frac{\dot{Y}_0}{\dot{\theta}}$$

$$(3.3–1)$$

$$Y_{ic} = Y_0 + \frac{\dot{X}_0}{\dot{\theta}}$$

The locus of such points in the fixed plane is the space centrode or the *herpolhode* curve. The locus of the instantaneous center on the moving xy plane is called the body centrode or the *polhode* curve.

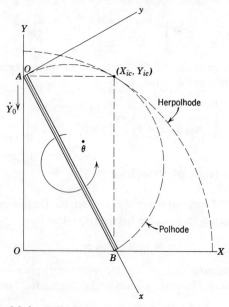

Fig. 3.3–I. Polhode curve rolls on the herpolhode curve.

As an illustration, consider a bar AB, shown in Fig. 3.3–1, the ends of which must move along the vertical and horizontal guides. Placing the moving coordinates x, y as shown with the origin coinciding with point A, $X_0 = \dot{X}_0 = 0$. We therefore have,

$$X_{ic} = -\frac{\dot{Y}_0}{\theta}$$

$$Y_{ic} = Y_0$$

which indicates that the locus of the instantaneous center (herpolhode) in the XY plane is a quarter-circle of radius equal to the length AB.

If for every position of the bar AB a hole is punched through the two planes at the instantaneous center, the set of holes on the moving plane will trace out the polhode curve, which in this case is a half-circle of diameter AB. It is evident, then, that the polhode curve rolls without slipping on the herpolhode curve.

3.4 Euler's Angles

A point on a rigid body can be defined in terms of body-fixed axes x, y, z. To determine the orientation of the body itself, we now introduce Euler's angles ψ, φ, θ which are three independent quantities capable of defining the position of the x, y, z, body axes relative to the inertial X, Y, Z axes, as shown in Fig. 3.4–1.

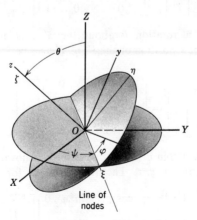

Fig. 3.4–1. Body axes x, y, z defined relative to inertial axes X, Y, Z by Euler's angles ψ, φ, θ.

The position of the body axes can be arrived at by three rotations which will also define other coordinates often encountered in rigid body dynamics.

With the x, y, z, axes coinciding with the X, Y, Z axes, allow the x, y, z coordinates to rotate about the Z axis through an angle ψ so as to take up the position $\xi'\eta'\zeta'$, shown in Fig. 3.4–2.

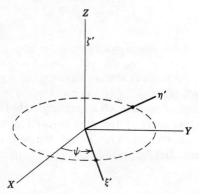

Fig. 3.4–2. Rotation about Z axis through angle ψ.

The relationship between the two coordinates is then given by the transfer matrix

$$\begin{bmatrix} \xi' \\ \eta' \\ \zeta' \end{bmatrix} = \begin{bmatrix} \cos\psi & \sin\psi & 0 \\ -\sin\psi & \cos\psi & 0 \\ 0 & 0 & 1 \end{bmatrix} \begin{bmatrix} X \\ Y \\ Z \end{bmatrix} \qquad (3.4\text{–}1)$$

We next allow a rotation θ about the ξ' axis as shown in Fig. 3.4–3, and let the new position of the ξ', η', ζ' axes be ξ, η, ζ with transfer matrix

$$\begin{bmatrix} \xi \\ \eta \\ \zeta \end{bmatrix} = \begin{bmatrix} 1 & 0 & 0 \\ 0 & \cos\theta & \sin\theta \\ 0 & -\sin\theta & \cos\theta \end{bmatrix} \begin{bmatrix} \xi' \\ \eta' \\ \zeta' \end{bmatrix} \qquad (3.4\text{–}2)$$

Finally we allow a spin φ about the axis ζ, as shown in Fig. 3.4–4, to arrive at the body axes x, y, z. The transfer matrix for this rotation is

$$\begin{bmatrix} x \\ y \\ z \end{bmatrix} = \begin{bmatrix} \cos\varphi & \sin\varphi & 0 \\ -\sin\varphi & \cos\varphi & 0 \\ 0 & 0 & 1 \end{bmatrix} \begin{bmatrix} \xi \\ \eta \\ \zeta \end{bmatrix} \qquad (3.4\text{–}3)$$

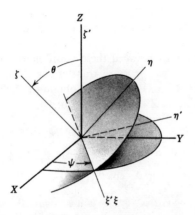

Fig. 3.4–3. Rotation about node axis $\xi' = \xi$ through angle θ.

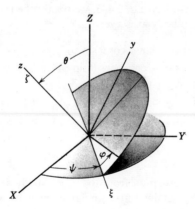

Fig. 3.4–4. Rotation about $z = \zeta$ axis through spin angle φ.

In arriving at the final position of the body axes, we have encountered four sets of orthogonal axes: X, Y, Z; ξ', η', ζ'; ξ, η, ζ; and x, y, z. Some of these axes coincide, such as the $Z\zeta'$, the ζz, and the $\xi'\xi$; however both letters will be retained to identify the coordinate system referred to. Of particular interest is the $\xi'\xi$ axis, called the *line of nodes*. It represents the intersection of the transverse body plane xy and the horizontal inertial plane XY.

Other transformations between these coordinates can be obtained by the multiplication of two or more transfer matrices. For instance, by substituting Eq. 3.4–1 into Eq. 3.4–2 we obtain the following transformation* between the X, Y, Z and the ξ, η, ζ axes:

$$
\begin{bmatrix} \xi \\ \eta \\ \zeta \end{bmatrix} = \begin{bmatrix} 1 & 0 & 0 \\ 0 & \cos\theta & \sin\theta \\ 0 & -\sin\theta & \cos\theta \end{bmatrix} \begin{bmatrix} \cos\psi & \sin\psi & 0 \\ -\sin\psi & \cos\psi & 0 \\ 0 & 0 & 1 \end{bmatrix} \begin{bmatrix} X \\ Y \\ Z \end{bmatrix}
$$

$$
= \begin{bmatrix} \cos\psi & \sin\psi & 0 \\ -\cos\theta\sin\psi & \cos\theta\cos\psi & \sin\theta \\ \sin\theta\sin\psi & -\sin\theta\cos\psi & \cos\theta \end{bmatrix} \begin{bmatrix} X \\ Y \\ Z \end{bmatrix} \tag{3.4-4}
$$

Substituting Eq. 3.4–4 into Eq. 3.4–3, we obtain the transformation from the XYZ axes to the body axes xyz:

$$
\begin{bmatrix} x \\ y \\ z \end{bmatrix} = \begin{bmatrix} (\cos\phi\cos\psi - \sin\phi\cos\theta\sin\psi) & (\cos\phi\sin\psi + \sin\phi\cos\theta\cos\psi) & (\sin\phi\sin\theta) \\ (-\sin\phi\cos\psi - \cos\phi\cos\theta\sin\psi) & (-\sin\phi\sin\psi + \cos\phi\cos\theta\cos\psi) & (\cos\phi\sin\theta) \\ (\sin\theta\sin\psi) & (-\sin\theta\cos\psi) & (\cos\theta) \end{bmatrix} \begin{bmatrix} X \\ Y \\ Z \end{bmatrix} \tag{3.4-5}
$$

The inverse transformation from the x, y, z body axes to the X, Y, Z inertial axes can be obtained in a similar manner by writing Eqs. 3.4–1, 3.4–2 and 3.4–3 in the inverse order, i.e.,

$$
\begin{bmatrix} X \\ Y \\ Z \end{bmatrix} = \begin{bmatrix} \cos\psi & -\sin\psi & 0 \\ \sin\psi & \cos\psi & 0 \\ 0 & 0 & 1 \end{bmatrix} \begin{bmatrix} \xi' \\ \eta' \\ \zeta' \end{bmatrix} \tag{3.4-6}
$$

* See Appendix A.

etc. Rules are also available for the direct inversion of matrices. The inverse of Eq. 3.4–5 is

$$
\begin{bmatrix} X \\ Y \\ Z \end{bmatrix} =
\begin{bmatrix}
(\cos \varphi \cos \psi - \sin \varphi \cos \theta \sin \psi) \\
(\cos \varphi \sin \psi + \sin \varphi \cos \theta \cos \psi) \\
(\sin \theta \sin \varphi)
\end{bmatrix}
$$

$$
\begin{bmatrix}
(-\sin \varphi \cos \psi - \sin \psi \cos \theta \cos \varphi) & (\sin \theta \sin \psi) \\
(-\sin \varphi \sin \psi + \cos \varphi \cos \theta \cos \psi) & (-\sin \theta \cos \psi) \\
(\sin \theta \cos \varphi) & (\cos \theta)
\end{bmatrix}
\begin{bmatrix} x \\ y \\ z \end{bmatrix} \quad (3.4\text{–}7)
$$

3.5 Transformation of Angular Velocities

Frequently we need to express the angular velocities $\omega_x\ \omega_y\ \omega_z$ about the body axes x, y, z in terms of Euler's angles. The transformation may be pursued as follows.

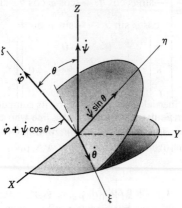

Fig. 3.5–1. Angular rates of Euler angles.

Resolve the angular velocity $\dot{\psi}$ along the ζ and η axes so that the orthogonal components of $\dot{\psi}$, $\dot{\varphi}$, and $\dot{\theta}$ are $\dot{\theta}$ along ξ, $\dot{\psi} \sin \theta$ along η, and $\dot{\varphi} + \dot{\psi} \cos \theta$ along ζ, as shown in Fig. 3.5–1.

Next resolve the components along the ξ and η axes to the x and y directions, the result being

$$
\begin{aligned}
\omega_x &= \dot{\psi} \sin \theta \sin \varphi + \dot{\theta} \cos \varphi \\
\omega_y &= \dot{\psi} \sin \theta \cos \varphi - \dot{\theta} \sin \varphi \\
\omega_z &= \dot{\varphi} + \dot{\psi} \cos \theta
\end{aligned}
\quad (3.5\text{–}1)
$$

or in its inverse form,

$$\dot{\psi} = \frac{1}{\sin\theta}(\omega_x \sin\varphi + \omega_y \cos\varphi)$$

$$\dot{\varphi} = \omega_z - \frac{\cos\theta}{\sin\theta}(\omega_x \sin\varphi + \omega_y \cos\varphi) \qquad (3.5\text{--}2)$$

$$\dot{\theta} = \omega_x \cos\varphi - \omega_y \sin\varphi$$

Arranged in matrix form, these equations become

$$\begin{bmatrix} \omega_x \\ \omega_y \\ \omega_z \end{bmatrix} = \begin{bmatrix} \sin\theta\sin\varphi & 0 & \cos\varphi \\ \sin\theta\cos\varphi & 0 & -\sin\varphi \\ \cos\theta & 1 & 0 \end{bmatrix} \begin{bmatrix} \dot{\psi} \\ \dot{\varphi} \\ \dot{\theta} \end{bmatrix} \qquad (3.5\text{--}3)$$

$$\begin{bmatrix} \dot{\psi} \\ \dot{\varphi} \\ \dot{\theta} \end{bmatrix} = \frac{1}{\sin\theta} \begin{bmatrix} \sin\varphi & \cos\varphi & 0 \\ -\sin\varphi\cos\theta & -\cos\varphi\cos\theta & \sin\theta \\ \cos\varphi\sin\theta & -\sin\varphi\sin\theta & 0 \end{bmatrix} \begin{bmatrix} \omega_x \\ \omega_y \\ \omega_z \end{bmatrix} \qquad (3.5\text{--}4)$$

PROBLEMS

1. A point P in the inertial space is defined by its components X_1, Y_1, and Z_1. From P a perpendicular PN is drawn to a line whose direction is specified by the angles α, β, and γ. Determine the lengths ON and PN.

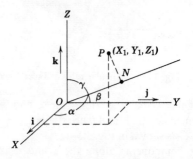

Prob. 1

2. Determine the inertial components of a point $(3, 4, 5)$ in the rotating co-ordinate system x, y, z, specified by the Euler angles $\theta = 30°$, $\psi = 45°$, and $\varphi = 30°$.

3. Using body-fixed axes along a bar of length l whose ends slide along a smooth vertical wall and horizontal floor, determine the acceleration of its midpoint.

4. One end A of a bar AB moves along a vertical wall while some intermediate point slides over the corner of a step a distance s from the wall. Derive the equation for the herpolhode and polhode curves. (Use fixed coordinates through the corner O, and let Z, Y, be coordinates of the space centrode.)

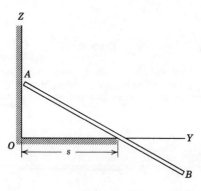

Prob. 4

5. The center of a wheel of radius R is moving to the right with velocity v_0 while the angular speed of the wheel is ω_0 in the counterclockwise direction. Using body-fixed axes x', y' and inertial axes X, Y as shown, determine the polhode and herpolhode curves.

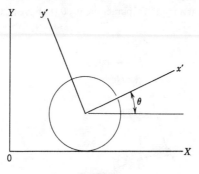

Prob. 5

6. The ends of a link of length $l = kR$ moves along the circumference and the diameter of a circle of radius R. Determine the instantaneous center as a function of the angle θ.

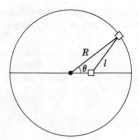

Prob. 6

7. A rod moves in a vertical plane, with the lower end along the horizontal and an intermediate point resting against a small pin at height h. Determine the instantaneous center as a function of θ and plot the polhode and herpolhode curves.

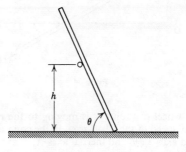

Prob. 7

8. At a given instant a man in a parachute is 60 ft above a 20° inclined hillside.

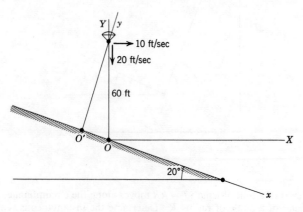

Prob. 8

If his horizontal and vertical velocities are constant and equal to 10 ft/sec and 20 ft/sec respectively, determine his coordinate as a function of time: (a) in the X, Y system; (b) in the x, y system. Determine the time and place of landing on the hillside.

9. A convenient coordinate system for surface navigation on earth is the longitude-latitude system with the origin coinciding with the moving vehicle shown in the sketch. The x, y axes lie in the horizontal plane along the latitude and longitude lines. Show that the angular velocities along the coordinates are

$$\boldsymbol{\omega} = -\dot{\lambda}\mathbf{i} + (\dot{\varphi} + \Omega)\cos\lambda\,\mathbf{j} + (\dot{\varphi} + \Omega)\sin\lambda\,\mathbf{k}$$

where the $\mathbf{i}$, $\mathbf{j}$, $\mathbf{k}$ vectors are along the x, y, z directions, and Ω is the earth's rotational speed.

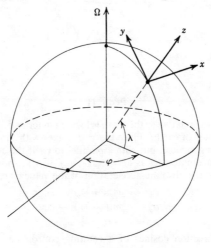

Prob. 9

10. For the system of Prob. 9, determine the x, y, z components of the acceleration.

11. A satellite s circles the earth with the orbit plane making an angle α with the earth's equatorial plane. The X axis is oriented so that it passes through the intersection of the orbit and equatorial planes. The position of the satellite at any time can be given in terms of r_s, the distance from the earth's center, ψ_s the angle of the meridian plane measured from the X axis, and λ_s the latitude; the corresponding coordinates of an observation station O are R_0, ψ_0, and λ_0.

(a) Determine the angle φ measured from the X axis to r_s in the plane of the orbit, in terms of ψ_s, λ_s, and α.

(b) Determine the cosine of the angle between R_0 and r_s, and the straight-line distance between O and S. Use h for altitude.

(c) Determine the direction cosine of the line OS relative to X, Y, Z.

(*d*) Determine the direction cosine of the line *OS* relative to a coordinate system x, y, z, with the origin at the observation station as shown.

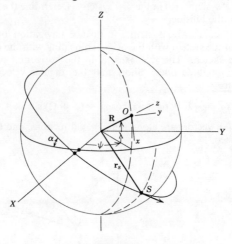

Prob. 11

12. A vehicle moving relative to the earth is located in terms of its position vector **r** from the center of earth. Place the x, y, z axes with the origin coinciding with the moving vehicle, but with z parallel to **r**. If the angular velocity of the x, y, z coordinate axes is $\boldsymbol{\omega} = \omega_x\mathbf{i} + \omega_y\mathbf{j} + \omega_z\mathbf{k}$, show that the x, y, z components of the absolute acceleration of the vehicle are,

$$a_x = \omega_x\omega_z r + \dot{\omega}_y r + 2\omega_y\dot{r}$$
$$a_y = \omega_y\omega_z r - \dot{\omega}_x r - 2\omega_x\dot{r}$$
$$a_z = \ddot{r} - (\omega_x^2 + \omega_y^2)r$$

13. Letting the direction cosines of a rotating coordinate system x, y, z be specified by

$$\begin{vmatrix} l_{Xx} & l_{Yx} & l_{Zx} \\ l_{Xy} & l_{Yy} & l_{Zy} \\ l_{Xz} & l_{Yz} & l_{Zz} \end{vmatrix}$$

where l_{Xy} is the cosine of the angle between X and y, determine the equations for the unit vectors $\mathbf{i}'$, $\mathbf{j}'$, $\mathbf{k}'$, along x, y, z, in terms of the unit vectors $\mathbf{i}$, $\mathbf{j}$, $\mathbf{k}$, along X, Y, Z.

14. If the x, y, z coordinate system is rotating with angular velocity $\boldsymbol{\omega} = \omega_x\mathbf{i}' + \omega_y\mathbf{j}' + \omega_z\mathbf{k}'$, show that the velocities of the unit vectors are

$$\frac{d}{dt}\mathbf{i}' = \omega_z\mathbf{j}' - \omega_y\mathbf{k}'$$

$$\frac{d}{dt}\mathbf{j}' = \omega_x\mathbf{k}' - \omega_z\mathbf{i}'$$

$$\frac{d}{dt}\mathbf{k}' = \omega_y\mathbf{i}' - \omega_x\mathbf{j}'$$

15. From Probs. 13 and 14, show that the rate of change of the direction cosines between the X, Y, Z and x, y, z axes is,

$$\dot{l}_{Xx} = \dot{\mathbf{i}}' \cdot \mathbf{i} = \omega_z \mathbf{j}' \cdot \mathbf{i} - \omega_y \mathbf{k}' \cdot \mathbf{i}$$

$$= \omega_z l_{Xy} - \omega_y l_{Xz}$$

$$\dot{l}_{Yx} = \omega_z l_{Yy} - \omega_y l_{Yz}$$

etc. Complete the other seven equations.

16. The following relationship for the unit vector $\mathbf{k}'$ can be written.

$$\mathbf{k}' = \mathbf{i}' \times \mathbf{j}' = \begin{vmatrix} \mathbf{i} & \mathbf{j} & \mathbf{k} \\ l_{Xx} & l_{Yx} & l_{Zx} \\ l_{Xy} & l_{Yy} & l_{Zy} \end{vmatrix} = l_{Xz}\mathbf{i} + l_{Yz}\mathbf{j} + l_{Zz}\mathbf{k}$$

This equation leads to the scalar equation

$$l_{Xz} = l_{Yx}l_{Zy} - l_{Yy}l_{Zx}$$

and two others. Complete the nine scalar relationships of this type.

17. Derive the matrix Eq. 3.4–7.

Particle Dynamics (Satellite Orbits)

CHAPTER 4

4.1 Force and Momentum

Newton's laws of motion were formulated for a single particle. If the mass m of the particle is multiplied by its velocity v, the resulting product is called the linear momentum.

$$\mathbf{p} = m\mathbf{v} \qquad (4.1\text{--}1)$$

The velocity $\mathbf{v}$ here is measured with respect to an inertial frame of reference so that, if the position of the particle is defined by its displacement vector $\mathbf{r}$, the velocity is $\mathbf{v} = \dot{\mathbf{r}}$.

Newton's second law states that the time rate of change of momentum is equal to the force producing it, and this change takes place in the direction in which the force acts.

$$\mathbf{F} = \dot{\mathbf{p}} \qquad (4.1\text{--}2)$$

With m a constant, this equation can also be written as,

$$\mathbf{F} = m\dot{\mathbf{v}} = m\ddot{\mathbf{r}} \qquad (4.1\text{--}3)$$

Newton's first law, which forms the basis for statics, is a special case of the second law when the force $\mathbf{F}$ is zero. It states that, if the forces acting on a particle balance to give a resultant force of zero ($\mathbf{F} = \dot{\mathbf{p}} = 0$), the

44

particle remains at rest ($\mathbf{p} = 0$), or continues to move in a straight line with constant velocity or momentum ($\mathbf{p} =$ constant).

4.2 Impulse and Momentum

If the force $\mathbf{F}$ is multiplied by the time dt and integrated, we obtain

$$\int_{t_1}^{t_2} \mathbf{F}\, dt = \int_{t_1}^{t_2} m\, \frac{d\mathbf{v}}{dt}\, dt = m\mathbf{v}_2 - m\mathbf{v}_1 \qquad (4.2\text{-}1)$$

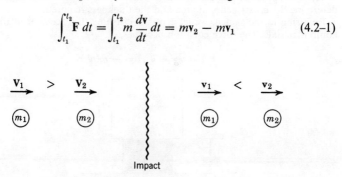

Impact

Fig. 4.2-I. Momentum before impact is equal to momentum after impact.

The time integral on the left side of the equation is called the impulse of the force, so that the above equation states that the change in momentum of a particle is equal to the impulse of the force acting on the particle.

When two bodies collide, a large force $\mathbf{f}(t)$ acts for a short time, and the impulse $\int \mathbf{f}\, dt$ exerted on the two bodies must be equal and opposite according to Newton's third law. Since impulse is equal to the change in momentum, for the two bodies considered together as a system, the impulses of collision cancel each other. Thus the change in momentum of the system is zero, and the momentum before impact must equal the momentum after impact. Energy however, is generally dissipated during impact, in which case the impulse during relaxation is less than the impulse during compression. For central impact we let this ratio be e, the coefficient of restitution, and it can be shown that e is also expressible in terms of the velocities as

$$e = \frac{\left(\int \mathbf{f}\, dt\right)_{\text{relax.}}}{\left(\int \mathbf{f}\, dt\right)_{\text{compr.}}} = \frac{v_2 - v_1}{V_1 - V_2} = \frac{\text{Velocity of separation}}{\text{Velocity of approach}} \qquad (4.2\text{-}2)$$

where the sequence of events is illustrated in Fig. 4.2-1. Thus when no energy is dissipated, the impact is elastic and $e = 1$, whereas for the completely plastic impact the relaxation impulse is zero and $e = 0$. In general e depends on the material, shape and the velocities of the two bodies.

Example 4.2–1

A honeycomb plastic has a crushing stress of σ_c lb/in.[2] If a package of mass m is to be dropped through a height h without exceeding a deceleration of ng, determine the cross-sectional area and the thickness required.

Referring to the sketch, we let ξ be the crushing displacement of the honeycomb material. The force equation becomes,

$$m\ddot{\xi} = mg - \sigma_c A = -m(ng) \qquad (a)$$

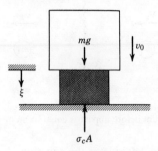

Ex. 4.2–1.

and its integral is

$$\dot{\xi} = \left(g - \frac{\sigma_c A}{m}\right)t + v_0 \qquad (b)$$

From (a) the required area is

$$A = \frac{mg}{\sigma_c}(1 + n) \qquad (c)$$

which substituted into (b) becomes,

$$\dot{\xi} = [g - (1 + n)g]t + v_0 \qquad (d)$$

Maximum crushing is attained when $\dot{\xi} = 0$, or

$$t = \frac{v_0}{ng} \qquad (e)$$

Integrating (d) and substituting (e), the material crushed is given by the equation,

$$\xi = -\frac{ng}{2}t^2 + v_0 t = \frac{1}{2}\frac{v_0^2}{ng} = \frac{h}{n} \qquad (f)$$

4.3 Work and Energy

If the force $\mathbf{F}$ acting on a particle moves through a distance $d\mathbf{r}$, the work done is equal to the scalar product $\mathbf{F} \cdot d\mathbf{r}$. The total work done in going from $\mathbf{r}_1$ to $\mathbf{r}_2$ is then

$$W = \int_{\mathbf{r}_1}^{\mathbf{r}_2} \mathbf{F} \cdot d\mathbf{r} \tag{4.3-1}$$

By substituting for $\mathbf{F}$ and changing the variable of integration to time by $d\mathbf{r} = \mathbf{v}\, dt$, the expression for work becomes,

$$\int_{\mathbf{r}_1}^{\mathbf{r}_2} \mathbf{F} \cdot d\mathbf{r} = \int_{t_1}^{t_2} m \frac{d\mathbf{v}}{dt} \cdot \mathbf{v}\, dt = \frac{1}{2} \int_{t_1}^{t_2} m \frac{d}{dt}(\mathbf{v} \cdot \mathbf{v})\, dt$$

$$= \frac{1}{2} \int_{t_1}^{t_2} m \frac{d}{dt} v^2\, dt = \tfrac{1}{2}mv_2{}^2 - \tfrac{1}{2}mv_1{}^2 \tag{4.3-2}$$

The scalar quantity $\tfrac{1}{2}mv^2$ is called the kinetic energy of the particle, so the work done on the particle by the force is equal to the change in kinetic energy of the particle.

We now define a conservative force system as one in which the work done is a function only of the position, and independent of the path taken by the force. It follows then that the work done by a conservative force system around any closed path must be zero.

$$\oint \mathbf{F} \cdot d\mathbf{r} = 0 \tag{4.3-3}$$

We will now define the potential energy $U(\mathbf{r}_1)$ as the work done by the conservative force in going from any point $\mathbf{r}_1$ to some reference point $\mathbf{r}_0$.

$$\int_{\mathbf{r}_1}^{\mathbf{r}_0} \mathbf{F} \cdot d\mathbf{r} = U(\mathbf{r}_1) \tag{4.3-4}$$

Thus every point in space can be assigned a scalar potential $U(\mathbf{r})$ which will depend on the reference point.

Consider next the work done by a conservative force in going from $\mathbf{r}_1$ to $\mathbf{r}_2$. Since the work done is independent of the path taken, we can go from $\mathbf{r}_1$ to $\mathbf{r}_0$ to $\mathbf{r}_2$ as follows:

$$\int_{\mathbf{r}_1}^{\mathbf{r}_2} \mathbf{F} \cdot d\mathbf{r} = \int_{\mathbf{r}_1}^{\mathbf{r}_0} \mathbf{F} \cdot d\mathbf{r} + \int_{\mathbf{r}_0}^{\mathbf{r}_2} \mathbf{F} \cdot d\mathbf{r}$$

$$= \int_{\mathbf{r}_1}^{\mathbf{r}_0} \mathbf{F} \cdot d\mathbf{r} - \int_{\mathbf{r}_2}^{\mathbf{r}_0} \mathbf{F} \cdot d\mathbf{r} = U(\mathbf{r}_1) - U(\mathbf{r}_2) \tag{4.3-5}$$

Thus the work done in going from $\mathbf{r}_1$ to $\mathbf{r}_2$ is the difference in the scalar potential $-[U(\mathbf{r}_2) - U(\mathbf{r}_1)]$, and it is evident that the result is independent of the reference point. In terms of the differential displacement, the above equation can be written as,

$$\mathbf{F} \cdot d\mathbf{r} = -dU \qquad (4.3-6)$$

which expresses the conservative force in terms of the potential or the potential energy. This discussion clearly indicates why the reference point for the potential energy is arbitrary in setting up the differential equations of motion, which are force or moment equations.

In a conservative system, the total energy is a constant. If we designate the kinetic energy by the letter T, Eq. 4.3-2 can be written as

$$\int_{\mathbf{r}_1}^{\mathbf{r}_2} \mathbf{F} \cdot d\mathbf{r} = T_2 - T_1 = -(U_2 - U_1) \qquad (4.3-7)$$

Rearranging, the total energies at 1 and 2 are seen to be equal,

$$T_2 + U_2 = T_1 + U_1 \qquad (4.3-8)$$

which illustrates the principle of conservation of energy for the conservative system.

As an example of a conservative force system, we have the gravitational attraction of the earth, which is inversely proportional to the square of the distance from the earth's center,

$$\mathbf{F} = -m\mathbf{g}\left(\frac{R}{r}\right)^2 \qquad (4.3-9)$$

where $\mathbf{g}$ and R are the acceleration of gravity at the earth's surface and radius of the earth respectively. If we use the earth's surface as the reference, the potential energy, or the potential of a mass m at height h above the earth's surface is,

$$U(h) = \int_{R+h}^{R} -m\mathbf{g}\left(\frac{R}{r}\right)^2 \cdot d\mathbf{r} = mgR^2\left(\frac{1}{R} - \frac{1}{R+h}\right)$$

$$= mg\frac{h}{\left(1 + \dfrac{h}{R}\right)} \qquad (4.3-10)$$

Thus, for the moderate heights h above the earth's surface, h/R is small, and we have for the potential energy the simple equation,

$$U(h) \cong mgh \qquad (4.3-11)$$

4.4 Moment of Momentum

The moment about an arbitrary point O of the momentum $\mathbf{p} = m\dot{\mathbf{R}}$ of a particle is

$$\mathbf{h}_0 = \mathbf{r} \times m\dot{\mathbf{R}} \qquad (4.4\text{-}1)$$

where $\dot{\mathbf{R}}$ is the absolute velocity of m and $\mathbf{r}$ is drawn from O as shown in Fig. 4.4–1. Differentiating this equation, we obtain

$$\dot{\mathbf{h}}_0 = \mathbf{r} \times m\ddot{\mathbf{R}} + \dot{\mathbf{r}} \times m\dot{\mathbf{R}} \qquad (4.4\text{-}2)$$

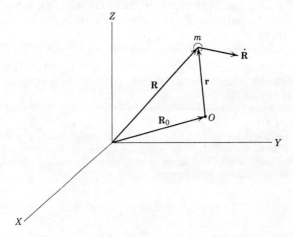

Fig. 4.4–1. Moment about O of momentum $m\dot{\mathbf{R}}$ is $\mathbf{r} \times m\dot{\mathbf{R}}$.

Substituting $\dot{\mathbf{R}} = \dot{\mathbf{R}}_0 + \dot{\mathbf{r}}$ and noting that $\dot{\mathbf{r}} \times \dot{\mathbf{r}} = 0$, this equation becomes

$$\dot{\mathbf{h}}_0 = \mathbf{r} \times m\ddot{\mathbf{R}} - \dot{\mathbf{R}}_0 \times m\dot{\mathbf{r}} \qquad (4.4\text{-}3)$$

To establish the relationship between $\dot{\mathbf{h}}_0$ and the moment $\mathbf{M}_0$ of the forces $\mathbf{F} = m\ddot{\mathbf{R}}$ acting on m, we have

$$\mathbf{M}_0 = \mathbf{r} \times m\ddot{\mathbf{R}} = \mathbf{r} \times m(\ddot{\mathbf{R}}_0 + \ddot{\mathbf{r}})$$

$$= \frac{d}{dt}(\mathbf{r} \times m\dot{\mathbf{r}}) - \ddot{\mathbf{R}}_0 \times m\mathbf{r} \qquad (4.4\text{-}4)$$

Substituting $\mathbf{M}_0 = \mathbf{r} \times m\ddot{\mathbf{R}}$ into Eq. 4.4–3, we can also write

$$M_0 = \dot{\mathbf{h}}_0 + \dot{\mathbf{R}}_0 \times m\dot{\mathbf{r}} \qquad (4.4\text{-}5)$$

Several interesting conclusions can be drawn from Eqs. 4.4–4 and 4.4–5 as follows:

a. If point O is fixed in space, then $\dot{\mathbf{R}}_0 = \ddot{\mathbf{R}}_0 = 0$ and $\dot{\mathbf{r}} = \dot{\mathbf{R}}$, which results in the simplified equation

$$\mathbf{M}_0 = \dot{\mathbf{h}}_0$$

b. If point O is moving with constant velocity, $\ddot{\mathbf{R}}_0 = 0$, and

$$\mathbf{M}_0 = \frac{d}{dt}(\mathbf{r} \times m\dot{\mathbf{r}})$$

which states that the moment is equal to the rate of change of the apparent moment of momentum expressed in terms of the relative velocity $\dot{\mathbf{r}}$.

c. If either $\ddot{\mathbf{R}}_0$ and $\mathbf{r}$ or $\dot{\mathbf{R}}_0$ and $\dot{\mathbf{r}}$ are parallel, again the simplified equations are valid.

d. If the system consists of more than one mass, then the second term of Eq. 4.4–4 becomes $-\ddot{\mathbf{R}}_0 \times \sum m\mathbf{r}$ which is zero ($\sum m\mathbf{r} = 0$) when the point O coincides with the center of mass. The moment equation is then the same as in case (b).

Example 4.4–1

A dumbbell idealized by two masses on a stiff, weightless rod of length l is dropped without rotation, and the left mass strikes a ledge with velocity v. Assuming the coefficient of restitution to be e, determine the angular rotation of the dumbbell immediately after impact.

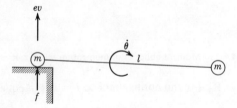

Ex. 4.4–I

The sketch shows the dumbbell immediately after impact. The velocity of the center of mass immediately after impact is $ev - \frac{l}{2}\dot{\theta}$, and the change in the linear momentum is,

$$\int f\,dt = 2m\left(ev - \frac{l}{2}\dot{\theta}\right) - (-2mv) \qquad (a)$$

The change in the moment of momentum about the center of mass is equal to the moment about the center of mass of the impulse,

$$\frac{l}{2}\int f\,dt = 2m\left(\frac{l}{2}\right)^2 \dot{\theta} \qquad (b)$$

Eliminating the impulse integral, the angular velocity immediately after impact is,

$$\dot\theta = \frac{v}{l}(1 + e) \qquad (c)$$

PROBLEMS

1. Assuming that meteors attracted by the earth start at infinity with zero speed, determine the speed with which they will strike the earth, neglecting friction (radius of earth = 3960 miles).

2. The acceleration $\ddot{\mathbf{r}}$ of a particle acted upon by a central force $f(r)\frac{\mathbf{r}}{r}$, where $\frac{\mathbf{r}}{r}$ is a unit vector along $\mathbf{r}$, is

$$m\ddot{\mathbf{r}} = -f(r)\frac{\mathbf{r}}{r}$$

Show by the vector method that equal areas are swept out by $\mathbf{r}$ per unit of time. *Hint:*

$$\mathbf{r} \times \frac{d^2\mathbf{r}}{dt^2} = \frac{d}{dt}\left(\mathbf{r} \times \frac{d\mathbf{r}}{dt}\right) = 0$$

3. A force field in a plane is defined by the equation $\mathbf{F} = -\frac{K}{r}\mathbf{1}_\theta$ where r and θ are polar coordinates. Show that the above force does not have a potential when the path encloses the origin. Alternatively, describe a closed path exclusive of the origin and show that a potential exists.

4. Show that the moment of momentum about an arbitrary point 0 for a system of particles is equal to

$$\mathbf{h}_0 = \sum_i \mathbf{r}_i \times m_i\dot{\mathbf{r}}_i - \dot{\mathbf{R}}_0 \times \sum_i m_i\mathbf{r}_i$$

where the notation is that of Fig. 4.4–1.

5. If a tempered-steel ball, weighing 0.01 lb will support a dead weight of 800 lb without being crushed, what is the greatest speed with which it can be projected without rupture, perpendicularly against a plane for which $e = 0.75$, assuming that the actual pressure is never greater than twice the average pressure, and the impact lasts 0.005 sec.

6. In Ex. 4.4–1 the dumbbell in the horizontal position just prior to impact has an angular velocity of 0.2 rad/sec counterclockwise. Determine the angular velocity immediately after impact.

7. In Ex. 4.4–1 determine the velocity of each mass before and after impact and prove that the change in the moment of momentum about the mass center is $2m(l/2)^2\dot\theta$.

8. During impact (see Fig. 4.2–1) the relative velocity of the two masses becomes equal to zero at the instant of maximum compression. Letting the common velocity of the masses at this instant be $\mathbf{V}_0$, it is possible to write for m_1 the equations

$$\left(\int \mathbf{f}\, dt\right)_{\text{compr.}} = m_1(\mathbf{V}_1 - \mathbf{V}_0)$$

$$\left(\int \mathbf{f}\, dt\right)_{\text{relax.}} = m_1(\mathbf{V}_0 - \mathbf{v}_1)$$

and likewise a similar set for m_2. Derive Eq. 4.2–2. *Hint:* When two fractions are equal, the two numerators and the two denominators may be added without altering their ratio.

4.5 Motion Under a Central Force

A force which is always directed towards a fixed point is called a central force. Choosing the origin O of polar coordinates as the fixed point, the moment of the central force about 0 must be zero.

$$\mathbf{M} = \mathbf{r} \times \mathbf{F} = \dot{\mathbf{h}} = 0 \qquad (4.5\text{–}1)$$

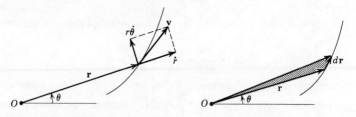

Fig. 4.5–1. Radial and transverse components of orbit velocity.

The moment of momentum about O must therefore be constant.

$$\mathbf{h} = \mathbf{r} \times m\mathbf{v} = \text{constant} \qquad (4.5\text{–}2)$$

As shown in Fig. 4.5–1, the magnitude of the cross product

$$|\mathbf{r} \times \mathbf{v}| = \frac{|\mathbf{r} \times d\mathbf{r}|}{dt}$$

is equal to twice the area swept out by the radial line per unit time. It is equal to the moment of momentum per unit mass, and we will designate it by the letter h.

$$h = |\mathbf{r} \times \mathbf{v}| = r^2\dot{\theta} \qquad (4.5\text{–}3)$$

We will now examine the motion under a central force $F(r)$ which is some arbitrary function of r for a unit mass. The acceleration in the radial and transverse directions are

$$\ddot{r} - r\dot{\theta}^2 = F(r) \qquad (4.5\text{–}4)$$

$$r\ddot{\theta} + 2\dot{r}\dot{\theta} = \frac{1}{r}\frac{d}{dt}r^2\dot{\theta} = 0 \qquad (4.5\text{–}5)$$

From the second equation we obtain the integral corresponding to Eq. 4.5–3

$$r^2\dot{\theta} = h = \text{constant}$$

Substituting $\dot\theta = h/r^2$ into the first equation, it can be written as

$$\ddot r - \frac{h^2}{r^3} = F(r) \tag{4.5-6}$$

or, since

$$\ddot r = \dot r \frac{d\dot r}{dr}$$

$$\dot r \frac{d\dot r}{dr} = \frac{h^2}{r^3} + F(r) \tag{4.5-7}$$

Integrating,

$$\dot r^2 = -\frac{h^2}{r^2} + 2\int F(r)\,dr + C \tag{4.5-8}$$

To eliminate the time, we note that $\dot r = (dr/d\theta)\dot\theta = (dr/d\theta)(h/r^2)$. Thus Eq. 4.5–8 can be written as

$$\left(\frac{dr}{d\theta}\right)^2 = -r^2 + 2\frac{r^4}{h^2}\int F(r)\,dr + C\frac{r^4}{h^2} \tag{4.5-9}$$

When $F(r)$ is specified, the orbit equation is obtained by the integration of the above equation.

Another equation of interest is the speed equation which can be determined from its two components as

$$v^2 = \dot r^2 + (r\dot\theta)^2 = 2\int F(r)\,dr + C \tag{4.5-10}$$

Since the direction of $\mathbf{h}$ as well as the magnitude must be a constant, the orbit plane, perpendicular to $\mathbf{h}$, must also remain fixed. Thus, the motion under a central force requires a constant area rate and a fixed orientation of the orbit plane. The motion of planets, as stated by Kepler's second law, closely conforms to the above requirements.

PROBLEMS

1. Determine the equation for the central attractive force $f(r)$ for which all circular orbits have the same areal rate λ.

2. Show that a particle in a central repulsive force field varying inversely as the square of the distance from the focus will move along a branch of a hyperbola.

3. If a particle describes a circle with the center of force on the circumference, determine the law of force.

4.6 The Two-Body Problem

Consider two bodies assumed as particles and moving under the influence of a mutual attractive force. Letting $\mathbf{r}_1$, $\mathbf{r}_2$, and $\mathbf{r}_c$ be the displacement vectors of each mass and their center of mass, as shown in Fig. 4.6–1, the vector $\mathbf{r} = \mathbf{r}_1 - \mathbf{r}_2$ will define their separation distance. The distance

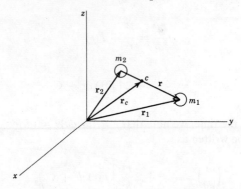

Fig. 4.6–1. Displacement vectors of two masses and their center of mass c.

of each mass from their center of mass is $[m_2/(m_1 + m_2)]\mathbf{r}$ and $[m_1/(m_1 + m_2)]\mathbf{r}$ so that $\mathbf{r}_1$ and $\mathbf{r}_2$ can be expressed in terms of $\mathbf{r}_c$ and $\mathbf{r}$ as

$$\mathbf{r}_1 = \mathbf{r}_c + \frac{m_2}{m_1 + m_2}\,\mathbf{r} \qquad (4.6\text{–}1)$$

$$\mathbf{r}_2 = \mathbf{r}_c - \frac{m_1}{m_1 + m_2}\,\mathbf{r}$$

We now let $\mathbf{F}_1$ and $\mathbf{F}_2$ be forces acting on m_1 and m_2 respectively, and write Newton's equations,

$$\mathbf{F}_1 = m_1\ddot{\mathbf{r}}_1 = m_1\ddot{\mathbf{r}}_c + \frac{m_1 m_2}{m_1 + m_2}\,\ddot{\mathbf{r}}$$

$$\mathbf{F}_2 = m_2\ddot{\mathbf{r}}_2 = m_2\ddot{\mathbf{r}}_c - \frac{m_1 m_2}{m_1 + m_2}\,\ddot{\mathbf{r}} \qquad (4.6\text{–}2)$$

In addition we can write the equation for the kinetic energy.

$$T = \frac{m_1 \dot{r}_1^{\,2}}{2} + \frac{m_2 \dot{r}_2^{\,2}}{2}$$

$$= \frac{(m_1 + m_2)}{2}\,\dot{r}_c^{\,2} + \frac{1}{2}\left(\frac{m_1 m_2}{m_1 + m_2}\right)\dot{r}^{\,2} \qquad (4.6\text{–}3)$$

If we specify that the system is isolated from external forces, then the resultant force $\mathbf{F} = \mathbf{F}_1 + \mathbf{F}_2 = 0$, which requires that the acceleration $\ddot{\mathbf{r}}_c$ of its center of mass must be zero. The force equations then reduce to

$$\mathbf{F}_1 = -\mathbf{F}_2 = \left(\frac{m_1 m_2}{m_1 + m_2}\right)\ddot{\mathbf{r}} \qquad (4.6\text{--}4)$$

Eqs. 4.6–3 and 4.6–4 indicate that the two-body problem can be reduced to that of a single body with an equivalent mass $(m_1 m_2)/(m_1 + m_2)$ at a distance $\mathbf{r}$ from the center of mass which is either stationary or in uniform motion along a straight line. It should be noted that the requirement $\mathbf{F}_1 = -\mathbf{F}_2$ does not restrict the forces to be collinear, so that the force system may be a couple as well as a collinear mutual attraction.

If one of the masses is very large compared to the other, the equivalent mass $(m_1 m_2)/(m_1 + m_2)$ reduces to that of the smaller mass moving relative to the center of the larger mass. This is essentially the condition encountered when a satellite is placed into an orbit around the earth. It is, however, of interest to recognize that we have a two-body problem which can be analyzed exactly in terms of an equivalent one-body problem.

Example 4.6–1
Assume that the ratio of the mass of the moon to that of the moon plus earth is known as

$$\mu = \frac{m_2}{m_1 + m_2}$$

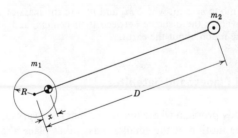

Fig. 4.6–2. Earth-moon system and their center of mass.

By observation relative to the fixed stars, the angular velocity ω of the line joining the centers of the earth and moon can be measured as $\omega = 2.66 \times 10^{-6}$ rad/sec. Show that the distance between the two bodies is

$$D^3 = \frac{gR^2}{\omega^2(1 - \mu)}$$

Referring to Fig. 4.6–2, the center of mass is given by the equation

$$m_1 x = (D - x)m_2 \qquad x = D\mu$$

Equating the force in Eq. 4.6–4 to Newton's gravitational force, we obtain

$$\frac{Gm_1 m_2}{D^2} = \frac{Km_2}{D^2} = \frac{m_1 m_2}{m_1 + m_2} D\omega^2$$

$$D^3 = \frac{K(m_1 + m_2)}{m_1 \omega^2}$$

Substituting $1 - \mu = m_1/(m_1 + m_2)$, and $K = gR^2$

$$D^3 = \frac{gR^2}{\omega^2(1 - \mu)}$$

PROBLEMS

1. If the mass of the moon is $\frac{1}{81}$ times that of the earth and its period is 27.32 days, determine the distance in miles between their centers, using $R = 3960$ miles for earth.
2. Determine the distance from the earth's center to the center of mass of the earth-moon system.
3. Determine the distance from the center of earth to the neutral point where the attractions of the earth and moon balance each other.
4. Derive the equation,

$$\frac{Gm_2}{r^3} = \frac{m_2}{m_1 + m_2}\left(\frac{2\pi}{\tau}\right)^2$$

for the earth-moon system, where m_2 and m_1 are the mass of the moon and earth respectively, r the distance between their centers, and τ the period of rotation of the moon about the earth.

4.7 Orbits of Planets and Satellites

In the two-body problem where one of the masses is very large compared to the other, the motion of the smaller mass takes place about the larger mass whose gravitational attraction is an inverse-square central force. For an artificial satellite moving around the earth as its focal center, the gravitational attraction is

$$F = -\frac{GMm}{r^2} \tag{4.7–1}$$

where M and m are the masses of the earth and satellite, G is a constant, and r is the distance of m from the center of the earth. Equation 4.7–1 also applies to the earth-sun and the moon-earth system. The constant GM

can be evaluated from a simple experiment of a falling body at the earth's surface. If the measured acceleration of the falling body at $r = R$ is g, then $F/m = -g = -GM/R^2$. We will now replace the constant $GM = gR^2$ by the letter K. The constant K can also be calculated from measured observations of earthbound satellite orbits.

Assuming the satellite to be successfully launched, its motion is governed by the following equations of force;

Radial force

$$\ddot{r} - r\dot{\theta}^2 = -\frac{K}{r^2} \qquad (4.7\text{--}2)$$

Transverse force

$$r\ddot{\theta} + 2\dot{r}\dot{\theta} = \frac{1}{r}\frac{d}{dt}r^2\dot{\theta} = 0 \qquad (4.7\text{--}3)$$

The second equation leads to the statement of conservation of moment of momentum per unit mass $r^2\dot{\theta} = h$.

Since our interest is centered about the shape of the orbit, it is advisable to eliminate the independent variable t in terms of θ as follows,

$$\dot{r} = \frac{dr}{d\theta}\dot{\theta} = \frac{h}{r^2}\frac{dr}{d\theta} = -h\frac{d}{d\theta}\frac{1}{r}$$

Thus by letting $1/r = u$, the following terms are converted to the new variables,

$$\dot{r} = -h\frac{du}{d\theta}$$

$$\ddot{r} = -h\frac{d^2u}{d\theta^2}\dot{\theta} = -h^2u^2\frac{d^2u}{d\theta^2}$$

Substituting these quantities into the radial force equation, the differential equation for the orbit becomes

$$\frac{d^2u}{d\theta^2} + u = \frac{K}{h^2} \qquad (4.7\text{--}4)$$

Equation 4.7–4, being a second-order differential equation, requires two arbitrary constants in the general solution. The general solution for this differential equation is,

$$u = \frac{K}{h^2} + C\cos(\theta - \theta_0) \qquad (4.7\text{--}5)$$

where K/h^2 is the particular integral.   The constant θ_0 can be made equal to zero by measuring θ from perigee (a point of minimum distance from the origin of r).

The evaluation of the constant C can be made from the energy equation. For a body at heights beyond the influence of the atmosphere, the system is conservative and the total energy $E = T + U$ of any orbit is a constant. In this equation it is convenient to consider the energies as those associated with a unit mass.

For the determination of the potential energy per unit mass, we choose the point at infinity for reference, and from Eq. 4.7–1 we obtain,

$$U(r) = -K\int_r^\infty \frac{dr}{r^2} = -\frac{K}{r} \tag{4.7–6}$$

Adding this to the kinetic energy per unit mass, the total energy becomes

$$E = \frac{v^2}{2} - \frac{K}{r} \tag{4.7–7}$$

In terms of u and θ,

$$v^2 = \dot{r}^2 + (r\dot\theta)^2 = h^2\left[\left(\frac{du}{d\theta}\right)^2 + u^2\right]$$

$$= h^2\left[C^2\sin^2\theta + \left(\frac{K}{h^2} + C\cos\theta\right)^2\right] \tag{4.7–8}$$

so that substituting Eqs. 4.7–8 and 4.7–5 into Eq. 4.7–7 results in

$$C^2 = \left(\frac{K}{h^2}\right)^2\left(1 + \frac{2Eh^2}{K^2}\right) \tag{4.7–9}$$

The equation for u can now be written as,

$$u = \frac{K}{h^2}(1 + e\cos\theta) \tag{4.7–10}$$

where

$$e = \sqrt{1 + \frac{2Eh^2}{K^2}} \tag{4.7–11}$$

Equations 4.7–10 and 4.7–11 apply to the general case for the motion under the inverse-square central force, and the type of orbit is established by the numerical value of e as follows:

Hyperbola   if $e > 1$
Parabola   if $e = 1$
Ellipse   if $0 < e < 1$ (perigee corresponding to $\theta = 0$)
Circle   if $e = 0$
Subcircular ellipse if $-1 < e < 0$ (apogee—point of maximum distance from the origin of r corresponding to $\theta = 0$)

4.8 Geometry of Conic Sections

Motion under central force results in an orbit which is one of the conic sections. The conic is the locus of a point whose distance from a fixed point F and a fixed line DD' have a constant ratio e. The fixed point F is called the focus, the fixed line DD' the directrix, and the ratio e the eccentricity.

Letting m be the distance from the focus to the directrix DD', the polar equation for the conic is

$$r = e(m - r \cos \theta)$$

or

$$r = \frac{em}{1 + e \cos \theta} \tag{4.8-1}$$

By letting $\theta = 0°, 90°, 180°$, and $\tan^{-1}(b/a)$, important distances are found. These are shown in Figs. 4.8-1, 4.8-2, and 4.8-3.

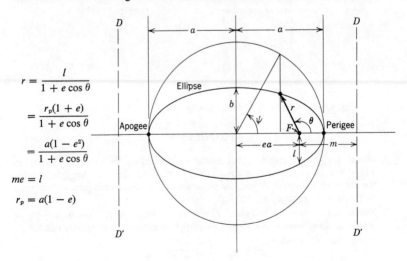

$$r = \frac{l}{1 + e \cos \theta}$$

$$= \frac{r_p(1 + e)}{1 + e \cos \theta}$$

$$= \frac{a(1 - e^2)}{1 + e \cos \theta}$$

$$me = l$$

$$r_p = a(1 - e)$$

Fig. 4.8-1. Geometry of the ellipse.

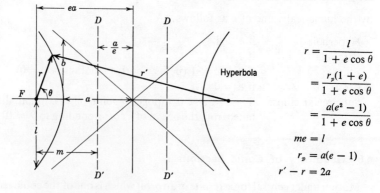

$$r = \frac{l}{1 + e \cos \theta}$$

$$= \frac{r_p(1 + e)}{1 + e \cos \theta}$$

$$= \frac{a(e^2 - 1)}{1 + e \cos \theta}$$

$$me = l$$

$$r_p = a(e - 1)$$

$$r' - r = 2a$$

Fig. 4.8–2. Geometry of the hyperbola.

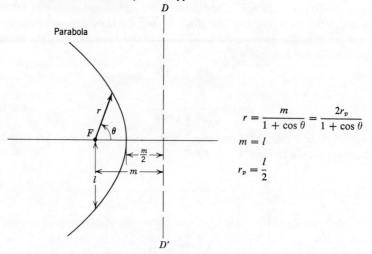

$$r = \frac{m}{1 + \cos \theta} = \frac{2r_p}{1 + \cos \theta}$$

$$m = l$$

$$r_p = \frac{l}{2}$$

Fig. 4.8–3. Geometry of the parabola.

PROBLEMS

1. If x and y are rectangular coordinates of a point on the ellipse as shown in Fig. 4.8–1, show that $x = a \cos \psi$ and $y = b \sin \psi$, where the angle ψ is called the *eccentric anomaly*. *Hint:* Use equation of ellipse in rectangular coordinates to relate y in terms of x.

2. Show that $\cos \psi = (a - r)/ae$. *Hint:* From Fig. 4.8–1, $r \cos \theta = -(ae - a \cos \psi)$. Combine with $r = [a(1 - e^2)]/(1 + e \cos \theta)$ to eliminate $\cos \theta$.

3. Prove the relationship

$$\tan \frac{\psi}{2} = \sqrt{\frac{1 - e}{1 + e}} \tan \frac{\theta}{2}$$

4.9 Orbit Established from Initial Conditions

The initial conditions at rocket burnout are:

$$r = r_0$$
$$v = v_0$$
$$\beta = \beta_0$$

where β is the heading angle, measured outward from the normal to r, as shown in Fig. 4.9–1. From this information we would like to determine

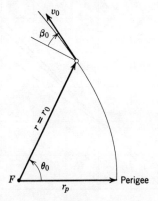

Fig. 4.9-1. Initial conditions at injection into orbit.

the value of the eccentricity e, which establishes the type of orbit, and θ_0, the angle between perigee and r_0.

We will let r_p be the perigee distance at $\theta = 0$ (when e is negative, $\theta = 0$ corresponds to apogee with distance r_a) in which case we have from Eq. 4.7–10

$$\frac{h^2}{K} = r_p(1 + e) \tag{4.9–1}$$

Equation 4.7–10 can then be written as

$$u = \frac{1 + e \cos \theta}{r_p(1 + e)} \tag{4.9–2}$$

The components of the initial velocity are,

$$v_0 \cos \beta_0 = r_0 \dot\theta_0 = \frac{h}{r_0} \tag{4.9–3}$$

$$v_0 \sin \beta_0 = \dot r_0 = -h\left(\frac{du}{d\theta}\right)_{\theta = \theta_0} = \frac{Ke}{r_0 v_0} \frac{\sin \theta_0}{\cos \beta_0} \tag{4.9–4}$$

Since from Eq. 4.7–10 we can write

$$\frac{1}{r_0} = \frac{K}{h^2}(1 + e \cos \theta_0) \tag{4.9-5}$$

substitution for h^2 from Eq. 4.9–3 results in the equation

$$\frac{r_0 v_0^2}{K} \cos^2 \beta_0 = 1 + e \cos \theta_0 \tag{4.9-6}$$

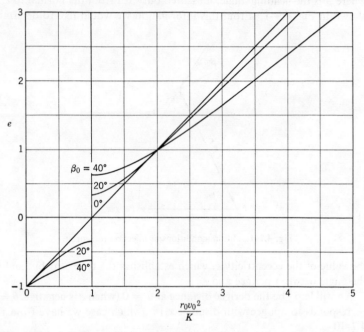

Fig. 4.9–2. Orbit eccentricity established from initial values of β and rv^2/K.

Solving for $e \sin \theta_0$ and $e \cos \theta_0$ in Eqs. 4.9–4 and 4.9–6 and dividing, the angular position from perigee is found

$$\tan \theta_0 = \frac{(r_0 v_0^2/K) \sin \beta_0 \cos \beta_0}{(r_0 v_0^2/K) \cos^2 \beta_0 - 1} \tag{4.9-7}$$

By adding the squares of $e \sin \theta_0$ and $e \cos \theta_0$, e^2 is obtained

$$e^2 = \left(\frac{r_0 v_0^2}{K} - 1\right)^2 \cos^2 \beta_0 + \sin^2 \beta_0 \tag{4.9-8}$$

Equations 4.9–7 and 4.9–8 completely establish the orbit for any initial conditions $r_0 v_0^2/K$ and β_0 arranged in nondimensional form. In addition, the total orbit energy established from Eq. 4.7–7 at burnout is

$$\frac{E r_0}{K} = \frac{1}{2} \frac{r_0 v_0^2}{K} - 1 \tag{4.9-9}$$

A plot of Eq. 4.9–8 showing e as a function of $r_0 v_0^2/K$ with β_0 as parameter is presented in Fig. 4.9–2. It is evident that, if $\beta_0 \neq 0$, e can never become zero, so that a circular orbit is not possible. Equation 4.9–7 indicates that, when $(r_0 v_0^2/K) \cos^2 \beta_0 = 1$, $\theta_0 = 90°$. For $(r_0 v_0^2/K) \cos^2 \beta_0 < 1$ and $\beta_0 > 0$, θ_0 is in the second quadrant.

PROBLEMS

1. Explorer No. 7 launched in October 1959 resulted in the following observations.
    Apogee distance above earth surface = 664 miles,
    Perigee distance above earth surface = 346 miles,
    Orbit period = 101.2 min.
    Using mean radius of earth to be 3960 miles, calculate K for earth.

2. For Prob. 1, determine the eccentricity and the perigee and apogee speeds.

3. Determine the circular orbit radius for which a satellite will remain stationary with respect to earth.

4. Explorer No. 6 launched in August 1959 is reported to have perigee and apogee heights above the earth's surface of 157 miles and 26,400 miles. Calculate the orbit period, its eccentricity, and the maximum speed.

5. If the initial conditions for a satellite at rocket burnout are β_0, r_0/R, and $r_0 v_0^2/K$, show that the perigee and apogee distances from the center of the earth are given by the equation

$$\frac{r_{p,a}}{r_0} = \frac{1}{2 - (r_0 v_0^2/K)} \left[1 \pm \sqrt{1 - \left(\frac{r_0 v_0^2}{K}\right)\left(2 - \frac{r_0 v_0^2}{K}\right) \cos^2 \beta_0} \right]$$

where $-$ corresponds to perigee and $+$ to apogee. Plot r_p/r_0 and r_a/r_0 versus $r_0 v_0^2/K \gtrless 1$ for $\beta = 1°$ and $5°$.

6. Assess the effect of the heading angle error β_0 on the perigee height when the velocity at rocket burnout is equal to the circular orbit value.

7. Plot θ_0 versus $r_0 v_0^2/K$, with β_0 as parameter. Use a range of $0 < r_0 v_0^2/K < 2$ and $\beta_0 = \frac{1}{2}°, 1°, 5°, 10°, 30°$.

4.10 Satellite Launched with $\beta_0 = 0$

The special case of a satellite launched with $\beta_0 = 0$ is instructive because of its simplicity of interpretation. From Eq. 4.9–7 it is evident that $\theta_0 = 0$, so that the launch point corresponds with perigee. Equation 4.9–8 now becomes

$$e = \frac{r_0 v_0^2}{K} - 1 \tag{4.10-1}$$

which is represented by the straight line for $\beta_0 = 0$ in Fig. 4.9–2. Equation 4.10–1 indicates that a circular orbit ($e = 0$) is obtained only when $r_0 v_0^2/K = 1$ and $\beta_0 = 0$. If v_0 or r_0 is increased so that $1 < r_0 v_0^2/K < 2$, the orbit will be an ellipse. For values of $r_0 v_0^2/K > 2$, the orbit will become a hyperbola and the satellite will escape from the earth. Thus $r_0 v_0^2/K = 2$ corresponds to the velocity of escape at height $r_0 = R + z$.

$$v_e = \sqrt{\frac{2K}{r_0}} = R\sqrt{\frac{2g}{R+z}} \tag{4.10–2}$$

Considering the geometry of the elliptic orbit, the semimajor and semiminor axes are:

$$\frac{a}{r_0} = \frac{1}{1-e} \tag{4.10–3}$$

$$\frac{b}{r_0} = \sqrt{\frac{1+e}{1-e}} \tag{4.10–4}$$

The apogee distance is:

$$\frac{r_a}{r_0} = \frac{1+e}{1-e} \tag{4.10–5}$$

and in terms of the altitude z above the earth's surface, the apogee and perigee altitudes are:

$$\frac{z_a}{R} = \frac{r_0}{R}\frac{1+e}{1-e} - 1 \tag{4.10–6}$$

$$\frac{z_p}{R} = \frac{r_0}{R} - 1 \tag{4.10–7}$$

Numerical values for small e are given in the following table to show the nearly circular shape of such elliptic orbits in spite of the large difference in the apogee and perigee heights.

Table 4.10–1. Calculations for Launching Altitude
$r_0/R = 1.10$

e	$\dfrac{1+e}{1-e}$	$\dfrac{z_a}{z_p}$	$\dfrac{a}{b}$	$\left(\dfrac{\text{Elliptic speed}}{\text{Circular speed}}\right)_{\text{at launch}}$
0.00	1.00	1.00	1.000	1.00
0.05	1.105	2.15	1.00125	1.025
0.10	1.22	3.40	1.0050	1.050
0.20	1.50	6.50	1.020	1.096

For an eccentricity of 0.20, the apogee height is 6.50 times the perigee height when the launch height is $r_0 = 1.10R$, or approximately 400 miles above the earth's surface.

We can next examine the case $r_0 v_0^2 / K < 1$. Equation 4.7–10 with negative e shows that we have an ellipse with the starting point corresponding to apogee, and perigee is at $\theta = 180°$. The speed is then not sufficient to balance the attractive force of the earth, and the satellite

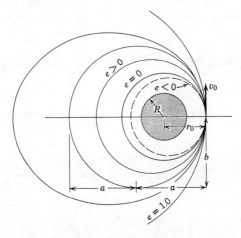

Fig. 4.10–1. Satellites launched with $\beta_0 = 0$.

distance r will diminish from its initial value r_0. With negative e, the center of the ellipse falls between the origin and the launching point. It is evident from the previous set of numbers that the satellite will fall into a region where atmospheric drag becomes important, even for small negative e. Figure 4.10–1 shows one such orbit along with orbits for positive e.

The period of closed orbits, ellipses, or circles can be found by dividing the enclosed area by the areal rate $h/2$. The area of the ellipse is πab. The semiminor axis, from Fig. 4.8–1, is $b = a \sqrt{1 - e^2}$, and h from Eq. 4.9–5 at $\theta = 0$ is $h = \sqrt{Kr_p(1 + e)} = \sqrt{Ka(1 - e^2)}$. Thus the equation for the period becomes

$$\tau = \frac{2\pi ab}{h} = \frac{2\pi}{\sqrt{K}} a^{3/2} \tag{4.10–8}$$

PROBLEMS

1. A satellite is launched parallel to the earth's surface at $r/R = 1.10$ with $rv^2/K = 1.20$. Determine the apogee distance and the ratio of the apogee to perigee heights above the earth's surface.

2. A satellite is launched parallel to the earth's surface with a velocity of 18,000 mph at a height of 400 miles. Calculate the apogee height above the earth's surface and the period.

3. Plot the escape velocity from the earth as a function of the altitude.

4. Determine and plot the orbit energy level Er_0/K of circular orbits as a function of the altitude z.

5. For Prob. 4, plot the period versus z.

6. For bodies launched with $\beta_0 = 0$ at height r_0/R, determine the equation for the apogee distance r_a as function of the velocity parameter rv^2/K at perigee.

4.11 Cotangential Transfer between Coplanar Circular Orbits

Transfer between coplanar circular orbits can be effected by an elliptic orbit with perigee and apogee distances equal to the radii of the respective circles, as shown in Fig. 4.11–1. The cotangential ellipse is known as the

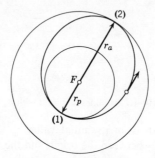

Fig. 4.11–1. Hohmann transfer orbit.

Hohmann transfer orbit, and it can be shown to be a minimum energy orbit for transfer between the coplanar circular orbits.

Assuming transfer to take place from 1 to 2, we can obtain the ratio r_a/r_p from Eq. 4.7–10. Letting $\theta = 180°$, $u = 1/r_a$ so that

$$\frac{r_a}{r_p} = \frac{1 + e}{1 - e} \qquad (4.11–1)$$

From Eq. 4.10–1, e can be eliminated in terms of $r_p v_p{}^2/K$ as

$$e = \frac{r_p v_p{}^2}{K} - 1 \qquad (4.11–2)$$

Substituting Eq. 4.11–1 into 4.11–2, we obtain

$$\frac{r_p v_p{}^2}{K} = \frac{2(r_a/r_p)}{1 + (r_a/r_p)} \tag{4.11-3}$$

which is plotted in Fig. 4.11–2.

In interpreting these results, we can assume the space vehicle to be initially orbiting around the inner circle of radius r_p. This requires $r_p v_p{}^2/K$ to be equal to 1.0. To escape from the inner circular orbit and

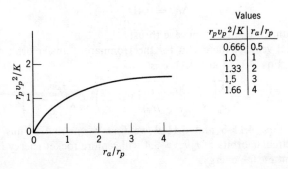

Values	
$r_p v_p{}^2/K$	r_a/r_p
0.666	0.5
1.0	1
1.33	2
1,5	3
1.66	4

Fig. 4.11–2. $r_p v_p{}^2/K$ necessary for Hohmann transfer between orbits r_a/r_p.

travel along the elliptic transfer orbit of ratio r_a/r_p, $r_p v_p{}^2/K$ must be increased to a value given by Eq. 4.11–3. This may be accomplished by firing a rocket in the tangential direction, the required increase in velocity being

$$\Delta v_p = \sqrt{\frac{K}{r_p}} \left[\sqrt{\frac{2(r_a/r_p)}{1 + (r_a/r_p)}} - 1 \right] \tag{4.11-4}$$

On reaching point 2, the apogee velocity, which can be found by equating the angular momentum at 1 and 2, i.e., $r_p v_p = r_a v_a$, becomes

$$\frac{r_a v_a{}^2}{K} = \frac{r_p}{r_a} \frac{r_p v_p{}^2}{K} = \frac{2}{1 + r_a/r_p} \tag{4.11-5}$$

Since the circular orbit velocity for radius r_a is $r_a v_a{}^2/K = 1$, and the apogee velocity as given by Eq. 4.11–5 is less than 1, another thrust in the forward direction is necessary. The increment in velocity required at point 2 to go into the circular orbit is then

$$\Delta v_a = \sqrt{\frac{K}{r_a}} \left(1 - \sqrt{\frac{2}{1 + (r_a/r_p)}} \right) \tag{4.11-6}$$

Thus the total impulse which must be applied in the direction of motion is determined by $\Delta v_p + \Delta v_a$, and the fuel energy corresponding to it is proportional to $(\Delta v_p + \Delta v_a)^2$.

It is of interest to compare the total velocity increment to transfer from orbit 1 to orbit 2 with that of the velocity increment for escape from orbit 1. The parabolic orbit velocity of escape from radius r_p is found from $r_p v_p{}^2/K = 2$ to be

$$v_{pe} = 1.414\sqrt{\frac{K}{r_p}} \qquad (4.11\text{--}7)$$

which requires a velocity increment of

$$\Delta v_p = 0.414\sqrt{\frac{K}{r_p}} \qquad (4.11\text{--}8)$$

acquired under a single impulsive thrust.

The total velocity increment for the Hohmann transfer orbit obtained by adding Eqs. 4.11–4 and 4.11–6 is

$$\Delta v_p = \sqrt{\frac{K}{r_p}}\left[\sqrt{\frac{2(r_a/r_p)}{1 + (r_a/r_p)}}\left(1 - \frac{r_p}{r_a}\right) + \sqrt{\frac{r_p}{r_a}} - 1\right] \qquad (4.11\text{--}9)$$

Equating Eqs. 4.11–8 and 4.11–9 we find $r_a/r_p = 3.4$. Thus transfer between circular orbits of $r_a/r_p > 3.4$ will require rocket energy in excess of the orbit escape energy.

Heliocentric orbits

In considering planetary orbits, the large mass of the sun (99.2% of the total mass of the solar system) enables one to ignore all other forces. Although planetary orbits are ellipses with their orbit planes inclined slightly from the ecliptic (earth's orbit plane), great simplification results from assuming the orbit to be circular and coplanar.

Assuming coplanar circular orbits, the equations for the Hohmann transfer orbit are applicable with the numerical values of K corresponding to the sun. K for the sun can be found from measured data pertaining to any planet. Assuming a circular orbit of radius r for the earth, we have

$$\frac{rv^2}{K} = 1 \qquad (4.11\text{--}10)$$

where $r = 490.5 \times 10^9$ ft $= 93 \times 10^6$ miles

       $K = GM_s$ for the sun

       $v = 2\pi r/\tau =$ velocity of earth

       $\tau = 365.25 \times 86{,}400$ sec $=$ period of the earth around the sun

Substituting these figures into Eq. 4.11–10, K for the heliocentric system is found to be 4.68×10^{21} ft³/sec².

Another convenient set of units for planetary and interplanetary orbits is one referenced to the earth's orbit, with $r = 1$ astronomical unit and

$\tau = 1$ year. Substituting unity for these quantities in the period equation $\tau = 2\pi\sqrt{(r^3/K)}$, the heliocentric constant K becomes equal to $4\pi^2$ astronomical units cubed per year squared.

PROBLEMS

1. Discuss how a space vehicle traveling around a circular orbit of radius r_2 can transfer to a coplanar circular orbit of radius r_1, where $r_1 < r_2$.

2. Show that, if $r_1/r_2 < 0.50$, the velocity increment necessary for the transfer to the inner orbit will exceed that of escape from the outer orbit.

3. Determine the time of flight for the Hohmann transfer orbit.

4. Determine the equation for the velocity v/v_c versus distance r/r_p for the Hohmann transfer orbit, where departure is from the inner orbit of radius r_p and circular orbit speed v_c.

5. A rocket traveling in a circular orbit $r_1 v_1^2/K = 1$ is given an impulsive thrust normal to the orbit so that the resultant velocity vector makes an angle β_0 outward from the trangent to the departing circular orbit. Determine the new orbit, specifying the perigee and apogee distances and the eccentricity. Determine θ_0 to perigee.

6. For the maneuver of Prob. 5, determine the areal rate and show that the area enclosed from the point of maneuver to apogee is given by the equation

$$A = \frac{r_p^2}{2(1-e)}\sqrt{\frac{1+e}{1-e}}\left(e\sqrt{1-e^2} + \sin^{-1}e + \frac{\pi}{2}\right)$$

7. A rocket traveling at 18,300 mph at perigee, fires a retrorocket at perigee height of 300 miles. What velocity change is necessary to reach minimum altitude of 100 miles during the first circuit?

8. The following table gives the distances of some of the planets from the sun.

Planet	Mean Distance from Sun
Mercury	0.39
Venus	0.72
Earth	$1.0 = 93 \times 10^6$ miles
Mars	1.52
Jupiter	5.2

Assuming the two orbits to be in the same plane, determine the Hohmann transfer orbit from earth to Mars and compute the time required for transit. Determine the position of Mars in its orbit relative to earth for interception to take place.

9. For the transfer orbit of Prob. 1, determine the velocity increments necessary on departure and on arrival.

10. Determine the spherical region around earth where the earth's gravitational attraction dominates over that of the sun.

11. Determine the equation for the escape velocity from the solar system. What is its value at the earth's orbit.

4.12 Transfer Between Coplanar Coaxial Elliptic Orbits

Figure 4.12–1 shows two coaxial elliptic orbits in the same plane. To transfer from the inner orbit 1 to the outer orbit 2, it can be shown that for minimum expenditure of energy, the thrust should be impulsive at perigee of the inner orbit and apogee of the outer orbit.

We will assume that the orbit parameters e and a of the two orbits are

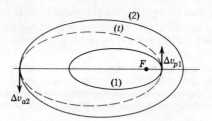

Fig. 4.12–1. Transfer between coplanar, coaxial, elliptic orbits.

given. The perigee and apogee distances are then known from the relationships $r_p = a(1 - e)$ and $r_a = a(1 + e)$.

Before impulse, the velocity at p_1 can be obtained from Eq. 4.9–6 by letting $\beta = 0$ and $\theta = 0$,

$$v_{p1} = \sqrt{\frac{K}{r_{p1}}(1 + e_1)} \tag{4.12–1}$$

For the transfer orbit, the necessary velocity at perigee can be found from Eq. 4.11–3 to be

$$v_{pt} = \sqrt{\frac{K}{r_{p1}}\left[\frac{2(r_{a2}/r_{p1})}{1 + (r_{a2}/r_{p1})}\right]} \tag{4.12–2}$$

The increment in velocity required at perigee of the inner orbit is then

$$\Delta v_{p1} = \sqrt{\frac{K}{r_{p1}}}\left[\sqrt{\frac{2(r_{a2}/r_{p1})}{1 + (r_{a2}/r_{p1})}} - \sqrt{1 + e_1}\right] \tag{4.12–3}$$

After departure, the vehicle proceeds along the transfer orbit until it reaches apogee. The velocity as it approaches apogee is

$$v_{at} = \frac{r_{p1}}{r_{a2}}v_{pt} = \sqrt{\frac{K}{r_{a2}}\frac{2}{1 + (r_{a2}/r_{p1})}} \tag{4.12–4}$$

The apogee velocity for orbit 2 can be found from Eq. 4.9–6, letting $\beta = 0$ and $\theta = 180°$,

$$v_{a2} = \sqrt{\frac{K}{r_{a2}}(1 - e_2)} \qquad (4.12\text{–}5)$$

The increment in velocity necessary to transfer from orbit (t) to orbit (2) at apogee is then

$$\Delta v_{a2} = \sqrt{\frac{K}{r_{a2}}}\left[\sqrt{1 - e_2} - \sqrt{\frac{2}{1 + (r_{a2}/r_{p1})}}\right] \qquad (4.12\text{–}6)$$

and the total increment in velocity in the tangential direction is

$$\Delta v_{p1} + \Delta v_{a2}$$

4.13   Orbital Change Due to Impulsive Thrust

In this section we will consider the general problem of changing an existing orbit to another of a given specification. Such changes may range from small corrections to an existing orbit, to large changes in the orbit for maneuvers. It will be assumed that the change will take place under impulsive thrust; e.g., a change in the direction and magnitude of the velocity vector takes place under negligible change in the displacement vector. This idealization is generally acceptable when the distance traveled during thrust is negligible in comparison to the radius vector.

In general, our concern is with elliptic and hyperbolic orbits, the circle and the parabola being special limiting cases. The relationship between the velocity v, the angular position θ, the heading angle β, and the eccentricity e, shown in Fig. 4.13–1, is already available from Eqs. 4.9–7 and 4.9–8, which are rewritten as follows:

$$\tan \theta = \frac{(rv^2/K)\sin \beta \cos \beta}{(rv^2/K)\cos^2 \beta - 1} \qquad (4.13\text{–}1)$$

$$e^2 = \left[\frac{rv^2}{K} - 1\right]^2 \cos^2 \beta + \sin^2 \beta \qquad (4.13\text{–}2)$$

By holding θ constant and varying β, rv^2/K can be computed from Eq. 4.13–1 rearranged as follows:

$$\frac{rv^2}{K} = \frac{1}{\cos^2 \beta - (\sin \beta \cos \beta)/(\tan \theta)} \qquad (4.13\text{–}3)$$

By holding e constant and varying β, the curve for rv^2/K versus β can be computed from Eq. 4.13-2 rearranged as Eq. 4.13-4:

$$\frac{rv^2}{K} = 1 \pm \sqrt{1 - \left(\frac{1 - e^2}{\cos^2 \beta}\right)} \qquad (4.13\text{-}4)$$

These results for the ellipse and the hyperbola are plotted as shown in Figs. 4.13-2 and 4.13-3.

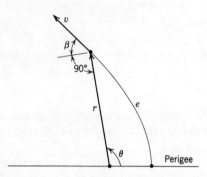

Fig. 4.13-1.   Displacement, velocity and heading angle β at any position θ.

In addition to these two equations we have the energy relationship of Eq. 4.7-7,

$$\frac{Er}{K} = \frac{1}{2}\frac{rv^2}{K} - 1 \qquad (4.13\text{-}5)$$

Since E is constant for a given orbit, we can evaluate it at perigee. Letting $\theta = \beta = 0$ in Eqs. 4.13-1 and 4.13-2,

$$e = \frac{r_p v_p{}^2}{K} - 1 \qquad (4.13\text{-}6)$$

which substituted into Eq. 4.13-5 with $r = r_p$ results in

$$\frac{2E}{K} = -\frac{1 - e}{r_p} \qquad (4.13\text{-}7)$$

Since $r_p = a(1 - e)$ for the ellipse and $r_p = a(e - 1)$ for the hyperbola (see Sec. 4.8), the energy E can be expressed in terms of a as follows.

$$\frac{2E}{K} = \begin{cases} -\dfrac{1}{a} & \text{for elliptic orbit} \\[2ex] +\dfrac{1}{a} & \text{for hyperbolic orbit} \end{cases} \qquad (4.13\text{-}8)$$

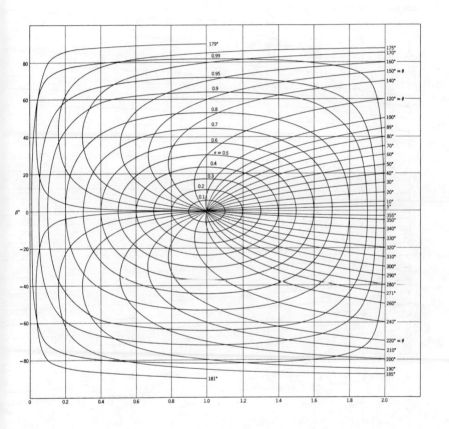

Fig. 4.13–2. Relation between β and rv^2/K with e and θ as parameters for elliptic orbits.

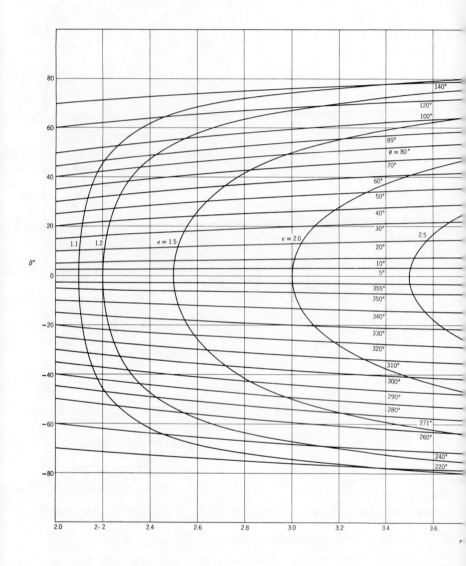

Fig. 4.13-3. Relation between β and rv^2/K with e and θ as parameters for hyperbolic orbits.

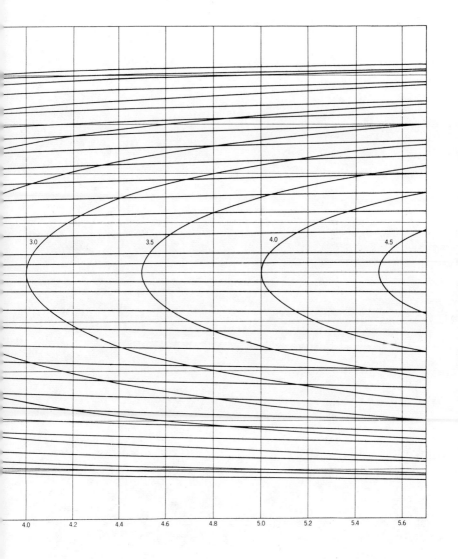

Substituting these values into Eq. 4.13–5, the energy equation can be written as

$$\frac{a}{r} = \frac{1}{2 - (rv^2/K)} \qquad \text{for elliptic orbit} \qquad (4.13\text{–}9)$$

$$\frac{a}{r} = \frac{1}{(rv^2/K) - 2} \qquad \text{for hyperbolic orbit} \qquad (4.13\text{–}10)$$

Finally, we need an equation from which the time elapsed during travel along an orbit can be computed. For this we examine the equation for the moment of momentum,

$$r^2\dot{\theta} = h = \sqrt{Kr_p(1 + e)} \qquad (4.13\text{–}11)$$

and rearrange it as follows:

$$\frac{d\theta}{(1 + e \cos \theta)^2} = \frac{\sqrt{Kr_p(1 + e)}}{r_p^2(1 + e)^2} dt$$

For $e < 1$, the integral of the left side is (see Peirce, Short Table of Integrals,[1] no. 308 and no. 300),

$$\int_0^\theta \frac{d\theta}{(1 + e \cos \theta)^2} = \frac{1}{1 - e^2}\left(\frac{-e \sin \theta}{1 + e \cos \theta} + \int_0^\theta \frac{d\theta}{1 + e \cos \theta}\right)$$

$$= \frac{1}{1 - e^2}\left[\frac{-e \sin \theta}{1 + e \cos \theta} + \frac{2}{\sqrt{1 - e^2}} \tan^{-1}\left(\frac{\sqrt{1 - e^2}}{1 + e} \tan \tfrac{1}{2}\theta\right)\right]$$

For $e > 1$,

$$\int_0^\theta \frac{d\theta}{(1 + e \cos \theta)^2} = \frac{1}{e^2 - 1}\left[\frac{e \sin \theta}{(1 + e \cos \theta)}\right.$$

$$\left. - \frac{1}{\sqrt{e^2 - 1}} \ln\left(\frac{\sqrt{e + 1} + \sqrt{e - 1} \tan \tfrac{1}{2}\theta}{\sqrt{e + 1} - \sqrt{e - 1} \tan \tfrac{1}{2}\theta}\right)\right]$$

Replacing r_p in terms of a and e as before, the equation for the time becomes:

For elliptic orbits ($e < 1$)

$$t_e = \frac{a^{3/2}}{\sqrt{K}}\left[2 \tan^{-1}\left(\sqrt{\frac{1 - e}{1 + e}} \tan \tfrac{1}{2}\theta\right) - \frac{e\sqrt{1 - e^2} \sin \theta}{1 + e \cos \theta}\right] \qquad (4.13\text{–}12)$$

For hyperbolic orbits ($e > 1$)

$$t_h = \frac{a^{3/2}}{\sqrt{K}}\left[\frac{e\sqrt{e^2 - 1} \sin \theta}{1 + e \cos \theta} - \ln\left(\frac{\sqrt{e + 1} + \sqrt{e - 1} \tan \tfrac{1}{2}\theta}{\sqrt{e + 1} - \sqrt{e - 1} \tan \tfrac{1}{2}\theta}\right)\right] \qquad (4.13\text{–}13)$$

[1] Third revised edition, Ginn & Co. 1929.

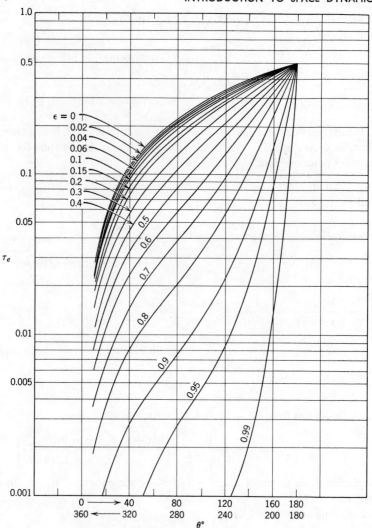

Fig. 4.13–4. Dimensionless time for elliptic orbits.

These equations in nondimensional form, $\tau_e = (t_e \sqrt{K})/(2\pi a^{3/2})$ and $\tau_h = (t_h \sqrt{K})/a^{3/2}$, have been computed and plotted by Augenstein[1] and are reproduced here as Figs. 4.13–4 and 4.13–5.

A somewhat simpler expression for the time along the elliptic orbit is available in terms of the eccentric anomaly ψ. For its derivation we need the following relationships:

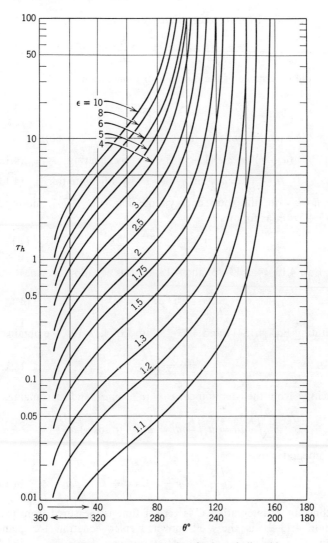

Fig. 4.13-5. Dimensionless time for hyperbolic orbits.

From the equation of the ellipse

$$\frac{1}{r} = \frac{K}{h^2}(1 + e \cos \theta) = \frac{1 + e \cos \theta}{a(1 - e^2)}$$

we obtain

$$h^2 = Ka(1 - e^2) \tag{4.13-14}$$

From Eq. 4.7–11 for the eccentricity, we have,

$$2E = -(1 - e^2)\frac{K^2}{h^2} \qquad (4.13\text{–}15)$$

From the invariance of the moment of momentum,

$$r^2\dot{\theta}^2 = \frac{h^2}{r^2} \qquad (4.13\text{–}16)$$

From the equation $\cos \psi = (a - r)/ae$ (see Prob. 2, p. 60),

$$(a - r)^2 = a^2e^2(1 - \sin^2 \psi) \qquad \text{by squaring} \qquad (4.13\text{–}17)$$

$$\dot{r} = ae\dot{\psi} \sin \psi \qquad \text{by differentiating} \qquad (4.13\text{–}18)$$

We now write the total energy equation, Eq. 4.7–7, noting that $v^2 = \dot{r}^2 + (r\dot{\theta})^2$, as follows:

$$\dot{r}^2 + (r\dot{\theta})^2 - \frac{2K}{r} = 2E \qquad (4.13\text{–}19a)$$

Using Eqs. 4.13–14, 4.13–15, and 4.13–16, this equation becomes

$$\frac{r^2\dot{r}^2}{K/a} = a^2e^2 - (a - r)^2 \qquad (4.13\text{–}19b)$$

Substituting Eqs. 4.13–17 and 4.13–18 into Eq. 4.13–19b, we obtain

$$r\dot{\psi} = \sqrt{\frac{K}{a}} \qquad (4.13\text{–}20)$$

Replacing r from the equation $\cos \psi = (a - r)/ae$, and rearranging,

$$\sqrt{\frac{K}{a}}\, dt = a\sqrt{\frac{K}{a^3}}\, dt = a(1 - e \cos \psi)\, d\psi$$

which integrates to

$$\sqrt{\frac{K}{a^3}}\, t = \psi - e \sin \psi + C \qquad (4.13\text{–}21)$$

The constant of integration C is zero if time is measured from perigee. Equation 4.13–21 is the well-known Kepler equation for planetary motion.

Example 4.13–I

A satellite is launched with the following initial conditions:

$$\frac{r_0 v_0^2}{K} = 1.40 \qquad \beta_0 = 20° \qquad \frac{r_0}{R} = 2.0$$

Determine the orbit parameters e and a/R, and establish the initial position with respect to perigee.

From Eqs. 4.13–2 and 4.13–1

$$e = \sqrt{(0.4)^2(0.939)^2 + (0.342)^2} = 0.508$$

$$\theta = \tan^{-1} \frac{(1.4)(0.939)(0.342)}{(1.4)(0.939)^2 - 1} = 62°23'$$

These values agree with those of the graph of Fig. 4.13–2.
From Eq. 4.13–9

$$\frac{a}{r_0} = \frac{1}{2 - 1.4} = 1.67 = \frac{a}{R}\frac{R}{r_0}$$

$$\frac{a}{R} = (1.67)(2.0) = 3.34$$

Example 4.13–2

The satellite orbit of Example 4.13–1 was characterized by $e = 0.508$ and $a/R = 3.34$, and its launch point was $r_0/R = 2.0$, $\theta = 62°23'$. If the satellite continues along this orbit to $\theta = 150°$, at which time the orbit is to be increased to a value $a/R = 3.60$ without rotating the apse line, determine the required increment in the velocity and its direction.

We first determine the value of rv^2/K and β before impulse for $\theta = 150°$ and $e = 0.508$. Using subscripts 1 and 2 for before and after impulse, we find from Fig. 4.13–2

$$\frac{r_1 v_1^2}{K} = 0.68 \qquad v_1 = 0.823\sqrt{\frac{K}{r_1}} \qquad \beta_1 = 24°$$

From Eq. 4.13–9 we have

$$\frac{a}{r_1} = \frac{1}{2 - 0.68} = 0.757 = \frac{a}{R}\frac{R}{r_1} = 3.34\frac{R}{r_1}$$

Therefore

$$\frac{r_1}{R} = \frac{3.34}{0.757} = 4.41$$

To maintain no rotation of the apse line, the new values of $r_2 v_2^2/K$ and β_2 after impulse must lie along the $\theta = 150°$ line in Fig. 4.13–2. (Note that $r_2 = r_1$ for the instantaneous impulse.) The value of a/R after impulse is specified as 3.60, so from Eq. 4.13–9 we have

$$\frac{a_2}{r_1} = \frac{a_2}{R}\frac{R}{r_1} = \frac{3.60}{4.41} = \frac{1}{2 - (r_1 v_2^2/K)}$$

Therefore

$$\frac{r_1 v_2^2}{K} = 0.780 \qquad v_2 = 0.882\sqrt{\frac{K}{r_1}}$$

The new eccentricity and heading angle corresponding to $r_1 v_2^2/K = 0.78$ and $\theta = 150°$ is, from Fig. 4.13–2,

$$e_2 = 0.30 \qquad \beta_2 = 11°$$

(Note $e_2 = 0.77$ and $\beta_2 = 49°$ is also a solution but one which requires a larger velocity increment.) Figure 4.13–6 shows a rough sketch of the two orbits.

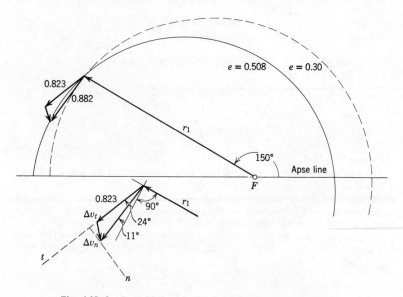

Fig. 4.13–6. Impulsive orbit change without changing apse line.

From the vector diagram of the velocities, the tangential and normal components of the required velocity increments are

$$\Delta v_t = (0.882 \cos 13° - 0.823)\sqrt{\frac{K}{r_1}} = 0.036\sqrt{\frac{K}{r_1}}$$

$$\Delta v_n = (0.882 \sin 13°)\sqrt{\frac{K}{r_1}} = 0.198\sqrt{\frac{K}{r_1}}$$

and the total velocity increment is

$$\Delta v = \sqrt{0.036^2 + 0.198^2}\sqrt{\frac{K}{r_1}} = 0.202\sqrt{\frac{K}{r_1}}$$

or 0.202 times the circular velocity at a radial distance r_1.

PROBLEMS

1. If in Example 4.13–2 the impulse of $\Delta v = 0.202\sqrt{K/r_1}$ is applied in the tangential direction, determine the new orbit parameters e and a/R and the rotation of the apse line.

2. In Example 4.13–2 determine the time required for the satellite to reach $\theta = 150°$ from the initial position of $\theta = 62°23'$.

3. In Prob. 2 determine the eccentric anomaly ψ corresponding to the two angles and check the time from Kepler's equation.

4. A satellite is launched at a height of 400 miles with $rv^2/K = 1.50$ and $\beta_0 = 10°$. Determine the eccentricity e, the orbit parameter a/R, and the position θ_0.

5. If in Prob. 4 the satellite is given an increment in velocity of $\Delta v = 2000$ ft/sec at apogee, determine the new orbit, e, a/R, and θ.

6. A satellite is placed into an orbit of $e = 0.60$ at perigee of height $r/R = 1.2$ with $v^2 = 1.6v_c^2$, where v_c is the circular velocity at this height. Determine a/R of the orbit and r/R at $\theta = 100°$.

7. If the satellite of Prob. 6 is to reduce the size of the orbit to $a/R = 2.23$ without rotating the line of apse, by an increment of velocity at the position $\theta = 100°$, determine the new eccentricity and the components of the velocity increment along the tangent and normal to it.

4.14 Perturbation of Orbital Parameters

The motion of a space vehicle moving along a specified orbit is completely defined by the following three equations:

$$\tan \theta = \frac{(rv^2/K) \sin \beta \cos \beta}{(rv^2/K) \cos^2 \beta - 1} \tag{4.14-1}$$

$$e^2 = \left(\frac{rv^2}{K} - 1\right)^2 \cos^2 \beta + \sin^2 \beta \tag{4.14-2}$$

$$\frac{a}{r} = \frac{\pm 1}{2 - \dfrac{rv^2}{K}} \qquad \begin{cases} + = \text{ellipse} \\ - = \text{hyperbola} \end{cases} \tag{4.14-3}$$

If at a specified position in the orbit a small impulsive thrust is imparted, in what way will the orbit parameters be affected? To answer this question we can examine each of the above equations separately.

Equation 4.14–1 indicates that the angular position of the apse line is a function of rv^2/K and β, so that

$$\theta = f\left(\frac{rv^2}{K}, \beta\right)$$

Differentiating,

$$d\theta = \frac{\partial f}{\partial \dfrac{rv^2}{K}} d\frac{rv^2}{K} + \frac{\partial f}{\partial \beta} d\beta \tag{4.14-4}$$

The first term of this equation represents a variation of θ due to a variation in the velocity (i.e., r is not changed during impulsive thrust), holding β constant. This is equivalent to moving the point in Fig. 4.13–2 along the horizontal line. Figure 4.13–2 shows that, if rv^2/K is increased along a horizontal line, then θ decreases, and vice versa. With larger velocities, the semimajor axis a will also increase according to Eq. 4.14–3. Figure 4.14–1

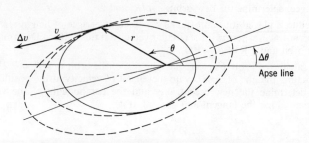

Fig. 4.14–1.   Orbit variation by tangential thrust.

illustrates how these orbits change with increasing velocity in the tangential direction. All orbits will be tangent to the velocity line.

To evaluate quantitatively the rotation of the apse line due to an increment in the tangential velocity, we can differentiate Eq. 4.14–1, holding β constant.

$$d\theta = \frac{-\sin \beta \cos \beta \cos^2 \theta \; d(rv^2/K)}{[(rv^2/K) \cos^2 \beta - 1]^2} \qquad (4.14\text{--}5)$$

To reduce this equation further, we first replace the denominator from Eq. 4.14–1

$$d\theta = \frac{-\sin^2 \theta \; d(rv^2/K)}{(rv^2/K)^2 \sin \beta \cos \beta}$$

then eliminate $\sin \beta \cos \beta = (K/rv^2)e \sin \theta$ (see Eq. 4.9–4) to obtain

$$d\theta = \frac{-\sin \theta}{e} \frac{d(rv^2/K)}{rv^2/K} = -\frac{2 \sin \theta}{e} \frac{dv}{v} \qquad (4.14\text{--}6)$$

If v is eliminated by Eq. 4.14–3, Eq. 4.14–6 can also be written as

$$d\theta = \frac{-2 \sin \theta}{e} \sqrt{\frac{r}{K[2 - (r/a)]}} \; dv \qquad (4.14\text{--}7)$$

If next the perturbation of the apse line is desired due to a small change in β, while holding the magnitude of the velocity constant, the change in θ can be found from Fig. 4.13–2 by moving the point along the vertical line. Such a change corresponds to the second term of Eq. 4.14–4, and the required increment in the velocity vector is $dv = v\, d\beta$.

The perturbation in the eccentricity e due to a small increment in the tangential velocity is again available from Fig. 4.13–2 by moving the point along the horizontal ($\beta = $ constant) line. It can be determined analytically by differentiating Eq. 4.14–2, holding β constant. The result is

$$de = \frac{2}{e}(1 - e^2)\left(\frac{a}{r} - 1\right)\frac{dv}{v} \qquad (4.14\text{–}8)$$

When the thrust is continuous over a finite length of time, it can be visualized as a series of small impulses, and the orbit change can be obtained by a succession of small changes.

PROBLEMS

1. Holding β constant, integrate the second form of Eq. 4.14–5 and compare the rotation of the apse line in Prob. 4.13–1 with this equation.

2. Show that Eq. 4.14–8 can be expressed in the form $\dfrac{de}{e} = 2\left(\dfrac{e + \cos\theta}{e}\right)\dfrac{dv}{v}$ which indicates that $\dfrac{de}{e} = 0$ for $\cos\theta = -e$. Verify points on Fig. 4.13–2 for which this is true.

3. According to Fig. 4.13–2, for θ greater than a certain value there is a value of rv^2/K, e, and β at which $d\theta/de$ and dv/de are zero. Determine the locus of rv^2/K, e, and β for such values.

4.15 Stability of Small Oscillations about a Circular Orbit

In a central force system, the circular orbit is always possible at a proper speed when the centrifugal force is balanced by the attractive force.

$$-r_0\dot\theta^2 = F(r_0) \qquad (4.15\text{–}1)$$

To determine the stability of such an orbit to a small radial disturbance r_1, we start with the general equation for the radial force

$$\ddot r - r\dot\theta^2 = F(r) \qquad (4.15\text{–}2)$$

and eliminate $\dot\theta$ from the condition that the moment of momentum $r^2\dot\theta = h$ must be a constant.

$$\ddot r - \frac{h^2}{r^3} = F(r) \qquad (4.15\text{–}3)$$

We now let $r = r_0 + r_1 = r_0[1 + (r_1/r_0)]$, so that $\ddot r = \ddot r_1$ and

$$\frac{h^2}{r^3} = \frac{h^2}{r_0^3}\left(1 + \frac{r_1}{r_0}\right)^{-3} = \frac{h^2}{r_0^3}\left(1 - \frac{3r_1}{r_0} + \frac{6r_1^2}{r_0^2}\cdots\right)$$

Also expand $F(r)$ about r_0 by the Taylor series

$$F(r) = F(r_0) + r_1 F'(r_0) + \tfrac{1}{2} r_1^2 F''(r_0) + \cdots$$

Substituting these expansions, ignoring higher-order terms, and noting that $-h^2/r_0^3 = F(r_0)$, we arrive at the differential equation for small oscillations about r_0.

$$\ddot{r}_1 - \left[\frac{3}{r_0} F(r_0) + F'(r_0) \right] r_1 = 0 \qquad (4.15\text{--}4)$$

This is a well known second order differential equation for harmonic oscillation provided $-\left[\dfrac{3}{r_0} F(r_0) + F'(r_0) \right]$ is a positive number; i.e., for stable oscillations we must have

$$\frac{3}{r_0} F(r_0) + F'(r_0) < 0 \qquad (4.15\text{--}5)$$

If $\dfrac{3}{r_0} F(r_0) + F'(r_0) > 0$, then the solution is an exponentially increasing function of time and the system is unstable.

Example 4.15–1

Determine the differential equation for small oscillations about a circular orbit when the attractive force is $-K/r^2$.

We have $F(r) = -K/r^2$. Differentiating,

$$F'(r) = \frac{2K}{r^3}$$

The differential equation of small oscillations is then

$$\ddot{r}_1 + \frac{K}{r_0^3} r_1 = 0$$

and the solution for an initial disturbance of $r_1(0)$ with $\dot{r}_1(0) = 0$ is

$$r_1(t) = r_1(0) \cos \sqrt{\frac{K}{r_0^3}} \, t$$

PROBLEMS

1. For a central force $-K/r^n$, show that a stable circular orbit is possible only for $n < 3$.

2. A body is moving in a circular orbit of radius r_0 under a central force $-K/r^2$. If the body is given a disturbance $r_1(0)$, show that the angular speed becomes

$$\dot{\theta} = \frac{h}{r_0^2} \left[1 - \frac{2r_1(0)}{r_0} \cos \sqrt{\frac{K}{r_0^3}} \, t \right]$$

4.16 Interception and Rendezvous

Problem 1. (*Circular orbits*)

We will consider first the problem of two vehicles moving in the same circular orbit r/R, one leading the other by a specified angle ϕ_{12} as shown in Fig. 4.16–1. We will let 1 and 2 be the pursuer and the pursued

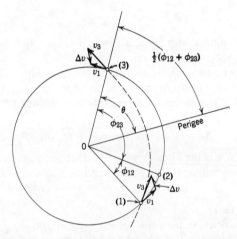

Fig. 4.16–1. Intercepting and rendezvous on circular orbit.

respectively. Since the orbit is circular, $rv_1^2/K = 1$, $\beta = 0$ for both vehicles, and ϕ_{12} remains unchanged until altered by thrust.

We wish now for 1 to overtake 2 at some position 3, indicated by angle ϕ_{23}, and to rendezvous with it along the circular orbit. What impulsive velocity increments are necessary at 1 and 3?

The problem is solved in the following manner. First the time required for 2 to travel to 3 is determined as

$$t_{23} = \frac{2\pi r^{3/2}}{\sqrt{K}} \frac{\phi_{23}}{360°} \qquad (4.16-1)$$

Vehicle 1 must travel to 3 on a new orbit which will require the same time. Due to equal radial distances 01 and 03, the perigee for the new orbit must bisect the angle $\phi_{12} + \phi_{23}$. Thus θ measured from perigee to 3 is $\frac{1}{2}(\phi_{12} + \phi_{23})$, as shown in Fig. 4.16–1.

We must now choose a value of e for the new orbit and, together with θ, determine $n = a/R$ for the time equation. If $e > 1$, we use the hyperbolic formula.

$$n = \frac{a}{R} = \left(\frac{r}{R}\right)\left(\frac{a}{r}\right) = \frac{r}{R}\left(\frac{1 + e \cos \theta}{e^2 - 1}\right) \tag{4.16-2}$$

From Fig. 4.13–5 we find $\tau_h = t_h(\sqrt{K/a^3})$ and compute the time for vehicle 1 to travel from $\theta = 0$ to point 3.

$$t_h = \tau_h\sqrt{\frac{a^3}{K}} = \tau_h\sqrt{n^3}\sqrt{\frac{R^3}{K}} = 806\tau_h\sqrt{n^3} \tag{4.16-3}*$$

If this value disagrees with $\tfrac{1}{2}t_{23}$, a new e is chosen and the procedure is repeated until agreement is found.

With e and θ known, rv^2/K and β are found from Fig. 4.13–3. Since β is zero for the circular orbit, the new β is the angle between the two velocity vectors at 3, and the increment in velocity is determined from the vector triangle as

$$\Delta v = \sqrt{(v_3 \cos \beta - v_1)^2 + (v_3 \sin \beta)^2} \tag{4.16-4}$$

where $v_1 = \sqrt{K/r}$ is the circular velocity. Due to symmetry the same Δv is applied at 1 to initiate the maneuver, and at 3 to rendezvous, as shown in Fig. 4.16–1.

Example 4.16–1

Given two vehicles on the same circular orbit of $r/R = 3.0$, with vehicle 1 lagging vehicle 2 by $80°$. It is desired for 1 to intercept and rendezvous with 2 at a position 3 which is $40°$ ahead of 2. Determine the transfer orbit and the required increments of velocity.

We have $\phi_{12} = 80°$, and $\phi_{23} = 40°$, so that perigee for the transfer orbit is $\theta = 60°$, bisecting angle 103. The time for 2 to travel to 3 is

$$t_{23} = \frac{40(2\pi)}{360}\sqrt{\frac{R^3}{K}}(3)^{3/2}$$

$$= (0.698)(806)(5.20) = 2930 \text{ sec}$$

and the half time is 1465 sec.

As an initial guess, we choose $e = 2.0$, and from Eq. 4.16–2 we find a/R.

$$n = \frac{a}{R} = 3.0\left(\frac{1 + 2 \cos 60°}{4 - 1}\right) = 2.0$$

From Fig. 4.13–5, $\tau_h = 0.80$ for $\theta = 60°$ and $e = 2.0$. The half time of flight from 1 to 3 is then

$$t_h = 0.80(2.0)^{3/2}(806) = 1825 \text{ sec.}$$

$$*\sqrt{\frac{R^3}{K}} = \sqrt{\frac{[(3960)5280]^3}{(1.407)10^{16}}} = 806 \text{ sec.}$$

Since this is larger than 1465, the orbit is too slow and we seek a faster one by choosing a larger e. A few trials result in

$$e = 3.0 \qquad \frac{a}{R} = 0.938 \qquad \frac{rv^2}{K} = 5.2$$

$$\tau_h = 2.0 \qquad t_h = 1465 \text{ sec.} \qquad \beta = 46°$$

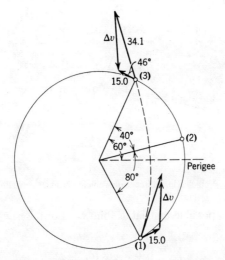

Fig. 4.16–2. Vehicle (1) intercepting vehicle (2) at (3).

The circular and hyperbolic velocities at 3 are,

$$v_1 = \sqrt{\frac{K}{r}} = \sqrt{\frac{K}{3R}} = \frac{1}{1.73}\sqrt{\frac{K}{R}} = \frac{25,930}{1.73} = 15,000 \text{ ft/sec*}$$

$$v_3 = \sqrt{\frac{5.2}{3}\frac{K}{R}} = 34,100 \text{ ft/sec}$$

and the required incremental velocity is

$$\Delta v = 10^3 \sqrt{(23.7 - 15)^2 + (24.58)^2} = 26,000 \text{ ft/sec}$$

The geometry of the maneuver is shown in Fig. 4.16–2.

Problem 2. (Elliptic orbits)

If the orbit on which the two vehicles are traveling is an ellipse, the problem becomes somewhat more complicated because the perigee for the transfer

---

$$* \sqrt{\frac{K}{R}} = \sqrt{\frac{(1.407)10^{16}}{(3960)5280}} = 25,930 \text{ ft/sec.}$$

orbit cannot be found by inspection as in the circular-orbit case. Using the same notation as in Prob. 1, the time required for 2 to reach 3 in Fig. 4.16–3 is shown by the shaded areas subtended by the angle 203. A maneuver at 1 must put vehicle 1 on a new orbit, and its subtended angle 103 must result in the same time. Although angle $103 = \phi$ is known, the perigee angle θ_1 is not known except in the special case $r_1 = r_3$.

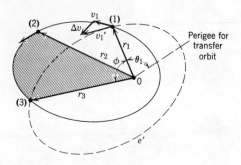

Fig. 4.16–3. Vehicle (1) intercepting vehicle (2) at (3) on elliptic orbit.

The solution is possible by trial as follows. For the new orbit we have

$$\frac{r_3}{r_1} = \frac{1 + e' \cos \theta_1}{1 + e' \cos (\theta_1 + \phi)} \tag{4.16-5}$$

Choosing a value of θ_1, the eccentricity e' can be found. $n = \dfrac{a}{R}$ can be found from $\pm \left(\dfrac{r_1}{R}\right)\left(\dfrac{1 + e' \cos \theta_1}{e'^2 - 1}\right)$ where $+$ is used for $e' > 1$ and $-$ for $e' < 1$. The angles θ_1, $\theta_1 + \phi$, and e' will establish τ_1 and τ_3 in Fig. 4.13–4 or 4.13–5. With the value of $a/R = n$, the elapsed time is found as in Prob. 1.

When agreement is established between the two elapsed times, the values of e' and θ_1 for the transfer orbit will result in rv^2/K and β', which can be found from Fig. 4.13–2 or 4.13–3. The remainder of the solution is then straightforward.

Problem 3. (Noncoplanar interception)

Vehicle 2 at $t = 0$ is at latitude 0 and longitude 0, traveling in a circular polar orbit of $r/R = 2.5$ and headed toward the north. Vehicle 1 at $t = 0$ is at latitude 0 and longitude 90° west, and traveling eastward in an equatorial elliptic orbit of $e = 0.50$, as shown in Fig. 4.16–4. The above position of vehicle 1 corresponds to perigee for which $r/R = 1.5$. Determine the impulsive velocity increment at 1 to intercept vehicle 2 at 3 when its latitude is 30° N.

The procedure for the solution of this problem is very much similar to that of Prob. 2. The transfer orbit 1, 3 is inclined 30° to the equatorial plane, $r_1/R = 1.5$, and $r_3/R = 2.5$, the angle between r_1 and r_3 being 90°. Perigee for the transfer orbit is again unknown and its position from r_1 is θ_1.

The elapsed time from 1 to 3 must equal that from 2 to 3, which is

$$t_{23} = \frac{30(2\pi)}{360} \frac{(2.5R)^{3/2}}{\sqrt{K}} = \frac{\pi}{6}(2.5)^{3/2}(806) = 1670 \text{ sec}$$

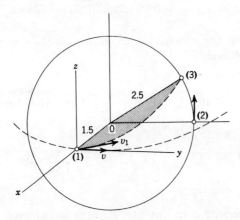

Fig. 4.16-4. Interception for noncoplanar orbits.

From the general equation of the orbit, we have for the two points on the transfer orbit,

$$\frac{r_3}{r_1} = \frac{2.5}{1.5} = 1.666 = \frac{1 + e' \cos \theta_1}{1 + e' \cos (\theta_1 + 90°)} \qquad (4.16\text{-}6)$$

or

$$0.666 = e'(\cos \theta_1 + 1.666 \sin \theta_1)$$

Choosing a value of θ_1, the eccentricity of the transfer orbit e' can be found from the above equation. With these two values of e' and θ_1, we can compute n from

$$n = \frac{a}{R} = \pm \frac{r_1}{R}\left(\frac{1 + e' \cos \theta_1}{e'^2 - 1}\right) \qquad (4.16\text{-}7)$$

where $+$ is used for $e' > 1$ and $-$ for $e' < 1$.

The nondimensional time τ_h is next found from Fig. 4.13–5 and, with a/R known, the elapsed time is computed and compared to the required time.

As a first choice of θ_1, try $340°$. Equation 4.16–6 gives $e' = 1.803$ and Eq. 4.16–7 gives $n = \dfrac{a}{R} = 1.795$. From Fig. 4.13–5, $\tau_h = 0.15$ for $\theta_1 = 340°$ (same as for $+20°$) and $\tau_h = 0.80$ for $\theta_3 = 70°$, making a total for the elapsed time of $\tau_h = 0.95$. The actual elapsed time is then

$$t_h = \tau_h \sqrt{\frac{a^3}{K}} = 0.95(1.795)^{3/2}(806) = 1845 \text{ sec}$$

Since this time is larger than 1670 sec, the orbit is too slow. A few trials result in the following:

$$\theta_1 = 338° \quad e' = 2.20$$

$$\frac{a}{R} = 1.19 \quad \tau_{\theta_1} = 0.29, \quad \tau_{\theta_3} = 1.31, \quad \tau_h = 1.60$$

$$t_h = 1.60(1.19)^{3/2}(806) = 1675 \text{ sec}$$

$$\left(\frac{rv^2}{K}\right)_1 = 3.26, \quad v_1 = 1.8\sqrt{\frac{K}{r_1}} = 1.8\sqrt{\frac{K}{1.5R}} = 1.47\sqrt{\frac{K}{R}}$$

$$\beta_1 = -15°$$

To find the velocity increment, let x, y, z be radial, transverse, and normal to the equatorial plane at point 1. Then the components of v_1 are (see Fig. 4.16–5):

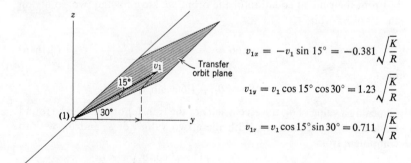

$$v_{1x} = -v_1 \sin 15° = -0.381\sqrt{\frac{K}{R}}$$

$$v_{1y} = v_1 \cos 15° \cos 30° = 1.23\sqrt{\frac{K}{R}}$$

$$v_{1z} = v_1 \cos 15° \sin 30° = 0.711\sqrt{\frac{K}{R}}$$

Fig. 4.16–5. Velocity increment required at (1) of Fig. 4.16–4.

The original velocity is entirely in the y direction, and since the initial orbit was an ellipse with $e = 0.50$, $r_p v_0^2/K = (1 + e) = 1.50$ and $v_0 = \sqrt{1.5K/r_p} = \sqrt{K/R}$.

The x, y, z components of the velocity increment at 1 are then,

$$\Delta v_x = -0.381\sqrt{\frac{K}{R}}$$

$$\Delta v_y = 0.23\sqrt{\frac{K}{R}}$$

$$\Delta v_z = 0.711\sqrt{\frac{K}{R}}$$

PROBLEMS

1. Two satellites 1 and 2 are in the same circular orbit of $r/R = a/R = 2$ in the same plane, but 2 is leading 1 by the angle $\phi_{12} = 30°$. What velocity increments are necessary to intercept and rendezvous when 2 has traveled through 45°.

2. Repeat Prob. 1 when 2 has traveled 90°.

3. Two satellites 1 and 2 are in the same circular orbit of $a/R = n_0$ in the same plane, but 2 is leading 1 by the angle ϕ_{12}. If 1 fires a retrorocket in the tangential direction, show that, in order for the two satellites to intercept after

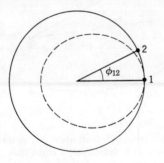

Prob. 3

1 has completed one revolution of its subcircular orbit, the necessary increment in the velocity is

$$\Delta v = v_c \left\{ 1 - \sqrt{2 - \frac{1}{[1 - (\phi_{12}/360)]^{2/3}}} \right\}$$

where v_c is the circular orbit velocity.

4. If in Prob. 3 the rocket is fired towards the rear so as to increase the velocity, determine the Δv necessary to intercept vehicle 2 on the Nth visit to 1. *Hint:* The time for the Nth visit of vehicle 2 at point 1 is

$$\left[N + \left(1 - \frac{\phi_{12}}{360} \right) \right] 2\pi \sqrt{\frac{R^3}{K}} \, n_0^{3/2}$$

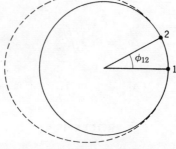

Prob. 4

5. Two satellites 1 and 2 are traveling in the same elliptic orbit in the same plane. The orbit is characterized by $e = 0.60$ and $a/R = 3.0$. When 1 is at $150°$, 2 is at $170°$. If interception and rendezvous are desired when 2 reaches $\theta = 210°$, determine the transfer orbit and the increments in velocity.

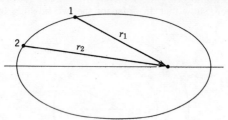

Prob. 5

6. Satellite 2 is leading satellite 1 by an angle ϕ_{12} in an elliptic orbit (see Fig. 4.16–3). To overtake 2 in a decreasingly short time, the eccentricity of the transfer orbit must increase to a large value. Show that in the limit as $e \to \infty$, the perigee of the transfer orbit can be determined from the equation

$$\cos \theta_1 - \frac{r_2}{r_1} \cos (\theta_1 + \phi_{12}) = 0$$

For fast transfer orbits, the actual θ_1 will be close to the above value.

7. Satellite 2 is traveling east in an equatorial circular orbit of $a/R = 2$, being at position longitude 0, latitude 0 at time $t = 0$. Satellite 1 at $t = 0$ is at latitude

90° and traveling in an elliptical orbit in the plane, longitude 0, with $a/R = 2$ and $e = 0.30$. If it is desired for 1 to intercept 2 at longitude 330°, determine the transfer orbit and the components of the velocity increment. The position of 1 at $t = 0$ corresponds to perigee for the elliptic orbit.

4.17 Long-Range Ballistic Trajectories

Since the shortest distance between two points on the surface of a sphere is along a great circle, ballistic trajectories are also considered in the great circle plane. Figure 4.17–1 shows the pertinent geometry of a

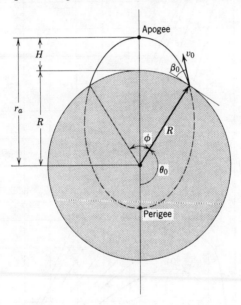

Fig. 4.17–1. Geometry of a ballistic trajectory.

ballistic trajectory which is an ellipse with the center of the earth as focus. Perigee is then inside the earth while the point of maximum height coincides with apogee.

Of interest here is the determination of the range $R\phi$, the height H, and the time t_b as function of the initial conditions which are $r_0 = R$, v_0, and β_0. We have at our disposal Eqs. 4.9–7 and 4.9–8 as developed in the initial-value problem of Sec. 4.9. The eccentricity is determined from the equation

$$e^2 = \left(\frac{Rv_0^2}{K} - 1\right)^2 \cos^2 \beta_0 + \sin^2 \beta_0 \qquad (4.17\text{–}1)$$

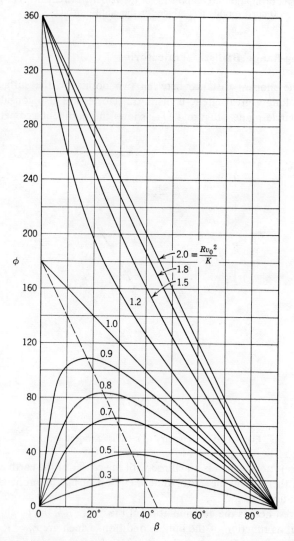

Fig. 4.17–2. Plot of Eq. 4.17–2 for the ballistic trajectory. (ϕ versus β_0 with Rv^2/K as parameter.

Since $\phi/2 = 180° - \theta_0$, $\tan\theta_0 = -\tan(\phi/2)$, and Eq. 4.9–7 can be written as

$$\tan\frac{\phi}{2} = \frac{-(Rv_0{}^2/K)\sin\beta_0\cos\beta_0}{(Rv_0{}^2/K)\cos^2\beta_0 - 1} \tag{4.17–2}$$

Figure 4.17–2 is a plot of ϕ versus β_0 with $Rv_0{}^2/K$ as parameter. The height H can be determined from its geometry.

$$r_a = a(1 + e) = H + R$$

$$\frac{H}{R} = \frac{a}{R}(1 + e) - 1 \tag{4.17–3}$$

From the equation of the ellipse, we have for $\theta_0 = 180° - \phi/2$, $r_0 = R$.

$$\frac{a}{R} = \frac{\left(1 - e\cos\dfrac{\phi}{2}\right)}{(1 - e)(1 + e)} \tag{4.17–4}$$

which, substituted into Eq. 4.17–3, results in

$$\frac{H}{R} = \frac{e}{1 - e}\left(1 - \cos\frac{\phi}{2}\right) \tag{4.17–5}$$

The time of flight is determined by subtracting the time required to go from perigee to $\theta = \theta_0$ from half the orbit period and doubling this figure, which from Eq. 4.13–12 is

$$t_b = 2\left(\frac{\pi a^{3/2}}{\sqrt{K}} - t_e\right)$$

$$= \frac{2a^{3/2}}{\sqrt{K}}\left\{\pi - \left[2\tan^{-1}\left(\sqrt{\frac{1-e}{1+e}}\tan\tfrac{1}{2}\theta_0\right) - \frac{e\sqrt{1-e^2}\sin\theta_0}{1 + e\cos\theta_0}\right]\right\} \tag{4.17–6}$$

PROBLEMS

1. For $Rv_0{}^2/K > 1$, show that the launching point corresponds to perigee if $\beta_0 = 0$.
2. For a given initial velocity $Rv_0{}^2/K < 1$, determine the angle β_0 for maximum range.
3. Relate the maximum range to the optimum heading angle β_0 and specified velocity $Rv_0{}^2/K$.
4. For a given range show that the minimum required velocity is related to β_0 by the equation $(Rv^2/K)_{min} = \dfrac{2\cos 2\beta_0}{1 + \cos 2\beta_0}$.

5. For a range of 5000 miles, determine the optimum angle β_0 the height H, and the velocity Rv_0^2/K.

6. For Prob. 5, determine the time of flight.

7. Discuss the effect of the earth's rotation on the motion of the ballistic missile.

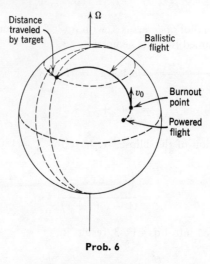

Prob. 6

4.18 Effect of the Earth's Oblateness

Due to the rotation of the earth from west to east, there is a speed advantage in launching a satellite in a direction with an easterly component. Such an orbit will precess in a westerly direction due to the earth's equatorial bulge, and thus a closed orbit is really not possible. The revolving satellite is like a gyroscope and, as shown in Fig. 4.18–1, its angular momentum vector $\mathbf{h}_s$, perpendicular to its orbit plane and directed towards the northern hemisphere, must slowly revolve about the north polar axis due to the moment exerted by the excess mass over the sphere near the equator. The rate of precession will depend on the orbit angle with respect to the equator and, to a somewhat smaller extent, on the altitude.

The moment due to the equatorial bulge responsible for the precession of the satellite orbit can be determined as follows:

Referring to Fig. 4.18–2 the satellite m_s is attracted towards the mass element dm of the earth according to the equation,

$$d\mathbf{F} = -\frac{Km_s\,dm}{mr^3}\mathbf{r} \qquad (4.18\text{–}1)$$

where $K = Gm$ and m is the mass of the earth.

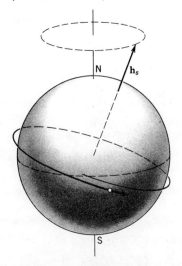

Fig. 4.18–1. Precession of orbit plane due to earth's oblateness.

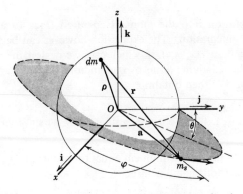

Fig. 4.18–2. Satellite m_s attracted by element dm of earth.

Resolving **r** into components,

$$\mathbf{r} = \mathbf{a} - \boldsymbol{\rho}$$
$$= (a \cos \varphi)\mathbf{i} + (a \sin \varphi \cos \theta)\mathbf{j} - (a \sin \varphi \sin \theta)\mathbf{k} - (x\mathbf{i} + y\mathbf{j} + z\mathbf{k})$$

$$(4.18\text{--}2)$$

where θ is the angle between the orbit plane and the equatorial plane and

x, y, z are the components of $\boldsymbol{\rho}$. The moment becomes,

$$d\mathbf{M} = \boldsymbol{\rho} \times d\mathbf{F}$$

$$= \frac{-Km_s\, dm}{mr^3} \begin{vmatrix} \mathbf{i} & \mathbf{j} & \mathbf{k} \\ x & y & z \\ (a\cos\varphi - x) & (a\sin\varphi\cos\theta - y) & (-a\sin\varphi\sin\theta - z) \end{vmatrix}$$

$$= \frac{Km_s a\, dm}{mr^3} \left[(y\sin\varphi\sin\theta + z\sin\varphi\cos\theta)\mathbf{i} - (x\sin\varphi\sin\theta + z\cos\varphi)\mathbf{j}\right.$$

$$\left. -(x\sin\varphi\cos\theta - y\cos\varphi)\mathbf{k}\right] \quad (4.18\text{--}3)$$

The quantity $1/r^3$ can be obtained by the following steps:

$$r^2 = (\mathbf{a} - \boldsymbol{\rho})\cdot(\mathbf{a} - \boldsymbol{\rho}) = a^2 + \rho^2 - 2\boldsymbol{\rho}\cdot\mathbf{a}$$

$$= a^2 + \rho^2 - 2a\,(x\cos\varphi + y\sin\varphi\cos\theta - z\sin\varphi\sin\theta)$$

$$\frac{1}{r^3} = \frac{1}{a^3}\left[1 + \left(\frac{\rho}{a}\right)^2 - \frac{2}{a}(x\cos\varphi + y\sin\varphi\cos\theta - z\sin\varphi\sin\theta)\right]^{-\frac{3}{2}}$$

$$(4.18\text{--}4)$$

and its substitution into the moment equation leads to a complicated expression for integration. The expression however can be simplified if a is much larger than $\rho(x, y, z)$, in which case we neglect the term $(\rho/a)^2$ in $1/r^3$, expand the remaining terms by the binomial theorem, and retain only the first terms. We then obtain,

$$\frac{1}{r^3} \cong \frac{1}{a^3}\left[1 + \frac{3}{a}(x\cos\varphi + y\sin\varphi\cos\theta - z\sin\varphi\sin\theta)\right] \quad (4.18\text{--}5)$$

and its substitution into the moment equation results in M_x, M_y, and M_z. The x component of the moment is,

$$M_x = \frac{Km_s}{ma^2}\left[\int(y\sin\varphi\sin\theta + z\sin\varphi\cos\theta)\, dm\right.$$

$$+ \frac{3}{a}\int(y\sin\varphi\sin\theta + z\sin\varphi\cos\theta)$$

$$\left. \times (x\cos\varphi + y\sin\varphi\cos\theta - z\sin\varphi\sin\theta)\, dm\right] \quad (4.18\text{--}6)$$

where θ and φ are held constant during integration. It is evident that the first integral is zero due to symmetry of the oblate spheroid. Also all cross products of the form xy, xz, yz will integrate to zero due to symmetry. We are then left with the integral,

$$M_x = \frac{3Km_s}{ma^3} \sin^2 \varphi \sin \theta \cos \theta \int (y^2 - z^2)\, dm$$

$$= \frac{3Km_s}{ma^3} \sin^2 \varphi \sin \theta \cos \theta \left[\int (x^2 + y^2)\, dm - \int (x^2 + z^2)\, dm \right]$$

$$= \frac{3Km_s}{ma^3} (C - A) \sin^2 \varphi \sin \theta \cos \theta \qquad (4.18\text{--}7)$$

where C and A are the moments of inertia of the earth about the polar and equatorial axes respectively.

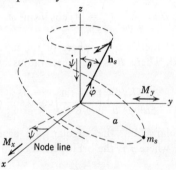

Fig. 4.18–3. Precession of vector $\mathbf{h}_s$ due to moments M_x and M_y.

Similarly, the moment about the y axis is,

$$M_y = \frac{-3Km_s}{ma^3} (C - A) \sin \theta \sin \varphi \cos \varphi \qquad (4.18\text{--}8)$$

and the moment about the z axis is zero.

$$M_z = 0 \qquad (4.18\text{--}9)$$

These equations indicate that the moment M_y is negative for $0 \le \varphi \le \pi/2$, and positive for $\pi/2 \le \varphi \le \pi$, the cycle repeating itself over π to 2π. Thus the net moment M_y over a complete cycle is zero. The moment M_x, however, is always positive and varying as $\sin^2 \varphi$.

Figure 4.18–3 shows the orbit plane, the satellite m_s, and the moments M_x and M_y exerted by the earth on the satellite. To determine the precession of the satellite orbit plane, we note that the angular momentum of the satellite has the value $h_s = m_s a^2 \dot{\varphi}$, which is normal to the orbit plane. The action of the moment M_y on $\mathbf{h}_s$ is oscillatory and zero over a complete cycle, but M_x requires the rate of change of $\mathbf{h}_s$ over the cycle to be cumulative in the x direction.

Measuring the regression of the node line ψ in the equatorial plane, the rate of precession $\dot{\psi}$ is directed along the $-z$ axis with components $\dot{\psi} \sin \theta$ in the plane of the orbit, and $\dot{\psi} \cos \theta$ normal to it. The component $\dot{\psi} \sin \theta$ rotates the vector $\mathbf{h}_s$ to give

$$h_s \dot{\psi} \sin \theta = M_x \qquad (4.18\text{--}10)$$

$$m_s a^2 \dot{\varphi} \dot{\psi} \sin \theta = \frac{3Km_s}{ma^3}(C - A) \sin \theta \cos \theta \sin^2 \varphi$$

from which the rate of precession of the orbit plane becomes,

$$\dot{\psi} = \frac{3K}{\dot{\varphi} ma^5}(C - A) \cos \theta \sin^2 \varphi$$

Since the moment of inertia of a sphere of radius R is $\frac{2}{5}mR^2$, we can introduce $C = \frac{2}{5}mR^2$, where R is the mean radius of earth, and rewrite $\dot{\psi}$ as

$$\dot{\psi} = \frac{6}{5a^3} \frac{K}{\dot{\varphi}} \left(\frac{R}{a}\right)^2 \left(\frac{C - A}{C}\right) \cos \theta \sin^2 \varphi \qquad (4.18\text{--}11)$$

Assuming a circular orbit, the angular rate around the orbit, is a constant, $\dot{\varphi} = 2\pi/\tau$, where τ is the orbit period equal to

$$\tau = \frac{2\pi a^{3/2}}{\sqrt{K}}$$

Thus φ can be replaced by $(2\pi/\tau)t$, and the precession angle measured in the equatorial plane per revolution of the satellite becomes,

$$\psi = \frac{6K}{5a^3}\left(\frac{C - A}{C}\right)\left(\frac{R}{a}\right)^2 \frac{\tau}{2\pi} \cos \theta \int_0^\tau \sin^2 \frac{2\pi}{\tau} t \, dt$$

$$= \frac{6K}{5}\left(\frac{C - A}{C}\right)\left(\frac{R}{a}\right)^2 \frac{\tau^2}{4\pi a^3} \cos \theta$$

$$= \frac{6\pi}{5}\left(\frac{C - A}{C}\right)\left(\frac{R}{a}\right)^2 \cos \theta \qquad (4.18\text{--}12)$$

The quantity $(C - A)/C$ for earth is 0.0032, so that the node line of the orbit regresses westward by the amount

$$\psi = 0.0121 \left(\frac{R}{a}\right)^2 \cos \theta \qquad (4.18\text{--}13)$$

for each revolution of the satellite around the earth.

We can compare this equation with the equation given by Blitzer,[2] which is

$$\psi = 2\pi J \left(\frac{R}{r}\right)^2 \cos i$$

Translated into our notation with $J = 1.637 \times 10^{-3}$ (see ref. 5) Blitzer's equation is

$$\psi = 0.01022 \left(\frac{R}{a}\right)^2 \cos \theta$$

which indicates fair agreement for our approximate equation, Eq. 4.18–13.

The reverse problem to the above is the precession of the earth's polar axis due to the moment exerted by the satellite on the earth. When the satellite is a sizeable mass, such as the moon, its influence is a measurable quantity. The problem of the precession of the earth's polar axis due to the sun and the moon is taken up in Chap. 5, Sec. 15.

PROBLEMS

1. Examination of Eq. 4.18–7 indicates that no restriction as to the density variation of earth was imposed; however in letting $\frac{2}{5}mR^2 = C$ just before Eq. 4.18–11, uniform density is implied. Indicate what would be changed in Eq. 4.18–12 if the density of the earth varied with the distance from its center.

2. If the term $(\rho/a)^2$ in Eq. 4.18–4 is retained to the first term of the binomial expansion, determine the correction to Eq. 4.18–5.

3. For a satellite launched southeasterly in a circular orbit at an angle of $35°$ with the equator and at an altitude of 400 miles, determine the regression of the node per revolution taking into account the rotation of the earth during the period.

4. Show that the attraction of a thin spherical shell of constant density is equal to that of a particle of the same mass concentrated at its center.

5. Assuming that $(C - A)/C = N$ differs from zero due to a narrow band around the equator of a perfect sphere, show that the mass of this narrow band must equal,

$$m = \frac{4N}{5 - 6N} m_0$$

where m_0 is the total mass of the sphere plus the narrow band.

REFERENCES

1. Augenstein, B. W., "Dynamics Problems Associated with Satellite Orbit Control," Trans. ASME, Series B (Nov. 1959), 281–288.
2. Blitzer, L., M. Weisfield, and D. Wheelon, "Perturbation of a Satellite Orbit Due to the Earth's Oblateness," J. Appl. Physics 27, No. 10 (Oct. 1956), 1141–1149.

3. Ehricke, Krafft A., "Interplanetary Operations," *Space Technology*, John Wiley and Sons, New York (1959), Chapter 8.
4. Ehricke, Krafft A., *Space Flight*, Vol. 1, D. van Nostrand, Princeton, N.J. (1960).
5. King-Hele, D. G., and D. M. C. Walker, "Methods for Predicting the Orbits of Near Earth-Satellites," *J. British Interplanetary Soc.*, **17**, No. 1 (Jan.–Feb. 1959), 2–14.
6. Moulton, F. R., *Celestial Mechanics*, Macmillan Co., New York (1914).
7. Ramsey, A. S., *Dynamics—Part II*, Cambridge University Press, New York (1956), Chapter 1.

Gyrodynamics

CHAPTER 5

5.1 Displacement of a Rigid Body

A rigid body can be viewed as a system of particles where the relative distances between particles are fixed. The position of a rigid body is defined by any three points on it, not in the same straight line.

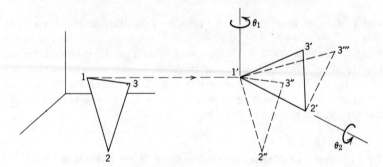

Fig. 5.1-1. Displacement of a rigid body.

The motion of a rigid body can be described by a translation of some reference point 0, plus a rotation about some axis through 0. Consider three arbitrary noncolinear points 1, 2, 3, in the initial and final positions 1', 2', 3', as shown in Fig. 5.1-1. The first point 1 can be brought to 1' by a translation so that the new position is 1', 2", 3". Next, rotate about an

101

axis through 1′ which is perpendicular to the plane 1′, 2″, 2′, bringing 2″ to coincide with 2′. Finally, rotate about an axis through 1′ and 2′, to bring 3‴ to 3′.

We will now show Euler's proof that the two individual rotations can be replaced by a single rotation. Draw a unit sphere about point 1′ and where the two rotation axes pierce it, connect the points by a great circle as shown in Fig. 5.1–2. Measure off $\frac{1}{2}\theta_1$ on each side of the great circle at

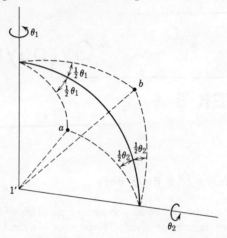

Fig. 5.1–2. Resultant rotation axis by Euler's proof.

axis 1, drawing two other great circles, and do likewise with angle θ_2 at axis 2. Now a rotation of θ_1 about axis 1 will bring point a to b, and a rotation θ_2 about axis 2 will bring b back to a. Thus $1'a$ is an undisturbed line during rotation θ_1 and θ_2, and therefore it must be the resultant axis of rotation. Note that $1'a$ is not in the plane containing the axes of rotation θ_1 and θ_2, which again points out the fact that finite rotations do not possess the properties of vectors.

5.2 Moment of Momentum of a Rigid Body (About a Fixed Point or the Moving Center of Mass)

Let body axes x, y, z be attached to the body with the origin O at any point. The velocity of any point i on the body is then,

$$\mathbf{v}_i = \mathbf{v}_0 + \boldsymbol{\omega} \times \mathbf{r}_i \qquad (5.2\text{–}1)$$

where $\boldsymbol{\omega}$ is the angular velocity of the body.

The moment of momentum about the origin O of the x, y, z, system is,

$$\mathbf{h}_0 = \sum_i \mathbf{r}_i \times m_i(\mathbf{v}_0 + \boldsymbol{\omega} \times \mathbf{r}_i)$$

$$= \sum_i \mathbf{r}_i \times (\boldsymbol{\omega} \times \mathbf{r}_i)m_i - \mathbf{v}_0 \times \sum_i m_i\mathbf{r}_i \qquad (5.2\text{-}2)$$

If the reference point O is stationary, $\mathbf{v}_0 = 0$, whereas if O coincides with the center of mass, $\sum_i m_i\mathbf{r}_i = 0$. Thus, if O is fixed, or a center of mass, the angular momentum is given by the first term of the above equation, which can be expressed by the following integral,

$$\mathbf{h}_0 = \int \mathbf{r} \times (\boldsymbol{\omega} \times \mathbf{r}) \, dm \qquad (5.2\text{-}3)$$

To evaluate this integral, we note that the first cross product $\boldsymbol{\omega} \times \mathbf{r}$ is

$$\boldsymbol{\omega} \times \mathbf{r} = \begin{vmatrix} \mathbf{i} & \mathbf{j} & \mathbf{k} \\ \omega_x & \omega_y & \omega_z \\ x & y & z \end{vmatrix} = (\omega_y z - \omega_z y)\mathbf{i} + (\omega_z x - \omega_x z)\mathbf{j} \\ + (\omega_x y - \omega_y x)\mathbf{k} \qquad (5.2\text{-}4)$$

Multiplying by dm, we have the x, y, z, components of the momentum of dm, as shown in Fig. 5.2–1.

The cross product $\mathbf{r} \times (\boldsymbol{\omega} \times \mathbf{r}) \, dm$ is,

$$\mathbf{r} \times (\boldsymbol{\omega} \times \mathbf{r}) \, dm = \begin{vmatrix} \mathbf{i} & \mathbf{j} & \mathbf{k} \\ x & y & z \\ (\omega_y z - \omega_z y) & (\omega_z x - \omega_x z) & (\omega_x y - \omega_y x) \end{vmatrix} dm$$

$$= \mathbf{i}[\omega_x(y^2 + z^2) - \omega_y(xy) - \omega_z(xz)] \, dm$$
$$+ \mathbf{j}[-\omega_x(xy) + \omega_y(x^2 + z^2) - \omega_z(yz)] \, dm$$
$$+ \mathbf{k}[-\omega_x(xz) - \omega_y(yz) + \omega_z(x^2 + y^2)] \, dm \qquad (5.2\text{-}5)$$

which represent the moment about the x, y, z, axes of the momentum vectors shown in Fig. 5.2–1. Integrating over the body, we arrive at the x, y, z, components of the moment of momentum of the body.

$$\mathbf{h}_0 = h_x\mathbf{i} + h_y\mathbf{j} + h_z\mathbf{k} \qquad (5.2\text{-}6)$$

We now define the moment of inertia of the body about the x, y, z, axes as

$$I_x = \int (y^2 + z^2) \, dm \quad I_y = \int (x^2 + z^2) \, dm \quad I_z = \int (x^2 + y^2) \, dm$$

and the products of inertia as,

$$I_{xy} = \int xy \, dm \quad I_{xz} = \int xz \, dm \quad I_{yz} = \int yz \, dm$$

in which case the moment of momentum components along the x, y, z, axes become,

$$h_x = I_x \omega_x - I_{xy} \omega_y - I_{xz} \omega_z$$
$$h_y = - I_{xy} \omega_x + I_y \omega_y - I_{yz} \omega_z \qquad (5.2\text{--}7)$$
$$h_z = - I_{xz} \omega_x - I_{yz} \omega_y + I_z \omega_z$$

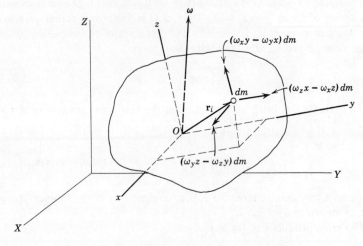

Fig. 5.2–I. Components of momentum ($\boldsymbol{\omega} \times \mathbf{r}$) dm.

The moments and products of inertia can be concisely presented by an inertia dyadic as follows:*

$$\mathscr{I} = \begin{vmatrix} \mathbf{ii}I_x & -\mathbf{ij}I_{xy} & -\mathbf{ik}I_{xz} \\ -\mathbf{ji}I_{xy} & \mathbf{jj}I_y & -\mathbf{jk}I_{yz} \\ -\mathbf{ki}I_{xz} & -\mathbf{kj}I_{yz} & \mathbf{kk}I_z \end{vmatrix} \qquad (5.2\text{--}8)$$

If we form the dot product of the inertia dyadic with the angular velocity vector

$$\boldsymbol{\omega} = \omega_x \mathbf{i} + \omega_y \mathbf{j} + \omega_z \mathbf{k}$$

we would obtain the moment of momentum

$$\mathbf{h} = \mathscr{I} \cdot \boldsymbol{\omega} \qquad (5.2\text{--}9)$$

The order of the dot product must be preserved with the following interpretation,

$$\mathbf{ij} \cdot \mathbf{i} = \mathbf{i}(\mathbf{j} \cdot \mathbf{i}) = 0$$
$$\mathbf{ji} \cdot \mathbf{i} = \mathbf{j}(\mathbf{i} \cdot \mathbf{i}) = \mathbf{j}$$

* See Appendix B.

5.3 Kinetic Energy of a Rigid Body

Consider a rigid body moving through space, and attach to it a set of body axes x, y, z, with its origin coinciding with its center of mass. Then any point $\mathbf{r}$ will have a velocity equal to,

$$\mathbf{v} = \mathbf{v}_0 + \boldsymbol{\omega} \times \mathbf{r} \tag{5.3-1}$$

The square of the velocity is obtained by the dot product of its vector,

$$v^2 = \mathbf{v} \cdot \mathbf{v} = v_0{}^2 + (\boldsymbol{\omega} \times \mathbf{r}) \cdot (\boldsymbol{\omega} \times \mathbf{r}) + 2\mathbf{v}_0 \cdot (\boldsymbol{\omega} \times \mathbf{r})$$

Thus, the kinetic energy of the body is given as,

$$T = \tfrac{1}{2} \int v^2 \, dm = \tfrac{1}{2} m v_0{}^2 + \tfrac{1}{2} \int (\boldsymbol{\omega} \times \mathbf{r}) \cdot (\boldsymbol{\omega} \times \mathbf{r}) \, dm + \mathbf{v}_0 \cdot \boldsymbol{\omega} \times \int \mathbf{r} \, dm$$

$$= \tfrac{1}{2} m v_0{}^2 + \tfrac{1}{2} \int (\boldsymbol{\omega} \times \mathbf{r}) \cdot (\boldsymbol{\omega} \times \mathbf{r}) \, dm \tag{5.3-2}$$

where $\int \mathbf{r} \, dm = 0$ for the origin of the body axes coinciding with the center of mass. We have thus found that the kinetic energy of translation is determined as if the entire mass is concentrated at the center of mass as a particle, and the second term is the kinetic energy of rotation about an axis $\boldsymbol{\omega}$ through the center of mass.

Focusing our attention to the kinetic energy of rotation, we examine the quantity $(\boldsymbol{\omega} \times \mathbf{r}) \cdot (\boldsymbol{\omega} \times \mathbf{r})$. Resolving $\boldsymbol{\omega} \times \mathbf{r}$ into components along the body axes, the dot product is given by the square of the $\mathbf{i}$, $\mathbf{j}$, $\mathbf{k}$ components.

$$
\begin{aligned}
(\boldsymbol{\omega} \times \mathbf{r}) \cdot (\boldsymbol{\omega} \times \mathbf{r}) &= (\omega_y z - \omega_z y)^2 + (\omega_z x - \omega_x z)^2 + (\omega_x y - \omega_y x)^2 \\
&= \omega_x{}^2(y^2 + z^2) + \omega_y{}^2(x^2 + z^2) + \omega_z{}^2(x^2 + y^2) \\
&\quad - 2\omega_x \omega_z xz - 2\omega_y \omega_z yz - 2\omega_x \omega_y xy
\end{aligned}
$$

Thus

$$
\begin{aligned}
2T_{\text{rot.}} &= \omega_x{}^2 I_x + \omega_y{}^2 I_y + \omega_z{}^2 I_z - 2\omega_x \omega_z I_{xz} - 2\omega_y \omega_z I_{yz} \\
&\quad - 2\omega_x \omega_y I_{xy}
\end{aligned} \tag{5.3-3}
$$

5.4 Moment of Inertia about a Rotated Axis

If $I_{\xi\xi}$ is the moment of inertia of a body about any axis ξ, with angular velocity $\boldsymbol{\omega}$, we can write

$$2T_{\text{rot.}} = I_{\xi\xi} \omega^2 \tag{5.4-1}$$

Substituting Eq. 5.3-3 into the above equation, we have

$$I_{\xi\xi} = \left(\frac{\omega_x}{\omega}\right)^2 I_x + \left(\frac{\omega_y}{\omega}\right)^2 I_y + \left(\frac{\omega_z}{\omega}\right)^2 I_z$$

$$- 2\left(\frac{\omega_x}{\omega}\right)\left(\frac{\omega_z}{\omega}\right)I_{xz} - 2\left(\frac{\omega_y}{\omega}\right)\left(\frac{\omega_z}{\omega}\right)I_{yz} - 2\left(\frac{\omega_x}{\omega}\right)\left(\frac{\omega_y}{\omega}\right)I_{xy} \quad (5.4-2)$$

By letting $l_{\xi x}$, $l_{\xi y}$, $l_{\xi z}$ be direction cosines of the vector $\boldsymbol{\omega}$ or axis ξ with respect to the x, y, z axes, the above equation can be written as

$$I_{\xi\xi} = l_{\xi x}{}^2 I_x + l_{\xi y}{}^2 I_y + l_{\xi z}{}^2 I_z - 2l_{\xi x}l_{\xi z}I_{xz} - 2l_{\xi y}l_{\xi z}I_{yz} - 2l_{\xi x}l_{\xi y}I_{xy} \quad (5.4-3)$$

Equation 5.4-3 can be concisely written in terms of a double summation as follows

$$I_{\xi\xi} = \sum_{\alpha}\sum_{\beta} l_{\xi\alpha}l_{\xi\beta}I_{\alpha\beta} \quad (5.4-4)$$

where α and β take on the letters x, y, z with $I_{\alpha\alpha}$ interpreted as $I_{xx} = I_x$, I_{yy}, etc., and $I_{\alpha\beta}$ as $-I_{xy}$, $-I_{yz}$, $-I_{xz}$. In fact Eq. 5.4-4 can be changed slightly to apply to products of inertia as well by the equation

$$-I_{\xi\eta} = \sum_{\alpha}\sum_{\beta} l_{\xi\alpha}l_{\eta\beta}I_{\alpha\beta} \quad (5.4-5)$$

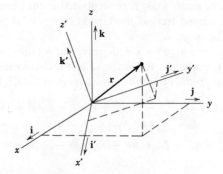

Fig. 5.4-1. Components of **r** in two coordinate systems.

The direction cosines $l_{\xi\alpha}$ to be used in Eq. 5.4-4 or 5.4-5 can be formed most conveniently from the transformation matrix between the two coordinate systems. If the two coordinate systems are x, y, z and x', y', z' with unit vectors **i**, **j**, **k** and **i′**, **j′**, **k′**, a point in space can be expressed in terms of either coordinate system as

$$\mathbf{r} = x\mathbf{i} + y\mathbf{j} + z\mathbf{k} = x'\mathbf{i'} + y'\mathbf{j'} + z'\mathbf{k'} \quad (5.4-6)$$

as shown in Fig. 5.4-1.

Scalar multiplication of Eq. 5.4–6 by **i**, **j**, and **k** results in

$$x = (\mathbf{i'} \cdot \mathbf{i})x' + (\mathbf{j'} \cdot \mathbf{i})y' + (\mathbf{k'} \cdot \mathbf{i})z'$$
$$y = (\mathbf{i'} \cdot \mathbf{j})x' + (\mathbf{j'} \cdot \mathbf{j})y' + (\mathbf{k'} \cdot \mathbf{j})z' \tag{5.4-7}$$
$$z = (\mathbf{i'} \cdot \mathbf{k})x' + (\mathbf{j'} \cdot \mathbf{k})y' + (\mathbf{k'} \cdot \mathbf{k})z'$$

which, in matrix notation, is

$$\begin{bmatrix} x \\ y \\ z \end{bmatrix} = \begin{bmatrix} l_{xx'} & l_{xy'} & l_{xz'} \\ l_{yx'} & l_{yy'} & l_{yz'} \\ l_{zx'} & l_{zy'} & l_{zz'} \end{bmatrix} \begin{bmatrix} x' \\ y' \\ z' \end{bmatrix} \tag{5.4-8}$$

Thus when the transfer matrix between the two coordinates is known, the elements of the matrix are the direction cosines.

We note further that the equation for the kinetic energy can be expressed by the dot product of the angular velocity and the moment of momentum,

$$2T = \boldsymbol{\omega} \cdot \mathbf{h}_0 \tag{5.4-9}$$

Since $\mathbf{h}_0$ can be expressed in terms of the momental dyadic (see Eq. 5.2–9), the above equation becomes

$$2T = \boldsymbol{\omega} \cdot \mathscr{I} \cdot \boldsymbol{\omega} \tag{5.4-10}$$

Again letting $\mathscr{I} = \mathbf{AB}$, the above double dot product is interpreted as

$$(\boldsymbol{\omega} \cdot \mathbf{A})(\mathbf{B} \cdot \boldsymbol{\omega})$$

which is a product of two scalars and, therefore, a pure number.

5.5 Principal Axes

We define the principal axes of the body, 1, 2, 3, as those about which the products of inertia vanish, and let A, B, C, be the moment of inertia about the 1, 2, 3 axes respectively. The moment of inertia about the instantaneous axes of rotation ξ, in terms of A, B, C then becomes,

$$I_\xi = A l_{\xi1}^2 + B l_{\xi2}^2 + C l_{\xi3}^2 \tag{5.5-1}$$

where $l_{\xi1}$, $l_{\xi2}$, $l_{\xi3}$ are the direction cosines of the vector $\boldsymbol{\omega}$, or axis ξ and the principal axes 1, 2, 3. Since the axes 1, 2, 3 are fixed in the body, A, B, C are constants. However, as the instantaneous axis ξ is moved, $l_{\xi1}$, $l_{\xi2}$, $l_{\xi3}$ change, and so will the value of I about the ξ axis.

If we let $\rho = 1/\sqrt{I_\xi}$, and lay off ρ and ξ, and do this for every orientation of the instantaneous axis ξ, we would obtain an ellipsoid of inertia.

Dividing the above equation by I_ξ and noting that $l_{\xi 1}/\sqrt{I_\xi} = l_{\xi 1}\rho = x$, $l_{\xi 2}\rho = y$, $l_{\xi 3}\rho = z$ are the principal coordinates of the equation for the ellipsoid of inertia, Eq. 5.5–1 becomes,

$$Ax^2 + By^2 + Cz^2 = 1 \qquad (5.5\text{–}2)$$

PROBLEMS

1. For the slender uniform bar of length l and mass m oriented in the position shown, determine the moment and product of inertia about each of the axes.

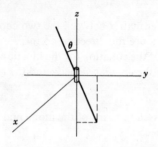

Prob. 1

2. If the bar of Prob. 1 is rotated about the z axis with angular velocity ω, determine the angular momentum about the three axes.

3. While getting into position for takeoff, a small aeroplane with two bladed propellers turns about its vertical axis as shown. Determine the x, y, z components of the angular momentum of the left propeller. Assume the propeller to be a uniform slender rod of length l.

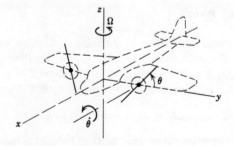

Prob. 3

4. Find the moments of inertia of six 10-lb weights arranged symmetrically on the x, y, z axes as follows: $x = \pm 5$, $y = \pm 2$, $z = \pm 6$.

5. Find the moment of inertia of the configuration of Prob. 4 about an axis through the origin and the point (1, 2, 2).

6. A thin disk of radius R is mounted on a shaft which deviates by an angle α from the normal. Choose orthogonal axes through the shaft and disk and determine the moment and product of inertia about each axis.

7. Determine the moment of inertia about the ξ, η, ζ axes for a cube of sides c, if ζ is placed along the diagonal and the ξ, η axes normal to ζ are rotated into a position such that $I_\xi = I_\eta$. What will be the direction cosines of ξ and η so placed?

8. Determine the length of a uniform cylinder of radius R such that its principal moments of inertia are equal.

9. If A, B, C are principal moments of inertia of a given body, show that $A + B$ is larger than C.

10. A space vehicle makes a landing on a hillside with velocities v_z and v_x just before impact as shown. If a leg strikes a boulder and the vehicle pivots about this point, show that the angular velocities immediately after impact are

$$\dot{\theta}_x = -\frac{m\bar{y}}{I_x} v_z$$

$$\dot{\theta}_y = \frac{m\bar{y}I_{yz} - m\bar{z}I_z}{I_yI_z - I_{yz}{}^2} v_x$$

$$\dot{\theta}_z = \frac{m\bar{y}}{I_z} v_x + \frac{I_{yz}}{I_z} \frac{m\bar{y}I_{yz} - m\bar{z}I_z}{I_yI_z - I_{yz}{}^2} v_x$$

where the x, y, z coordinates are oriented with z vertical, x parallel to the horizontal velocity v_x, and y horizontal.

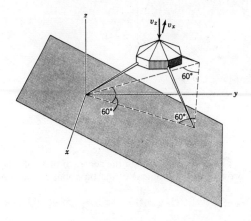

Prob. 10

11. Assume that the moments and products of inertia about the x, y, z axes are known. Determine the equation for the moment of inertia about the ξ, η, ζ axes by noting that $\rho_\xi{}^2 = \rho^2 - \xi^2 = \rho^2 - (\mathbf{\rho} \cdot \mathbf{i}')^2$, $\mathbf{\rho} = x\mathbf{i} + y\mathbf{j} + z\mathbf{k}$, and $I_{\xi\xi} = \int \rho_\xi{}^2 \, dm$, etc.

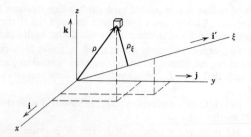

Prob. 11

12. In the same manner as in Prob. 11, determine the equation for the product of inertia $I_{\xi\eta}$, $I_{\xi\zeta}$, and $I_{\eta\zeta}$.

13. An airplane of mass M and moments of inertia A, B, C about principal axes drops at a uniform rate of V ft/sec, and at the same time spins at the rate of 10 rpm about an axis which makes equal angles with the three principal axes. Determine its total kinetic energy.

14. Show that the moment of inertia of three equal masses 120° apart as shown in the sketch is equal to $\frac{3}{2}ml^2$ about any axis in its plane.

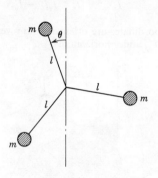

Prob. 14

15. With the x, y, z axes passing through the center of mass of a body shown in the sketch, prove that the moment of inertia about any axis n through the origin is

$$I_n = \sum_i m_i[x_i^2(1 - n_x^2) + y_i^2(1 - n_y^2) + z_i^2(1 - n_z^2)$$

$$- 2(x_iy_in_xn_y + x_iz_in_xn_z + y_iz_in_yn_z)]$$

where $\mathbf{n} = n_x\mathbf{i} + n_y\mathbf{j} + n_z\mathbf{k}$ is a unit vector along the axis $\mathbf{n}$. *Hint:* Start with

$$I_n = \sum_i m_i(\mathbf{r}_i \times \mathbf{n}) \cdot (\mathbf{r}_i \times \mathbf{n}) = \sum_i m_i[r_i^2 - (\mathbf{r}_i \cdot \mathbf{n})^2]$$

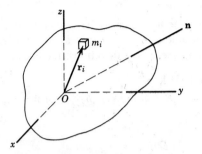

Prob. 15

16. Starting with $I_{n'} = \sum_i m_i[(\mathbf{r_0} + \mathbf{r}_i) \times \mathbf{n}] \cdot [(\mathbf{r_0} + \mathbf{r}_i) \times \mathbf{n}]$, derive the equation for the moment of inertia about an axis n' parallel to n, and displaced by the vector $\mathbf{r_0}$.

17. By actually multiplying, using the rule $\mathbf{ij} \cdot \mathbf{i} = 0$, $\mathbf{ij} \cdot \mathbf{j} = \mathbf{i}$, etc., show that $\mathbf{h} = \mathscr{I} \cdot \boldsymbol{\omega}$.

18. Show that

$$\dot{\mathscr{I}} = [\dot{\mathscr{I}}] + \boldsymbol{\omega} \times \mathscr{I} + \begin{vmatrix} \mathbf{i}\omega \times \mathbf{i}I_x & -\mathbf{i}\omega \times \mathbf{j}I_{xy} & -\mathbf{i}\omega \times \mathbf{k}I_{xy} \\ -\mathbf{j}\omega \times \mathbf{i}I_{xy} & \mathbf{j}\omega \times \mathbf{j}I_y & -\mathbf{j}\omega \times \mathbf{k}I_{yz} \\ -\mathbf{k}\omega \times \mathbf{i}I_{xz} & -\mathbf{k}\omega \times \mathbf{j}I_{yz} & \mathbf{k}\omega \times \mathbf{k}I_z \end{vmatrix}$$

where

$$[\dot{\mathscr{I}}] = \begin{vmatrix} \mathbf{ii}\dot{I}_x & -\mathbf{ij}\dot{I}_{xy} & -\mathbf{ik}\dot{I}_{xz} \\ -\mathbf{ji}\dot{I}_{xy} & \mathbf{jj}\dot{I}_y & -\mathbf{jk}\dot{I}_{yz} \\ -\mathbf{ki}\dot{I}_{xz} & -\mathbf{kj}\dot{I}_{yz} & \mathbf{kk}\dot{I}_z \end{vmatrix}$$

19. Determine the total time derivative

$$\frac{d\mathbf{h}}{dt} = \mathscr{I} \cdot \dot{\boldsymbol{\omega}} + \dot{\mathscr{I}} \cdot \boldsymbol{\omega}$$

with respect to inertial coordinates, and show that it is equal to

$$\frac{d\mathbf{h}}{dt} = \mathscr{I} \cdot \dot{\boldsymbol{\omega}} + [\dot{\mathscr{I}}] \cdot \boldsymbol{\omega} + \boldsymbol{\omega} \times \mathbf{h}$$

5.6 Euler's Moment Equation

We have shown previously that the moment about the mass center is equal to the time derivative of the moment of momentum about this point. With $\mathbf{h}_c = h_x\mathbf{i} + h_y\mathbf{j} + h_z\mathbf{k}$, we can differentiate, noting that $\mathbf{i}$, $\mathbf{j}$, $\mathbf{k}$ rotate with the body.

$$\mathbf{M}_c = [\dot{\mathbf{h}}_c] + \boldsymbol{\omega} \times \mathbf{h}_c$$
$$= (\dot{h}_x\mathbf{i} + \dot{h}_y\mathbf{j} + \dot{h}_z\mathbf{k}) + \boldsymbol{\omega} \times \mathbf{h}_c \qquad (5.6\text{--}1)$$

The cross product $\boldsymbol{\omega} \times \mathbf{h}_c$ can easily be visualized as the rotation of the vectors $h_x\mathbf{i}$, $h_y\mathbf{j}$, $h_z\mathbf{k}$, due to ω_x, ω_y, ω_z, as shown in Fig. 5.6–1. Thus, by adding vectors along the x, y, z directions, the above equation becomes,

$$\mathbf{M}_c = M_x\mathbf{i} + M_y\mathbf{j} + M_z\mathbf{k}$$
$$= (\dot{h}_x + \omega_y h_z - \omega_z h_y)\mathbf{i} + (\dot{h}_y + \omega_z h_x - \omega_x h_z)\mathbf{j} + (\dot{h}_z + \omega_x h_y - \omega_y h_x)\mathbf{k}$$

$$(5.6\text{–}2)$$

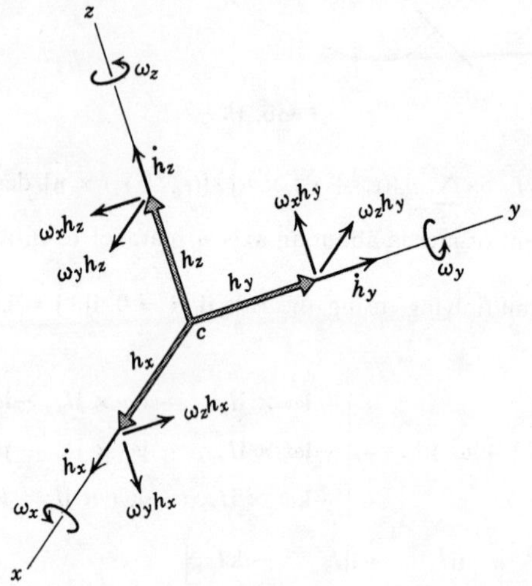

Fig. 5.6–1. Components of moment of momentum and their rate of change.

The component equations, known as Euler's moment equations, are

$$M_x = \dot{h}_x + \omega_y h_z - \omega_z h_y$$
$$M_y = \dot{h}_y + \omega_z h_x - \omega_x h_z \qquad (5.6\text{–}3)$$
$$M_z = \dot{h}_z + \omega_x h_y - \omega_y h_x$$

where the x, y, z axes with the origin coinciding with the center of mass rotate with angular velocity $\boldsymbol{\omega}$.

Equation 5.6–1 or 5.6–3 for the moment is applicable to any coordinate system with a fixed origin or a moving origin coinciding with the center of mass. The angular velocity $\boldsymbol{\omega}$ is that of the coordinate system and, if the axes are fixed in the body, the moments and products of inertia are constant.

For a body of revolution with moments of inertia A, A, C about principal axes, A about any transverse axis is the same, and so we might choose a set of transverse axes rotating at a speed different from that of the body without introducing a variable moment of inertia with time. We can for instance use the node axis system ξ, η, ζ, in which case the moments are,

$$M_\xi = \dot{h}_\xi + h_\zeta \omega_\eta - h_\eta \omega_\zeta$$
$$M_\eta = \dot{h}_\eta + h_\xi \omega_\zeta - h_\zeta \omega_\xi \qquad (5.6\text{--}4)$$
$$M_\zeta = \dot{h}_\zeta + h_\eta \omega_\xi - h_\xi \omega_\eta$$

These equations are known as Euler's modified equations.

5.7 Euler's Equation for Principal Axes

With the origin of the body axes coinciding with the center of mass, we can orient the x, y, z axes to coincide with the principal axes 1, 2, 3 of the body to eliminate the products of inertia terms in the moment of momentum expressions. We then have,

$$h_1 = A\omega_1 \qquad h_2 = B\omega_2 \qquad h_3 = C\omega_3$$

where A, B, C, are principal moments of inertia which are constant since 1, 2, 3 are fixed in the body. The moment equations are then,

$$M_1 = A\dot{\omega}_1 + \omega_2\omega_3(C - B)$$
$$M_2 = B\dot{\omega}_2 + \omega_1\omega_3(A - C)$$
$$M_3 = C\dot{\omega}_3 + \omega_1\omega_2(B - A)$$

which are called the Euler's equations for principal axes. The general solution of these equations is difficult, and in the sections to follow we will consider some special cases which enable an analytical solution.

5.8 Body with $A = B$ and Zero External Moment (Body Coordinates)

We will consider here a cylindrical disk with axis 3 normal to the circular face, as shown in Fig. 5.8–1.
The moment of inertia about the three body axes are,

$$B = A \qquad \text{about 1 and 2}$$
$$\text{and} \quad C \qquad \text{about 3}$$

Euler's equation then becomes,

$$A\dot{\omega}_1 + (C - A)\omega_2\omega_3 = 0$$
$$A\dot{\omega}_2 - (C - A)\omega_1\omega_3 = 0 \qquad (5.8\text{--}1)$$
$$C\dot{\omega}_3 = 0$$

From the third of these equations, we conclude that ω_3 must be a constant.

$$\omega_3 = n \qquad (5.8\text{-}2)$$

Making the substitution,

$$n\left(\frac{C-A}{A}\right) = \lambda \qquad (5.8\text{-}3)$$

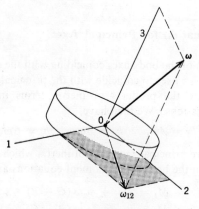

Fig. 5.8–1. Body of revolution with principal axes 1, 2, 3.

the first two equations can be written as,

$$\dot{\omega}_1 + \lambda\omega_2 = 0$$
$$\dot{\omega}_2 - \lambda\omega_1 = 0 \qquad (5.8\text{-}4)$$

Multiplying the first equation by ω_1 and the second by ω_2, and adding, we obtain the equation,

$$\omega_1\dot{\omega}_1 + \omega_2\dot{\omega}_2 = 0$$

or

$$\omega_1^2 + \omega_2^2 = \omega_{12}^2 = \text{constant} \qquad (5.8\text{-}5)$$

Thus the magnitude of the resultant angular velocity vector $\boldsymbol{\omega}$ is a constant.

$$\omega = \sqrt{\omega_1^2 + \omega_2^2 + \omega_3^2} = \sqrt{\omega_{12}^2 + n^2} = \text{constant} \qquad (5.8\text{-}6)$$

Since there is no moment acting on the body, we have,

$$\mathbf{M} = \dot{\mathbf{h}} = 0 \qquad (5.8\text{-}7)$$

which requires that the angular momentum vector $\mathbf{h}$ be a constant, fixed in space. It is evident that **h** must lie in the plane containing axis 3 and **ω**, since the component of **h** in the 1, 2-plane is

$$A(\omega_1 \mathbf{i} + \omega_2 \mathbf{j}) = A\omega_{12}\mathbf{1}_{12} \qquad (5.8\text{–}8)$$

which has the same direction as the component of **ω** in the 1, 2-plane. Thus, the plane containing axis 3 and the **ω** vector rotates about the fixed **h** vector, as shown in Fig. 5.8–2. The motion can be visualized by the rolling of the body cone on the space cone, which is fixed in space by the vector **h**.

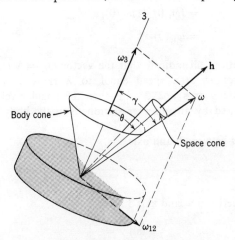

Fig. 5.8–2. Body cone rolls on space cone.

The speed of rotation of the plane containing **ω** and axis 3 about the line **h** can be found as follows. Differentiating the first of Eq. 5.8–4, and substituting from the second, we have,

$$\ddot{\omega}_1 + \lambda^2 \omega_1 = 0 \qquad (5.8\text{–}9)$$

Letting $\omega_1(0)$ and $\dot{\omega}_1(0)$ be the initial conditions at $t = 0$, the solution of this equation is,

$$\omega_1 = \omega_1(0) \cos \lambda t + \frac{\dot{\omega}_1(0)}{\lambda} \sin \lambda t \qquad (5.8\text{–}10)$$

Also, from this set of equations,

$$\omega_2 = -\frac{\dot{\omega}_1}{\lambda} = \omega_1(0) \sin \lambda t - \frac{\dot{\omega}_1(0)}{\lambda} \cos \lambda t \qquad (5.8\text{–}11)$$

From the last equation we obtain for $t = 0$,

$$\omega_2(0) = -\frac{\dot{\omega}_1(0)}{\lambda}$$

We note further that

$$\omega_{12} = \omega_1 + i\omega_2 \tag{5.8-12}$$

where $i = \sqrt{-1}$ so that, another form of the solution is obtained by adding ω_1 and ω_2 in quadrature

$$\omega_{12} = \left[\omega_1(0) - i\frac{\dot\omega_1(0)}{\lambda} \right](\cos \lambda t + i \sin \lambda t)$$

$$= [\omega_1(0) + i\omega_2(0)]\, e^{i\lambda t} \tag{5.8-13}$$

$$= \omega_{12}(0)\, e^{i\lambda t}$$

These equations all indicate that the vector $\omega_{12} = \sqrt{\omega_1{}^2 + \omega_2{}^2}$ in the 1, 2-plane, rotates at a speed equal to λ rad/sec, with respect to the rotating body axes 1, 2, 3. λ is then the angular velocity of the $\boldsymbol{\omega}$ vector as viewed by an observer stationed on the body at the axis of symmetry.

The angle θ between $\mathbf{h}$ and axis 3 is,

$$\tan \theta = \frac{A\omega_{12}}{Cn} \tag{5.8-14}$$

The angle γ between $\boldsymbol{\omega}$ and axis 3 is,

$$\tan \gamma = \frac{\omega_{12}}{n} \tag{5.8-15}$$

By comparison we have,

$$\tan \theta = \frac{A}{C} \tan \gamma \tag{5.8-16}$$

If $C > A$, then γ will be greater than θ, as shown in Fig. 5.8–2, and $\mathbf{h}$ will lie between axis 3 and $\boldsymbol{\omega}$. If $C < A$, γ will be smaller than θ and $\boldsymbol{\omega}$ will lie between $\mathbf{h}$ and axis 3.

The equations derived in this section all refer to the body axes which are in motion with the body. They do not tell us how the body moves in space. However, the following conclusions were obtained.

1. The vectors $\boldsymbol{\omega}$, $\mathbf{h}$, and axis 3 lie in the same plane.

2. In the plane of $\boldsymbol{\omega}$, $\mathbf{h}$ and axis 3, the angle θ between 03 and $\mathbf{h}$ is constant.

3. The angle γ between 03 and $\boldsymbol{\omega}$ is constant.

4. The vector $\mathbf{h}$ is constant or fixed in direction and magnitude.

5. The vector $\boldsymbol{\omega}$ has constant magnitude $\sqrt{\omega_{12}{}^2 + n^2}$.

6. The plane containing the three vectors rotates relative to the body axes at an angular speed λ.

5.9 Body of Revolution with Zero Moment, in Terms of Euler's Angles

We will consider the previous problem of Section 5.8, by introducing the Euler angles ψ, θ, φ, in order to establish the motion with respect to fixed axes. Since $\mathbf{h}$ is fixed in space, we will orient it along the OZ axis, in which case $\dot\psi$ becomes the angular velocity of the node line and the spin axis about the OZ direction. We also had $\dot\theta = 0$ from the previous section, so that Eq. 3.5–1 becomes,

$$\begin{aligned}
\omega_1 &= \dot\psi \sin\theta \sin\varphi \\
\omega_2 &= \dot\psi \sin\theta \cos\varphi \\
\omega_3 &= \dot\varphi + \dot\psi \cos\theta = n
\end{aligned} \tag{5.9-1}$$

Differentiating with $\dot\theta = 0$, the angular accelerations are,

$$\begin{aligned}
\dot\omega_1 &= \dot\psi\dot\varphi \sin\theta \cos\varphi \\
\dot\omega_2 &= -\dot\psi\dot\varphi \sin\theta \sin\varphi \\
\dot\omega_3 &= 0
\end{aligned} \tag{5.9-2}$$

Substituting into the first of Eq. 5.8–1, we obtain,

$$A\dot\psi\dot\varphi \sin\theta \cos\varphi + (\dot\varphi\dot\psi \sin\theta \cos\varphi + \dot\psi^2 \sin\theta \cos\theta \cos\varphi)(C - A) = 0 \tag{5.9-3}$$

Thus, the precession velocity in terms of the spin velocity $\dot\varphi$, C, A, and θ becomes,

$$\dot\psi = \frac{C\dot\varphi}{(A - C)\cos\theta} \tag{5.9-4}$$

Equation 5.9–4 states that the roll axis (also plane containing z and $\boldsymbol{\omega}$) rotates about a fixed OZ or $\mathbf{h}$ axis with a speed of $\dot\psi$ proportional to the angular velocity of spin $\dot\varphi$. If $C > A$, the spin must be negative, since $\dot\psi$ was drawn in the positive direction. Thus if $C > A$, the spin is opposite in sense to the precession, and we call the case *retrograde precession*. If $C < A$, then $\dot\varphi$ and $\dot\psi$ have the same sense, and we have *direct precession*. Thus a flat disk spinning about an axis perpendicular to its plane will have retrograde precession, whereas a slender rod spinning about its longitudinal axis will have direct precession. The space and body cones for the two cases are shown in Figs. 5.9–1 and 5.9–2. It should be noted also that, when $C > A$,

$$\left| \frac{C}{(A - C)\cos\theta} \right| > 1$$

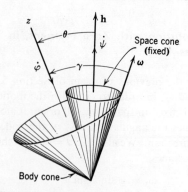

Fig. 5.9–1. Retrograde precession $C > A$.

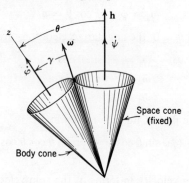

Fig. 5.9–2. Direct precession $C < A$.

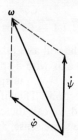

Fig. 5.9–3. For steady precession $\boldsymbol{\omega}$ is the vector sum of $\dot{\psi}$ and $\dot{\varphi}$.

Therefore,

$$\dot{\psi} > \dot{\varphi}$$

As $\theta \rightarrow 90°$, $\dot{\psi}$ becomes very large compared to $\dot{\varphi}$. Since $\dot{\theta} = 0$, the resultant angular velocity $\boldsymbol{\omega}$ is the vector sum of $\dot{\psi}$ and $\dot{\varphi}$, as shown in Fig. 5.9–3.

PROBLEMS

1. Show that a thin disk thrown spinning with its plane nearly parallel will make two wobbles to every cycle of spin.

2. A thin disk thrown into the air is seen to wobble so that its normal generates a cone of 20° twice per second. Determine the rate of spin and the total angular velocity vector with respect to the inertial coordinates.

3. A thin disk is spun about an axis making an angle γ with the normal to the disk and then released. Assuming the disk to be moment free, find the half-angle of the cone generated by the disk normal, and the time required for one complete rotation of the normal around the cone. Is the precession direct or retrograde?

4. A cylindrical disk has a thickness equal to $R/2$ where R is the radius. If it is spinning and precessing with its normal generating a cone angle of 15° with a fixed direction in space, determine the cone angle in space generated by the angular velocity vector. If the magnitude of ω is 10π rad/sec, determine its component ω_{12} in the plane of the disk.

5. In Prob. 2, determine the equations for the angular velocities about the body axes 1, 2, 3, where 1 and 2 are in the plane of the disk and 3 is normal to it.

6. In Prob. 2, determine the equations for the angular acceleration about the axes 1, 2, and 3.

7. Letting m be the mass of the disk of Prob. 4, write the equations for the vectors ω and $\mathbf{h}$, and determine the kinetic energy of rotation.

8. A uniform disk of radius R and mass m is mounted through its center to a shaft so that its normal makes an angle α with the shaft. If the shaft rotates at speed ω between bearings a distance l apart, determine the bearing reactions.

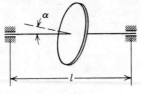

Prob. 8

9. A thin rectangular plate of sides a and b is mounted on a shaft in the plane of the plate through its center. The shaft makes an angle α with the long

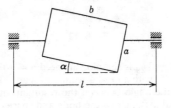

Prob. 9

side b. Determine the bearing reactions when the shaft rotates at speed ω. The distance between bearings is l.

10. Let x, y, z be the principal axes through the center of gravity of an airplane, with A, B, C as principal moments of inertia. Show that in order for the airplane to make a turn of radius R about a vertical line, with speed V, it must bank at an angle $\theta = \tan^{-1} V^2/Rg$ and supply a rolling moment in the same direction as θ, equal to

$$\frac{1}{2}(C - A)\frac{V^2}{R^2}\sin 2\theta$$

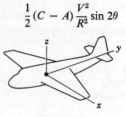

Prob. 10

11. Determine the gyroscopic moment necessary for the airplane of Prob. 10 to spin with speed $\dot{\psi}$ about the vertical OZ axis, when the nose of the airplane is inclined at an angle α below the horizontal.

12. If the spin $\dot{\psi}$ takes place about an axis through the center of gravity making angles α, β, γ, with respect to the x, y, z axes of the airplane, write the differential equations for the moments M_x, M_y, and M_z.

13. A slender rod of mass m and length l is welded to a shaft at its $\frac{1}{3}$ point at an angle θ. Determine the bearing reactions when the shaft rotates with speed ω, and the distance between bearings is $2c$.

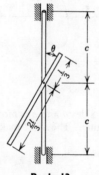

Prob. 13

14. A missile with $C/A = \frac{1}{20}$ is spinning at speed $n = 10\pi$ rad/sec and precessing at an angle $\theta = 5°$ with a fixed direction in space. Determine the precession speed $\dot{\psi}$ and the angle between the resultant angular velocity vector and the longitudinal axis. Draw the body and space cones. What can you conclude regarding the precessional speed of slender bodies?

5.10 Unsymmetrical Body with Zero External Moment (Poinsot's Geometric Solution)

In the general case of a body without an axis of revolution, the principal moments of inertia are unequal, so that Euler's equations for no external torque become,

$$A\dot{\omega}_1 + (C - B)\omega_2\omega_3 = 0$$
$$B\dot{\omega}_2 + (A - C)\omega_1\omega_3 = 0 \qquad (5.10\text{--}1)$$
$$C\dot{\omega}_3 + (B - A)\omega_1\omega_2 = 0$$

where 1, 2, 3 are body-fixed axes coinciding with the principal axis. The solution of these equations involve elliptic functions, and is taken up in Sec. 5.11. The following is a geometric discussion of the solution due to Poinsot.

With no external torque, the kinetic energy and the moment of momentum about the center of mass C are constants.

$$\mathbf{h} = \text{constant} \qquad (5.10\text{--}2)$$

No work is done on the body, therefore,

$$T = \text{constant} \qquad (5.10\text{--}3)$$

The expression for $\boldsymbol{\omega}$ and $\mathbf{h}$ are,

$$\boldsymbol{\omega} = \omega_1\mathbf{i} + \omega_2\mathbf{j} + \omega_3\mathbf{k} \qquad (5.10\text{--}4)$$

$$\mathbf{h} = A\omega_1\mathbf{i} + B\omega_2\mathbf{j} + C\omega_3\mathbf{k} \qquad (5.10\text{--}5)$$

from which we obtain,

$$\boldsymbol{\omega} \cdot \mathbf{h} = A\omega_1{}^2 + B\omega_2{}^2 + C\omega_3{}^2 = 2T = \text{constant} \qquad (5.10\text{--}6)$$

$$h^2 = A^2\omega_1{}^2 + B^2\omega_2{}^2 + C^2\omega_3{}^2 = \text{constant} \qquad (5.10\text{--}7)$$

With $\mathbf{h}$ drawn in a specified direction, its unit vector is $\mathbf{h}/h$, and the component of $\boldsymbol{\omega}$ along $\mathbf{h}$ will be given by

$$ON = \boldsymbol{\omega} \cdot \frac{\mathbf{h}}{h} = \frac{2T}{h} = \text{constant} \qquad (5.10\text{--}8)$$

Thus N is a fixed point along the direction of $\mathbf{h}$, as shown in Fig. 5.10–1, and the end of the $\boldsymbol{\omega}$ vector must lie in a plane through N perpendicular to ON. The line ON is referred to as the *invariable line*, and the perpendicular plane the *invariable plane*. Thus the end of the vector $\boldsymbol{\omega}$ must move in this invariable plane.

Letting the coordinates of $\boldsymbol{\omega}$ be ξ', η', ζ', we have $\omega_1 = \xi'$, $\omega_2 = \eta'$, and $\omega_3 = \zeta'$. The equation for $2T$ may then be written as

$$\frac{\xi'^2}{(\sqrt{2T/A})^2} + \frac{\eta'^2}{(\sqrt{2T/B})^2} + \frac{\zeta'^2}{(\sqrt{2T/C})^2} = 1 \qquad (5.10\text{-}9)$$

which is called the Poinsot ellipsoid. Figure 5.10–2 is an illustration of the ellipsoid.

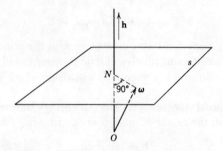

Fig. 5.10–1. Invariable line ON and invariable plane s.

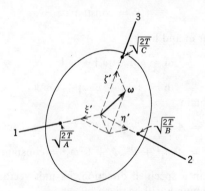

Fig. 5.10–2. Poinsot ellipsoid.

We next examine the inertia ellipsoid which was previously described as the locus of $\rho = 1/\sqrt{I}$ laid off along the $\boldsymbol{\omega}$ vector. Letting the coordinates of ρ be ξ, η, ζ, the equation for the inertia ellipsoid from Eq. 5.5–2 becomes,

$$\frac{\xi^2}{(\sqrt{1/A})^2} + \frac{\eta^2}{(\sqrt{1/B})^2} + \frac{\zeta^2}{(\sqrt{1/C})^2} = 1 \qquad (5.10\text{-}10)$$

We thus see that the Poinsot ellipsoid is proportional to the inertia ellipsoid with its coordinates equal to $\sqrt{2T}$ times the coordinates of the inertia ellipsoid.

The Poinsot ellipsoid can be shown to roll on the invariable plane with its center a distance $ON = 2T/h$ from it. Starting with the equation,

$$2T = \boldsymbol{\omega} \cdot \mathbf{h} \qquad (5.10\text{–}11)$$

and remembering that $\mathbf{h}$ and T are constant, we can obtain the following differential relationship.

$$d(2T) = d\boldsymbol{\omega} \cdot \mathbf{h} = 0 \qquad (5.10\text{–}12)$$

Since the vanishing of the dot product of two vectors $d\boldsymbol{\omega}$ and $\mathbf{h}$ requires that the cosine of the angle between them be zero, we conclude that $d\boldsymbol{\omega}$ and $\mathbf{h}$ are perpendicular to each other. With the end point of $\boldsymbol{\omega}$ moving in the invariable plane, any change $d\boldsymbol{\omega}$ of $\boldsymbol{\omega}$ is perpendicular to $\mathbf{h}$, and since the locus of $\boldsymbol{\omega}$ (ξ', η', ζ') corresponds to the Poinsot ellipsoid, it must be tangent to the invariable plane. Thus to an observer stationed to a fixed coordinate system, the motion of the body is described by the rolling of the Poinsot ellipsoid (or inertia ellipsoid with ξ, η, ζ coordinates increased by $\sqrt{2T}$) on the invariable plane.

If we wish to examine the motion from the point of view of an observer stationed on the moving body, the invariable plane will now appear to move with respect to the body. From the angular momentum equation, Eq. 5.10 7, we can rearrange the terms to form the angular momentum ellipsoid as follows.

$$\frac{\xi'^2}{(h/A)^2} + \frac{\eta'^2}{(h/B)^2} + \frac{\zeta'^2}{(h/C)^2} = 1 \qquad (5.10\text{–}13)$$

The curve traced by the end of the $\boldsymbol{\omega}$ vector is then defined by the intersection of the Poinsot, ellipsoid Eq. 5.10–9, and the angular momentum Eq. 5.10–13, ellipsoid.

The instantaneous axis $\boldsymbol{\omega}$ passes through the point of contact between the Poinsot ellipsoid and the invariable plane. It therefore generates simultaneously two cones, one in the fixed space, and the other in the body or the Poinsot ellipsoid. These cones, called the herpolhode and the polhode cones respectively, are shown in Fig. 5.10–3.

Since $\boldsymbol{\omega}$ is common to both the Poinsot ellipsoid and the momentum ellipsoid, the equation for the body cone can be obtained by subtracting Eq. 5.10–9 from 5.10–13 and multiplying by h^2.

$$A\left(A - \frac{h^2}{2T}\right)\xi'^2 + B\left(B - \frac{h^2}{2T}\right)\eta'^2 + C\left(C - \frac{h^2}{2T}\right)\zeta'^2 = 0$$

$$(5.10\text{–}14)$$

This equation indicates that for the body cone to be real, $h^2/2T$ must lie between the greatest and least values of A, B, and C. For instance, if we let $A > B > C$, then $h^2/2T$ must satisfy the relation,

$$A \geq \frac{h^2}{2T} \geq C$$

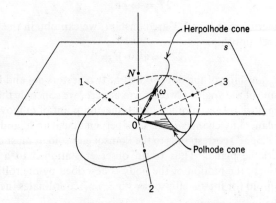

Fig. 5.10–3. Poinsot ellipsoid rolls on the invariable plane s.

To find the polhode curves, we let the Poinsot ellipsoid intersect the polhode cone. The form of these curves is best visualized by looking at their principal plane projections, obtained by eliminating in turn one of the coordinates between Eqs. 5.10–14 and 5.10–9. The three equations so obtained are,

$$A(A - C)\xi'^2 + B(B - C)\eta'^2 = 2T\left(\frac{h^2}{2T} - C\right)$$

$$A(A - B)\xi'^2 - C(B - C)\zeta'^2 = 2T\left(\frac{h^2}{2T} - B\right)$$

$$B(A - B)\eta'^2 + C(A - C)\zeta'^2 = 2T\left(A - \frac{h^2}{2T}\right) \qquad (5.10\text{–}15)$$

If $h^2/2T = C$, the first equation can be satisfied for $\xi' = \eta' = 0$. The polhode curve then degenerates to a point on the ζ' axis.

If $h^2/2T = B$, we obtain from the second of the three equations

$$\frac{\zeta'}{\xi'} = \sqrt{\frac{A(A - B)}{C(B - C)}} \qquad (5.10\text{–}16)$$

which indicates two planes passing through η' axis. The $\xi'\eta'$ and the $\eta'\zeta'$ projections from the other two equations, are ellipses.

If $h^2/2T = A$, the third of the three equations can only be satisfied if $\eta' = \zeta' = 0$, and the polhode curve degenerates to a point on the ξ' axis.

If $h^2/2T$ lies between B and C, the polhodes lie between the planes of Eq. 5.10–16 and the ζ' axis. Their $\xi'\eta'$ projections are ellipses.

If $h^2/2T$ lies between A and B, the polhodes lie in the central part of the ellipsoid between the planes of Eq. 5.10–16. Their $\eta'\zeta'$ projections are

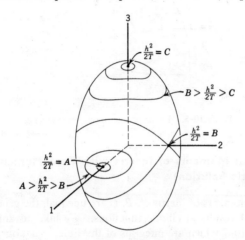

Fig. 5.10–4.   Polhode curves.

ellipses and their $\xi'\zeta'$ projections are hyperbolas. The general nature of the polhode curve for the various cases are shown in Fig. 5.10–4.

If we write Eqs. 5.10–6 and 5.10–7 in terms of momentum components, h_1, h_2, h_3, they become

$$\frac{h_1^2}{A} + \frac{h_2^2}{B} + \frac{h_3^2}{C} = 2T \tag{5.10--17}$$

$$h_1^2 + h_2^2 + h_3^2 = h^2 \tag{5.10--18}$$

and their simultaneous solution give the intersection of the Poinsot ellipsoid, Eq. 5.10–17, and the momentum sphere, Eq. 5.10–18, in terms of the momentum coordinates. Since Eqs. 5.10–14 and 5.10–15 are also solutions to the same problem, we need only to rewrite these equations in terms of h_1, h_2, and h_3. Thus Eq. 5.10–14 rewritten in terms of momentum coordinates is

$$\left(1 - \frac{1}{A}\frac{h^2}{2T}\right)h_1^2 + \left(1 - \frac{1}{B}\frac{h^2}{2T}\right)h_2^2 + \left(1 - \frac{1}{C}\frac{h^2}{2T}\right)h_3^2 = 0 \tag{5.10--19}$$

Figure 5.10–5 is a plot of Eq. 5.10–19 showing the locus of the **h** vector on the momentum sphere for various values of $h^2/2T$, where $A > B > C$, and $A \geq h^2/2T \geq C$.

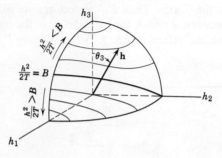

Fig. 5.10–5.   Locus of **h** on momentum sphere.

5.11   Unequal Moments of Inertia with Zero Moment (Analytic Solution)

When the moments of inertia A, B, C are unequal, the general solution of Eqs. 5.10–1 results in elliptic functions. We take advantage of Eqs. 5.10–6 and 5.10–7 which are integrals of the Euler equations, and obtain

$$h^2 - 2TA = B(B - A)\omega_2{}^2 + C(C - A)\omega_3{}^2 \qquad (5.11\text{–}1)$$

$$h^2 - 2TB = A(A - B)\omega_1{}^2 + C(C - B)\omega_3{}^2 \qquad (5.11\text{–}2)$$

$$h^2 - 2TC = A(A - C)\omega_1{}^2 + B(B - C)\omega_2{}^2 \qquad (5.11\text{–}3)$$

If we assume $A > B > C$, Eq. 5.11–1 is always negative, Eq. 5.11–3 is always positive, and Eq. 5.11–2 may be either positive or negative. From Eqs. 5.11–1 and 3, we write

$$\omega_1{}^2 = \frac{h^2 - 2TC}{A(A - C)} \left\{ 1 - \frac{B(B - C)}{h^2 - 2TC}\, \omega_2{}^2 \right\}$$

$$\omega_3{}^2 = \frac{2TA - h^2}{C(A - C)} \left\{ 1 - \left(\frac{A - B}{B - C}\right)\left(\frac{h^2 - 2TC}{2TA - h^2}\right)\frac{B(B - C)}{(h^2 - 2TC)}\, \omega_2{}^2 \right\}$$

$$(5.11\text{–}4)$$

We now let

$$y = \sqrt{\frac{B(B - C)}{h^2 - 2TC}}\, \omega_2$$

$$(5.11\text{–}5)$$

$$k = \sqrt{\left(\frac{A - B}{B - C}\right)\left(\frac{h^2 - 2TC}{2TA - h^2}\right)} \qquad 0 \leq k \leq 1$$

so that

$$\omega_1 = \sqrt{\frac{h^2 - 2TC}{A(A - C)}} \sqrt{1 - y^2} \qquad (5.11\text{--}6)$$

$$\omega_3 = \sqrt{\frac{2TA - h^2}{C(A - C)}} \sqrt{1 - k^2 y^2} \qquad (5.11\text{--}7)$$

Substituting these into the second of Eq. 5.10–1, we obtain

$$\dot{\omega}_2 = \sqrt{\frac{h^2 - 2TC}{B(B - C)}} \; \dot{y}$$

$$= \frac{(C - A)}{B} \sqrt{\frac{(h^2 - 2TC)(2TA - h^2)}{AC(C - A)^2}} \sqrt{(1 - y^2)(1 - k^2 y^2)} \qquad (5.11\text{--}8)$$

which is recognized as an elliptic integral of the first kind,

$$Nt = u = \sqrt{\frac{(B - C)(2TA - h^2)}{ABC}} \; t = \int_0^y \frac{dy}{\sqrt{(1 - y^2)(1 - k^2 y^2)}} \qquad (5.11\text{--}9)$$

where t is measured from the instant when $\omega_2 = 0$. On letting $y = \sin \phi$

$$u = \int_0^\phi \frac{d\phi}{\sqrt{1 - k^2 \sin^2 \phi}} = F(\phi, k) \qquad (5.11\text{--}10)$$

and u becomes a function of the modulus k and the amplitude ϕ. Conversely, y is a function of $u = Nt$ and k, and is available as a tabulated function for $0 \le k \le 1$, where N is defined by Eq. 5.11–9.

$$y = \mathscr{S}n(u, k) \qquad (5.11\text{--}11)$$

Thus from Eq. 5.11–5 the solution for ω_2 is

$$\omega_2 = \sqrt{\frac{h^2 - 2TC}{B(B - C)}} \; \mathscr{S}n(Nt, k) \qquad (5.11\text{--}12)$$

Likewise, it can be shown that the solution for the angular velocities about the other two axes are

$$\omega_1 = \sqrt{\frac{h^2 - 2TC}{A(A - C)}} \, \mathscr{C}n(Nt, k)$$

$$\omega_3 = -\sqrt{\frac{2TA - h^2}{C(A - C)}} \, \mathscr{D}n(Nt, k)$$

(5.11–13)

where the $\mathscr{C}n$ and $\mathscr{D}n$ functions are related to the $\mathscr{S}n$ function by the equations,

$$\mathscr{C}n^2x = 1 - \mathscr{S}n^2x$$

$$\mathscr{D}n^2x = 1 - k^2\mathscr{S}n^2x$$

(5.11–14)

These solutions correspond to the case $h^2 < 2TB$ which is required for $0 \le k \le 1$.

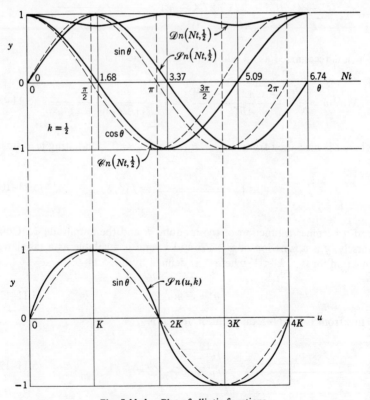

Fig. 5.11–1. Plot of elliptic functions.

Fig. 5.11–2. Plot of elliptic functions.

For small values of $k \to 0$ (i.e., A approaching B), the elliptic functions approach the trigonometric functions,

$$\mathcal{S}n(Nt, 0) = \sin (Nt)$$
$$\mathcal{C}n(Nt, 0) = \cos (Nt)$$
$$\mathcal{D}n(Nt, 0) = 1.0$$

Thus when $A = B$, the angular velocities reduce to the form:

$$\omega_2 = \beta \sin (Nt)$$
$$\omega_1 = \beta \cos (Nt)$$
$$\omega_3 = n$$

which agree with the results of Sec. 5.8.

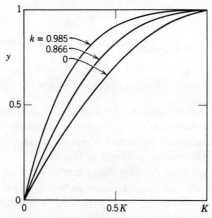

Table of $y = \sin \phi = \mathcal{S}n(u, k)$ Taken from Peirce's Table of Integrals
3rd Revised Ed. p. 122

	Ordinate, $\mathcal{S}n(u, k) =$	Abscissa $= u = Nt = \int_0^\phi \dfrac{d\phi}{\sqrt{1 - k^2 \sin^2 \phi}}$				
ϕ	$\sin \phi$	$k = 0$	$k = 0.50$	$k = 0.707$	$k = 0.866$	$k = 0.9848$
0°	0	0	0	0	0	0
10°	0.1736	0.111	0.1037	0.0943	0.0814	0.0556
20°	0.3420	0.222	0.2080	0.190	0.1646	0.113
30°	0.500	0.3333	0.3140	0.289	0.2515	0.174
45°	0.707	0.500	0.477	0.445	0.3945	0.278
60°	0.866	0.666	0.645	0.616	0.563	0.413
75°	0.9563	0.834	0.821	0.802	0.765	0.617
90°	1.000	1.000	1.000	1.000	1.000	1.000
		K	$= 1.686$	$= 1.854$	$= 2.156$	$= 3.153$

Fig. 5.11–3. Plot of elliptic function $\mathcal{S}n(u, k)$.

To show the effect of $A \neq B$, a plot of the elliptic function for $k = \frac{1}{2}$ is compared to the trigonometric functions in Fig. 5.11–1. If $y = \sin \phi = \mathscr{S}n(u, k)$ is plotted against $u = Nt$ for various values of k, with u normalized to unity at $\phi = 90°$, the plot of $\mathscr{S}n(u, k)$ will appear as in Figs. 5.11–2 and 5.11–3.

5.12 Stability of Rotation about Principal Axes

If we write Euler's equations for no external moment in the form,

$$A\dot{\omega}_1 = (B - C)\omega_2\omega_3$$
$$B\dot{\omega}_2 = (C - A)\omega_1\omega_3 \qquad (5.12\text{–}1)$$
$$C\dot{\omega}_3 = (A - B)\omega_1\omega_2$$

we find that,

$$\omega_1 = \text{constant} \quad \text{if} \quad \omega_2 = \omega_3 = 0$$
$$\omega_2 = \text{constant} \quad \text{if} \quad \omega_1 = \omega_3 = 0$$
$$\omega_3 = \text{constant} \quad \text{if} \quad \omega_1 = \omega_2 = 0$$

indicating that permanent rotations are possible about each of the principal axes. We will now show that these permanent rotations are stable about the axes of maximum and minimum moments of inertia, and unstable about the axis of intermediate moment of inertia.

We will assume constant rotation about one of the axes, say axis 1, and allow a small perturbation to determine its stability. We then have an initial condition $\omega_1 = \omega_0$, $\omega_2 = \omega_3 = 0$, and a perturbed condition, $\omega_1 = \omega_0 + \epsilon$, with ω_2 and ω_3 small. The linearized equations are,

$$A\dot{\epsilon} = 0$$

$$B\dot{\omega}_2 = (C - A)\omega_0\omega_3 \qquad (5.12\text{–}2)$$

$$C\dot{\omega}_3 = (A - B)\omega_0\omega_2$$

Differentiating the second and third of Eq. 5.12–2, we obtain,

$$\ddot{\omega}_2 + \frac{(A - B)(A - C)}{BC}\,\omega_0{}^2\omega_2 = 0$$

$$\qquad (5.12\text{–}3)$$

$$\ddot{\omega}_3 + \frac{(A - B)(A - C)}{BC}\,\omega_0{}^2\omega_3 = 0$$

which are stable provided $(A - B)(A - C)$ is positive. We see that this condition is satisfied provided $A > B$, and $A > C$ (A a major principal axis), or $B > A$, and $C > A$ (A a minor principal axis). When A is an intermediate value between B and C, $(A - B)(A - C)$ is negative, and small values of ω_2 and ω_3 will increase. The first of Eq. 5.12–2 requires that the small perturbation ϵ remain constant, and, hence, if axis 1 is either a major or minor principal axis, the rotation is stable. If axis 1 is an intermediate axis, ω_2 and ω_3 will increase, resulting in an unstable oscillation. The above conclusions can be simply demonstrated by dropping an eraser spinning about each of the principal axes. Situations requiring modification of these conclusions are discussed in Sec. 7.6.

PROBLEMS

1. Determine the ellipsoid of inertia for the configuration of Prob. 5.5–4.
2. Determine the ellipsoid of inertia for a solid uniform cylinder of radius R and length $2R$. What is its moment of inertia about a line passing through its geometric center and the perifery of one end.
3. Consider a spinning body with $A > B > C$. With initial conditions $\omega_1 = p_0$, $\omega_2 = 0$, $\omega_3 = r_0$, find ω_1, ω_2, and ω_3 for a motion characterized by one of the plane polhodes

$$\frac{r_0}{p_0} = \pm \sqrt{\frac{A(A - B)}{C(B - C)}}$$

Show that, with t approaching ∞, the motion tends to a permanent rotation about axis 2.

4. Show that, for a body with principal moments of inertia A, B, C, the polhodes are closed curves while the herpolhodes generated on the invariable plane are generally open curves.
5. Derive the equations for ω_1, ω_2, ω_3, when $h^2 > 2TB$ and $A > B > C$.
6. Discuss the stability of rotation of an unsymmetric body in the moment-free case in terms of the polhode curves of Fig. 5.10–4.
7. Solve for the exact solution of a simple pendulum oscillating through large amplitudes. Express the equation for the period in terms of an elliptical integral and its series approximation.
8. For $h^2/2T < B$, where $A > B > C$, the angle θ_3 made by the $\mathbf{h}$ vector and the h_3 axis is defined by the equation (see Fig. 5.10–5)

$$\sin \theta_3 = \frac{1}{h} \sqrt{h_1{}^2 + h_2{}^2}$$

Show that the value of θ_3 varies between

$$\sin \theta_{3\text{max}} = \sqrt{\frac{BC}{B - C}\left(\frac{1}{C} - \frac{1}{h^2/2T}\right)}$$

$$\sin \theta_{3\text{min}} = \sqrt{\frac{AC}{A - C}\left(\frac{1}{C} - \frac{1}{h^2/2T}\right)}$$

Hint: Rewrite Eqs. 5.10–15 in terms of h_1, h_2, h_3.

9. Derive the equations for ω_1, ω_2, ω_3 when $C > B > A$ and examine the conditions $h^2 < 2TB$ and $h^2 > 2TB$.

5.13   General Motion of a Symmetric Gyro or Top

Figure 5.13–1 shows a symmetric gyro spinning about axis ζ and supported by two gimbals. The inner gimbal allows a pitching rotation of the spin axis about the horizontal bearings ξ, while the outer gimbal is free

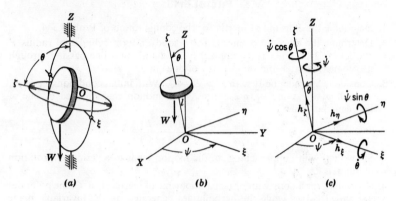

(a)   (b)   (c)

Fig. 5.13–1.   Symmetric gyro-angular momentum about gimbal axes.

to rotate about the vertical Z axis. The gyro is then pivoted about the stationary geometric center of the gimbal system, and the center of mass does not coincide with the fixed center O.

We will now define precession and nutation in the following manner. The rotation $\dot{\psi}$ of the horizontal axis ξ (node axis) about the vertical Z axis is called precession. If the angle θ is held constant, the spin axis will generate a cone due to precession. The rotation $\dot{\theta}$ of the inner gimbal about the node axis ξ is called nutation. The term signifies a nodding of the spin axis. In the general case, both precession and nutation may exist simultaneously.

We will at first neglect the mass of the gimbals, in which case we would have only the spinning wheel free to rotate in any manner about the stationary geometric center O, as shown in Fig. 5.13–1b. The system is then identical to that of a spinning top pivoted about a fixed point O, and subject to a gravity torque $Wl \sin \theta$ about the axis ξ. If $l = 0$, we have the special case of a gyro free to rotate about the center of gravity.

It is convenient here to write the moment equations about the cartesian coordinates through the geometric center with the line of nodes ξ as one of the axes. We will let the moment of inertia of the wheel about the node axis system ξ, η, and ζ be A, A, C. The angular velocities of the ξ, η, ζ axes and the angular momentum about them are,

$$
\begin{aligned}
\omega_\xi &= \dot\theta & h_\xi &= A\dot\theta \\
\omega_\eta &= \dot\psi \sin\theta & h_\eta &= A\dot\psi \sin\theta \\
\omega_\zeta &= \dot\psi \cos\theta & h_\zeta &= C(\dot\varphi + \dot\psi \cos\theta)
\end{aligned}
\tag{5.13-1}
$$

where φ is the spin angle of the body axes 1, 2, referenced to the node axis ξ.

The components of the moment equation,

$$
\mathbf{M} = [\dot{\mathbf{h}}] + \boldsymbol\omega \times \mathbf{h}
\tag{5.13-2}
$$

can be immediately written down by examination of Fig. 5.13-1c.

$$
\begin{aligned}
M_\xi &= \dot h_\xi + h_\zeta \dot\psi \sin\theta - h_\eta \dot\psi \cos\theta \\
M_\eta &= \dot h_\eta + h_\xi \dot\psi \cos\theta - h_\zeta \dot\theta \\
M_\zeta &= \dot h_\zeta + h_\eta \dot\theta - h_\xi \dot\psi \sin\theta
\end{aligned}
\tag{5.13-3}
$$

Substituting for the components of $\mathbf{h}$ and $\mathbf{M}$, these equations become,

$$
Wl \sin\theta = A\ddot\theta + C(\dot\varphi + \dot\psi \cos\theta)\dot\psi \sin\theta - A\dot\psi^2 \sin\theta \cos\theta
$$

$$
0 = A\frac{d}{dt}(\dot\psi \sin\theta) + A\dot\theta\dot\psi \cos\theta - C\dot\theta(\dot\varphi + \dot\psi \cos\theta)
$$

$$
0 = C\frac{d}{dt}(\dot\varphi + \dot\psi \cos\theta)
\tag{5.13-4}
$$

The last equation indicates that $\dot\varphi + \dot\psi \cos\theta$ is a constant, and we will let it equal n.

$$
n = \dot\varphi + \dot\psi \cos\theta
\tag{5.13-5}
$$

With this substitution, the first two equations of Eq. 5.13-4 can be integrated with proper integrating factors. We will, however, consider an alternative approach based on certain integrals of the equations of motion which are constant. These are the conservation of total energy, and the conservation of angular momentum about the vertical Z axis which is moment-free.

If we examine the equations for the angular velocities about the body axes,

$$
\begin{aligned}
\omega_1 &= \dot\theta \cos\varphi + \dot\psi \sin\theta \sin\varphi \\
\omega_2 &= -\dot\theta \sin\varphi + \dot\psi \sin\theta \cos\varphi \\
\omega_3 &= \dot\varphi + \dot\psi \cos\theta = n
\end{aligned}
\tag{5.13-6}
$$

we would find that by adding the squares of the first two, that the square of the resultant velocity in the equatorial plane, in terms of Euler's angles is

$$\omega_1{}^2 + \omega_2{}^2 = \dot\theta^2 + \dot\psi^2 \sin^2\theta \qquad (5.13\text{--}7)$$

Since ω_3 is constant and equal to n, the kinetic energy can now be written as,

$$T = \tfrac{1}{2}Cn^2 + \tfrac{1}{2}A(\dot\theta^2 + \dot\psi^2 \sin^2\theta) \qquad (5.13\text{--}8)$$

Referencing the potential energy to the level of the origin of the coordinate system,

$$U = Wl \cos\theta \qquad (5.13\text{--}9)$$

and the total energy E, which must be a constant, becomes

$$E = \tfrac{1}{2}Cn^2 + \tfrac{1}{2}A(\dot\theta^2 + \dot\psi^2 \sin^2\theta) + Wl \cos\theta \qquad (5.13\text{--}10)$$

which is one of the first integrals of the differential equations of motion.

With the moment about the Z axis equal to zero, the momentum $h_Z = h_\zeta \cos\theta + h_\eta \sin\theta$ must be a constant,

$$h_Z = Cn \cos\theta + A\dot\psi \sin^2\theta \qquad (5.13\text{--}11)$$

This equation could also be obtained from the Lagrangian approach since the generalized coordinate ψ is a cyclic coordinate with $M_Z = 0$ (see Chap. 9) Solving for $\dot\psi$ from the above equation,

$$\dot\psi = \frac{h_Z - Cn \cos\theta}{A \sin^2\theta} \qquad (5.13\text{--}12)$$

and substituting into Eq. 5.13–10, we obtain the following form of the energy equation,

$$E - \frac{Cn^2}{2} = \frac{A\dot\theta^2}{2} + \frac{(h_Z - Cn \cos\theta)^2}{2A \sin^2\theta} + Wl \cos\theta \qquad (5.13\text{--}13)$$

Equation 5.13–13 is entirely in terms of θ, and its solution substituted into Eq. 5.13–10 completely describes the motion of the system.

We now make the following substitution of symbols:

$$\alpha = \frac{2}{A}\left(E - \frac{Cn^2}{2}\right) \qquad \text{a constant}$$

$$\beta = \frac{2Wl}{A} \qquad \text{a constant}$$

$$\gamma = \frac{h_Z}{A} \qquad \text{a constant}$$

$$N = \frac{Cn}{A} \qquad \text{a constant}$$

$$u = \cos\theta \qquad \text{a variable}$$

which enables Eq. 5.13–13 to be written as,

$$\alpha \sin^2 \theta = \dot{\theta}^2 \sin^2 \theta + (\gamma - N \cos \theta)^2 + \beta \cos \theta \sin^2 \theta \quad (5.13-14)$$

In terms of u, the equation becomes,

$$\dot{u}^2 = (\alpha - \beta u)(1 - u^2) - (\gamma - Nu)^2 \quad (5.13-15)$$

The solution of the above equation is given by the following integral which can be evaluated in terms of elliptic functions.

$$t = \int_{u(0)}^{u(t)} \frac{du}{\sqrt{(\alpha - \beta u)(1 - u^2) - (\gamma - Nu)^2}} \quad (5.13-16)$$

The mathematical solution resulting from the above elliptical integral is difficult to interpret, however; fortunately it is not necessary to carry out the above solution to evaluate the behavior of the gyro. If we let,

$$\dot{u}^2 = f(u) \quad (5.13-17)$$

Equation 5.13–15 can be written as,

$$f(u) = (\alpha - \beta u)(1 - u^2) - (\gamma - Nu)^2 \quad (5.13-18)$$

and the roots of this equation will tell us a great deal about the motion of the gyro.

Although $u = \cos \theta$ is limited between ± 1 for the physical problem, mathematically u can extend outside this region. For large values of u, the dominant term in $f(u)$ is βu^3, thus $f(u)$ must be positive for large positive u, and negative for large negative u, as shown in Fig. 5.13–2. Also at $u = 1$, the first term drops out, leaving,

$$f(\pm 1) = -(\gamma \mp N)^2$$

It is evident then that $f(u)$ at $u = \pm 1$ must always be negative. Looking at the expression for $f(u)$ in terms of θ,

$$f(u) = \dot{u}^2 = \dot{\theta}^2 \sin^2 \theta \quad (5.13-19)$$

we find that, for real values of θ and $\dot{\theta}$, $f(u)$ must be positive. We therefore conclude that, for the physical problem, $u = \cos \theta$, must always lie between u_1 and u_2 for which $f(u)$ is positive.

We note next that for $\theta > 0$, $\dot{\theta}$ must be zero at u_1 and u_2, which requires that the spin axis $O\zeta$ move between the bounding circles $u_1 = \cos \theta_1$ and $u_2 = \cos \theta_2$, as shown in Fig. 5.13–3.

The type of curve traced by the spin axis in the region θ_1 and θ_2, depends on the relative values of γ and N. For example, Eq. 5.13–12 can be written as,

$$\dot{\psi} = \frac{\gamma - Nu}{1 - u^2}$$

and since u is less than 1 between u_1 and u_2, the sign of $\dot\psi$ depends on $\gamma - Nu$. If $\gamma = Nu_1$, then $\dot\psi$ must be zero at the upper bounding circle and positive for θ greater than θ_1, as shown in Fig. 5.13–4. To obtain the curve of Fig. 5.13–5, $\dot\psi$ must change sign for some value of u between u_1 and u_2. Thus for this case, $\gamma - Nu_i = 0$ for $u_2 < u_i < u_1$.

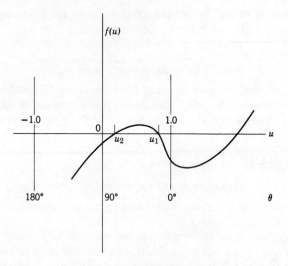

Fig. 5.13–2. Cubic equation representing motion of symmetric gyro.

Initial conditions

If a gyro or top is started at $t = 0$ with $\theta = \theta_0$ and $\dot\theta = \dot\psi = 0$, the values of the two constants of the system, E and h_Z are found from Eqs. 5.13–12 and 5.13–13, to be,

$$h_Z = Cn \cos \theta_0$$
$$E - \tfrac{1}{2}Cn^2 = Wl \cos \theta_0$$

Substituting into Eqs. 5.13–12 and 5.13–13, the equations for the precession and nutation are,

$$\dot\psi = \frac{N(\cos \theta_0 - \cos \theta)}{\sin^2 \theta}$$

$$\dot\theta^2 = (\cos \theta_0 - \cos \theta)\left[\beta - \frac{N^2}{\sin^2 \theta}(\cos \theta_0 - \cos \theta)\right] \quad (5.13\text{–}20)$$

These equations both agree with the initial conditions imposed on the system. The second of the above equations indicates that the right side of the equation must always be positive, therefore θ_0 must correspond to the

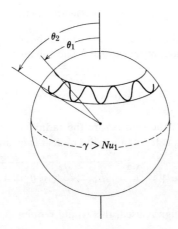

Fig. 5.13–3. Oscillation bounded between θ_1 and θ_2 with $\gamma > Nu_1$.

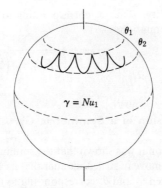

Fig. 5.13–4. Oscillation bounded between θ_1 and θ_2 with $\gamma = Nu_1$.

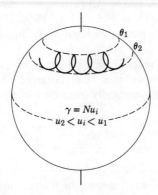

Fig. 5.13–5. Oscillation bounded between θ_1 and θ_2 with $\gamma = Nu_i$, where $u_2 < u_i < u_1$.

highest elevation of the spin axis, or the upper bounding circle. The lower bounding circle is found by letting $\dot\theta = 0$ and solving for $\cos\theta$. The result is

$$\cos\theta = \frac{N^2}{2\beta} \pm \sqrt{1 - \frac{N^2}{\beta}\cos\theta_0 + \left(\frac{N^2}{2\beta}\right)^2} \qquad (5.13\text{--}21)$$

It can be shown that the sign before the radical must be negative. For instance, since $\cos\theta_0$ is less than 1, the radical is greater than

$$\sqrt{1 - \frac{N^2}{\beta} + \left(\frac{N^2}{2\beta}\right)^2} = 1 - \frac{N^2}{2\beta} < \sqrt{1 - \frac{N^2}{\beta}\cos\theta_0 + \left(\frac{N^2}{2\beta}\right)^2} = \cos\theta - \frac{N^2}{2\beta}$$

Thus if the positive sign is used, this would require that,

$$\cos\theta > 1$$

which is an impossibility. Thus θ_2 corresponding to the lower bounding circle is given by the equation

$$\cos\theta_2 = \frac{N^2}{2\beta} - \sqrt{1 - \frac{N^2}{\beta}\cos\theta_0 + \left(\frac{N^2}{2\beta}\right)^2} \qquad (5.13\text{--}21)$$

5.14 Steady Precession of a Symmetric Gyro or Top

In the previous section it was shown that the spinning top pivoted about a fixed point is able to move in such a manner that its axis of symmetry $O\zeta$, occupies a zone between θ_1 and θ_2 corresponding to the roots of $f(u) = 0$ at u_1 and u_2. It is evident then that as u_1 and u_2 approach each other, the annular zone between θ_1 and θ_2 will narrow until eventually they merge to a single value θ_s, as shown in Fig. 5.14–1. Physically, this is the angle of steady precession which can be initiated by the initial conditions, $\theta = \theta_s$, $\dot\theta = 0$, and $\dot\psi = \dot\psi_s$.

The analysis for this special case is probably one of the simplest cases of the spinning top, which will now be investigated. Although the problem can be approached mathematically from the two equations,

$$f(u) = 0$$

$$\frac{df(u)}{du} = 0$$

it will be more instructive to examine the problem from a physical basis as follows.

For steady precession, $\dot\theta = 0$, and the angular velocities about the ξ, η, ζ axes, shown in Fig. 5.13–1c, are

$$\omega_\xi = 0$$
$$\omega_\eta = \dot\psi \sin\theta$$
$$\omega_\zeta = \dot\psi \cos\theta$$

The angular momentum about the corresponding axes are,

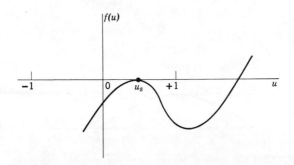

Fig. 5.14–1. Steady precession corresponding to $u_1 = u_2 = u_s$.

$$h_\xi = 0$$
$$h_\eta = A\dot\psi \sin\theta \qquad\qquad (5.14\text{–}1)$$
$$h_\zeta = C(\dot\varphi + \dot\psi \cos\theta) = Cn$$

and the moment about the ξ axis becomes

$$M_\xi = h_\zeta \dot\psi \sin\theta - h_\eta \dot\psi \cos\theta$$
$$= Cn\dot\psi \sin\theta - A\dot\psi^2 \sin\theta \cos\theta = Wl \sin\theta \qquad (5.14\text{–}2)$$

or

$$\dot\psi^2 - \left(\frac{Cn}{A\cos\theta}\right)\dot\psi + \frac{Wl}{A\cos\theta} = 0 \qquad (5.14\text{–}3)$$

The two precessional speeds are then given by the equation,

$$\dot\psi_{1,2} = \frac{Cn}{2A\cos\theta} \pm \sqrt{\left(\frac{Cn}{2A\cos\theta}\right)^2 - \frac{Wl}{A\cos\theta}} \qquad (5.14\text{–}4)$$

provided the spin is great enough to keep the radical of the above equation positive. This requirement is satisfied if,

$$n^2 > \frac{4AWl \cos\theta}{C^2} \qquad (5.14\text{–}5)$$

Further insight to the problem is obtained by plotting M_ξ versus $\dot\psi$ from Eq. 5.14–2, as shown in Fig. 5.14–2. The curve is a parabola and for any value of M_ξ there correspond two precessional speeds. When $M_\xi = 0$, the precessional speeds are zero and $\dot\psi_0$. This value of the precessional speed is given by the equation,

$$\dot\psi_0 = \frac{Cn}{A \cos \theta} = \frac{C\dot\varphi}{(A - C) \cos \theta} \tag{5.14–6}$$

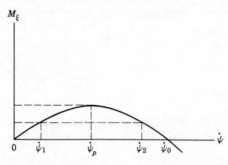

Fig. 5.14–2. Constant moment M_ξ results in two possible precessional speeds $\dot\psi_1$ and $\dot\psi_2$.

which agree with that of the moment-free gyro (see Eq. 5.9–4). The precessional speed corresponding to the peak moment is

$$\dot\psi_p = \frac{Cn}{2A \cos \theta} \tag{5.14–7}$$

and the corresponding peak moment is

$$M_{\xi\text{max}} = \frac{1}{4} \frac{C^2 n^2}{A} \tan \theta \tag{5.14–8}$$

For intermediate moment M_ξ, the two precessional speeds $\dot\psi_1$ and $\dot\psi_2$ are referred to as the slow and the fast precession. In general, the fast precession is not attained due to the high kinetic energy required, and the precession of a spinning top is usually slow precession.

Limiting cases

For $\dot\psi$ to be real, it is necessary for the terms under the radical of Eq. 5.14–4 to be greater than zero, or,

$$C^2 n^2 \geq 4WlA \cos \theta$$

If $\theta = 90°$, Eq. 5.14–5 indicates that the minimum required n is zero. However, Eq. 5.14–4 indicates that $\dot\psi$ then is indeterminate. This limiting case can be resolved for n greater than zero by the elementary consideration of Fig. 5.14–3. The rate of change of the angular momentum vector $h_\zeta = Cn$, is equal to the moment $M_\xi = Wl$,

$$M_\xi = \dot h_\zeta$$
$$Wl = Cn\dot\psi$$

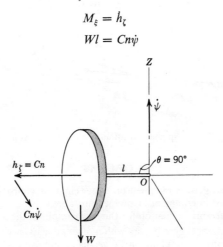

Fig. 5.14–3.   Limiting case $\theta = 90°$.

The whirling speed $\dot\psi$ for $\theta = 90°$ is therefore equal to,

$$\dot\psi = \frac{Wl}{Cn} \tag{5.14–9}$$

and as long as n is finite, $\dot\psi$ will also remain finite.

The same result can also be obtained from Eq. 5.14–3, for as $\cos \to 0$, the term $\dot\psi^2$ is negligible compared to the other two.

For $\theta = 0°$, the required value of n from Eq. 5.14–5 must satisfy the equation,

$$n \ge \frac{2}{C} \sqrt{WlA} \tag{5.14–10}$$

in which case we have the sleeping top. Equation 5.14–10 is often used to determine the spin of a missile or a projectile necessary for stability. Referring to Fig. 5.14–4, as long as the spin axis $O\zeta$ coincides with the velocity vector of the missile, the drag force will also coincide with $O\zeta$. However, if the spin axis deviates slightly from the velocity vector by a small angle θ, the drag force, which acts at the center of pressure, a distance l ahead of the center of mass, will have a moment $Rl \sin \theta$ about the center of mass. Thus this problem is identical to that of the sleeping top with Wl

replaced by Rl. The missile or projectile will hence be stable if the spin is great enough as established by,

$$n > \frac{2}{C} \sqrt{RlA},$$

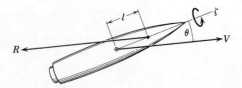

Fig. 5.14–4. Spin stabilization of missiles and projectiles.

PROBLEMS

1. A symmetrical gyro is mounted on two weightless gimbals. What are the expressions for the moment of momentum about the ξ, η, ζ, and Z axes. Write the differential equation for the moment about the above axes.

2. For the symmetrical gyro or top, with moment due to gravity, determine the equation for the phase plane plot $\dot{u}$ versus u, and express the equation for the period of nutation in terms of $f(u)$.

3. The condition for the sleeping top, $u = \cos 0° = 1$, can be satisfied by either of the curves shown below. Show that (a) is stable and (b) is unstable by considering a small variation of the curves as indicated by the dotted lines.

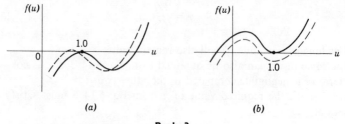

Prob. 3

4. If the stability of the sleeping top is indicated by (a) in the Fig. for Prob. 3, deduce the relationship $n \geq (2/C) \sqrt{WlA}$ by noting that,

$$u = 1 \qquad\qquad f(u) = (-\dot{\theta} \sin \theta)^2 = 0$$

$$\frac{df(u)}{du} = 0 \qquad \text{and} \qquad \frac{d^2f(u)}{du^2} = \text{negative}$$

5. Show that the equation for the steady precession of a top can be written as

$$\dot{\psi}_{1,2} = \frac{h_Z}{2A} \pm \sqrt{\left(\frac{h_Z}{2A}\right)^2 - \frac{Wl \cos \theta}{A}}$$

6. Show that for large n, the lower and higher precessional speeds become

$$\dot{\psi}_1 = \frac{Wl}{Cn} \qquad \dot{\psi}_2 = \frac{Cn}{A \cos \theta}$$

Why is the lower precessional speed independent of θ?

7. For steady precession, show that the angle between the momentum vector **h** and the spin axis is given by the equation,

$$\tan \alpha_{1,2} = \frac{1}{2} \tan \theta \left[1 \pm \sqrt{1 - \frac{4WAl \cos \theta}{C^2 n^2}} \right]$$

where the minus sign must be used for the slow precession.

8. Show that for large n, the angle α corresponding to the slow and fast precessional speeds are,

$$\tan \alpha_1 = \frac{AWl \sin \theta}{C^2 n^2} \qquad \tan \alpha_2 = \tan \theta$$

9. What is the expression for α corresponding to the minimum value of n consistent with steady precession?

10. Determine the conditions necessary for small oscillation of the spin axis of a gyro, between two annular circles θ_1 and θ_2, such that $\dot{\psi}$ is the same at the two circles.

11. The end of the spin axis of a top spinning about a fixed pivot describes on a sphere the curve shown. If $Cn/A = N$ is given, determine the following quantities in terms of N.

$$\gamma = \frac{h_Z}{A} \tag{a}$$

$$\beta = \frac{2Wl}{A} \tag{b}$$

$$\alpha = \frac{2}{A} (E - \tfrac{1}{2}Cn^2) \tag{c}$$

$$\dot{\psi}_2 \tag{d}$$

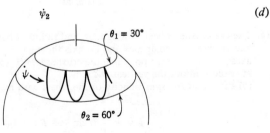

Prob. II

Show in general that a motion with the sharp cusp at $\theta_2 > \theta_1$ cannot take place.

12. A thin disk of radius r and mass m has a stem of length a which is pinned to a vertical shaft as shown. Show that if $r < 2a$, it will be unstable if $\dot{\psi}^2 > \dfrac{4ga}{4a^2 - r^2}$. Show also that it will be stable at all $\dot{\psi}$ if $r > 2a$.

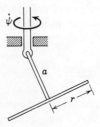

Prob. 12

13. A flywheel with axes x, y, z has moment of inertia I_x, $I_y = I_z$, and is spinning with high speed ω. If it is oscillating in pitch and yaw with the center of gravity stationary, set up the equations of motion with restraining moments equal to $K_p\theta$ in pitch and $K_y\psi$ in yaw.

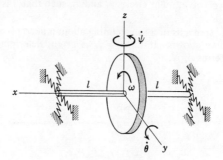

Prob. 13

14. The ore crusher wheel of weight W and radius R has moments of inertia C and A about its polar axis and an axis through O parallel to the diametric axes. The spin axis z on which the crusher wheel is free to rotate, precesses at speed $\dot{\psi}$ about the vertical OZ axis.

(a) show that the spin velocity of the wheel with respect to the axle is

$$\dot{\varphi} = -\frac{1}{R}(l \sin \theta + R \cos \theta)\dot{\psi}$$

(b) Show that the velocity of the center of the wheel is equal to

$$v_c = -R(\dot{\varphi} + \dot{\psi} \cos \theta)$$

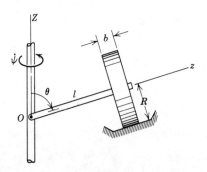

Prob. 14

15. For the ore crusher of Prob. 14, write the equation for the moment about the pin O, and show that the normal force between the wheel and the inclined track is

$$F = W \sin \theta + \frac{1}{2} \dot\psi^2 \left[\frac{C}{R} (1 - \cos 2\theta) + \frac{A}{l} \sin 2\theta \right]$$

16. From the result of Prob. 15, show that the crushing force F has a maximum or minimum between $\theta = 45°$ and $90°$, or between $\theta = 135°$ and $180°$. Since the force due to gravity is a maximum at $\theta = 90°$, we can reason that the maximum occurs for $\theta = 45°$ to $90°$, or actually between $0°$ and $45°$ with the horizontal. Determine the optimum axle tilt above the horizontal for a crusher of the following dimensions.

$$R = 18 \text{ in.} \qquad l = 3 \text{ ft (from center plane of wheel}$$
$$\text{to pin } O)$$
$$b = 12 \text{ in.} \qquad \dot\psi = 100 \text{ rpm}$$

17. Determine the velocity and acceleration of the top of the crusher wheel for the dimensions given in Prob. 16.

18. The general equation for the symmetrical gyro is given by Eq. 5.13–18. Show that

$$f'(u) = \frac{df(u)}{du} = 2\ddot{u} = -2(\dot\theta^2 \cos \theta - \ddot\theta \sin \theta)$$

19. For steady precession $\dot\theta = \ddot\theta = 0$, which requires that $f(u) = f'(u) = 0$. From these two equations show that the steady precession is defined by the quadratic,

$$\dot\psi^2 - \frac{N}{u} \dot\psi + \frac{\beta}{2u} = 0$$

with the following restriction

$$N \ge \sqrt{2\beta u}$$

20. A gyro of weight mg, spinning with angular speed n is tied to a string of length l_1 and precesses around the vertical axis with constant speed Ω. If C and A are moments of inertia about the spin axis and its normal through the center of mass, determine the three equations of motion.

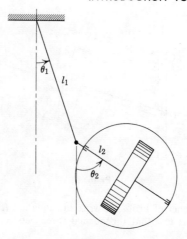

Prob. 20

21. The outer frame of a toy gyro is symmetric about the three axes through its geometric center, and has a moment of inertia C_2. If when the rotor is spinning with speed $\dot{\psi}$, the outer frame also spins at a different speed $\dot{\psi}_2$ in the same direction, write the new momentum and moment equations for steady precession.

Prob. 21

22. If the outer frame of the toy gyro is a single ring in one plane, the moment of inertia about an axis normal to the plane of the ring will be $2C_2$. How does this affect the equations of Prob. 21?

5.15 Precession and Nutation of the Earth's Polar Axis

In Sec. 4.18, we derived the equations for the moment exerted by the earth, due to its oblateness, on a satellite revolving around the earth. According to Newton's third law, the satellite must exert an equal and opposite moment on the earth, but, due to the large mass of the earth in comparison to that of the satellite, its effect is measurable only on the satellite motion.

The moment equations from Sec. 4.18, with opposite signs for moment exerted by the satellite on earth, are,

$$M_x = -\frac{3Km_s}{ma^3}(C-A)\sin^2\varphi\sin\theta\cos\theta$$

$$M_y = \frac{3Km_s}{ma^3}(C-A)\sin\theta\sin\varphi\cos\varphi \qquad (5.15\text{--}1)$$

$$M_z = 0$$

These equations apply to any two bodies oriented as in Sec. 4.18, and, therefore, are applicable to the sun and earth, or the moon and earth. The fact that the earth's polar axis is inclined from the normal to the plane of the ecliptic by 23° 27′ results in a moment exerted by the sun on the earth as shown in Fig. 5.15–1.

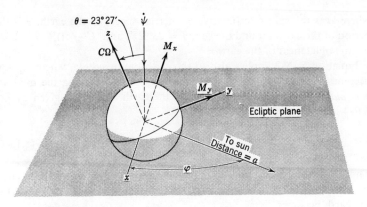

Fig. 5.15–1. Moments M_x and M_y exerted by sun on earth.

The angular momentum vector of the spinning earth is $C\Omega$, and the moment M_y causes the end of the vector $C\Omega$ to move in the same direction and, therefore, change the angle θ. Since M_y is oscillatory with net result per cycle equal to zero, its effect is to produce an oscillatory nutation of zero net angle per cycle.

The moment M_x is oscillatory but cumulative along the negative x axis, which requires a net precession per cycle. Letting $\dot{\psi}$ be the precession rate normal to the plane of the ecliptic, the component $\dot{\psi}\sin\theta$ along the negative y direction will rotate the angular momentum vector $C\Omega$ in the direction M_x to give

$$C\Omega\dot\psi\sin\theta = M_x \tag{5.15-2}$$

The precession rate from Eq. 5.15-1 is then

$$\dot\psi = -\frac{3Km_s}{m\Omega a^3}\left(\frac{C-A}{C}\right)\cos\theta\sin^2\varphi \tag{5.15-3}$$

The quantity $Km_s/m = Gm_s$ in this equation, where m_s and m are the mass of the sun and earth respectively, can be eliminated from the equation of the central force between the sun and the earth as follows.

$$\frac{Gm_sm}{a^2} = \frac{Km_s}{a^2} = ma\left(\frac{2\pi}{\tau}\right)^2 \tag{5.15-4}$$

Also the average value of $\sin^2\varphi$ for $0 \leq \varphi \leq 2\pi$ is $\frac{1}{2}$, so that the average precession rate per year due to the sun is given by the equation,

$$\dot\psi_{av} = -\frac{3}{2\Omega}\left(\frac{2\pi}{\tau}\right)^2\left(\frac{C-A}{C}\right)\cos\theta \tag{5.15-5}$$

where Ω is the spin rate (relative to inertial space) of the earth, τ is the period of the earth around the sun, $\theta = 23°\,27'$, and $(C-A)/C = 0.0032$ for the oblateness of the earth.

Equation 5.15-3 applies also to the earth-moon system, where m_s now becomes the mass of the moon. However, in eliminating the quantity Km_s/ma^3, the attractive force between the earth and moon given by Eq. 4.6-4 must be used as follows.

$$\frac{Gmm_s}{a^2} = \left(\frac{mm_s}{m+m_s}\right)a\left(\frac{2\pi}{\tau}\right)^2 \tag{5.15-6}$$

Thus $(Km_s/ma^3) = [m_s/(m+m_s)](2\pi/\tau)^2$ and the equation for the average precession rate of the earth's polar axis per revolution of the moon around the earth becomes,

$$\dot\psi_{av} = -\frac{3}{2\Omega}\frac{m_s}{m+m_s}\left(\frac{2\pi}{\tau}\right)^2\left(\frac{C-A}{C}\right)\cos\theta \tag{5.15-7}$$

PROBLEMS

1. The earth's spin rate is $\Omega = 2\pi \times 366.25$ rad/year. Show that the earth's polar axis precesses 0.0000765 rad/year due to the sun.

2. The mass of the moon is $\frac{1}{81}$ that of earth, and its period is 27.32 days. Its orbit plane θ varies between $18°\,19'$ and $28°\,35'$, with an average value of $23°\,27'$. Determine the precession of the earth's polar axis per year due to

the moon. Determine the combined precession due to the sun and moon and show that the period of the precession of the earth's axis is approximately 26,000 years.

3. Consider the earth and moon alone with the moon's orbit plane inclined from the Earth's equatorial plane by an average of 23° 27'. Determine the period of regression of the moon's node and show that its period is approximately 18 yrs.

5.16 General Motion of a Rigid Body

So far we have considered problems in which the motion about the center of mass or a fixed point can be determined independently by the moment equation

$$\mathbf{M} = \dot{\mathbf{h}} + \boldsymbol{\omega} \times \mathbf{h} \qquad (5.16\text{--}1)$$

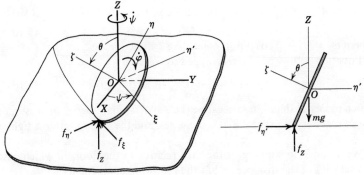

Fig. 5.16–1. Rolling of a thin disk on a plane.

However in the more general type of motion the force

$$\mathbf{F} = m(\dot{\mathbf{v}} + \boldsymbol{\omega} \times \mathbf{v}) \qquad (5.16\text{--}2)$$

may contribute to the moment $\mathbf{M}$ and the two equations are no longer independent. With each of the above equations resolved into three components, we have in general six interrelated equations which must be solved simultaneously. Such problems are, of course, more difficult to solve; however, the general procedure can be illustrated by the following simple problems.

Rolling of a thin circular disk on a rough horizontal plane

The configuration of the rolling disk is shown in Fig. 5.16–1. We will use two sets of axes with origin coinciding with the center of the disk; the X, Y, Z axes with fixed direction, and the ξ, η, ζ axes moving with the disk.

The ζ axis is normal to the plane of the disk, and the ξ, η axes are in the plane of the disk with ξ along the horizontal. It is convenient to introduce the η' axis so that ξ, η', Z form a third set rotated from the X, Y, Z axes about the Z axis. The η' axis so defined is the horizontal projection of axis η. The orientation of the ζ and ξ axes are defined by the rotation θ about the ξ axis for ζ, and a rotation ψ about the Z axis for ξ.

The constraint force of the floor consists of the normal reaction f_Z, and a friction force f, which is resolved into components f_ξ and $f_{\eta'}$ parallel and perpendicular to the ξ axis.

With A, A, C as moments of inertia about the ξ, η, ζ, axes respectively, and R as the radius of the disk, the moment equations are;

$$M_\zeta = -C\dot{\omega}_\zeta = -f_\xi R \tag{5.16-3}$$

$$M_\eta = A(\ddot{\psi}\sin\theta + \dot{\psi}\dot{\theta}\cos\theta) + A\dot{\theta}\dot{\psi}\cos\theta + C\omega_\zeta\dot{\theta} = 0 \tag{5.16-4}$$

$$M_\xi = A\ddot{\theta} - C\omega_\zeta\dot{\psi}\sin\theta - A\dot{\psi}^2\sin\theta\cos\theta = -f_Z R\cos\theta + f_{\eta'}R\sin\theta \tag{5.16-5}$$

where $\omega_\zeta = \dot{\varphi} - \dot{\psi}\cos\theta$ in the negative ζ direction.

Equation 5.16–4 can be rearranged as follows,

$$A\ddot{\psi}\sin\theta = -\dot{\theta}(C\omega_\zeta + 2A\dot{\psi}\cos\theta) \tag{5.16-6}$$

which indicates that, if $\dot{\theta}$ is negative (i.e., disk falling), then $\ddot{\psi}$ increases, or the spin about the vertical Z axis increases, and the disk rolls into a tighter circle.

We assume no slipping of the disk relative to the floor, in which case the velocity of the disk center has the following components.

$$v_\xi = R\omega_\zeta \tag{5.16-7}$$

$$v_\zeta = R\dot{\theta} \tag{5.16-8}$$

$$v_Z = R\dot{\theta}\cos\theta \tag{5.16-9}$$

$$v_{\eta'} = -R\dot{\theta}\sin\theta \tag{5.16-10}$$

which are shown in Fig. 5.16–2. The force equation can then be written as,

$$f_\xi = -m(R\dot{\omega}_\zeta + R\dot{\theta}\dot{\psi}\sin\theta) \tag{5.16-11}$$

$$f_{\eta'} = -m(R\ddot{\theta}\sin\theta + R\dot{\theta}^2\cos\theta - R\omega_\zeta\dot{\psi}) \tag{5.16-12}$$

$$f_Z - mg = m(R\ddot{\theta}\cos\theta - R\dot{\theta}^2\sin\theta) \tag{5.16-13}$$

Equations 5.16–3, 5.16–4, 5.16–5 and 5.16–11, 5.16–12, 5.16–13 are the six component equations which must be solved simultaneously. However

these equations have simple solutions only for certain special conditions
which are;

$$\theta \cong 90° \quad \text{and } \omega_\zeta \text{ large}$$
$$\theta \cong 90° \quad \text{and } \omega_\zeta \to 0,\ \dot\psi \text{ large}$$
$$\theta \cong 0° \quad \text{and } \omega_\zeta \text{ small},\ \dot\psi \text{ large}$$

The first case represents a rolling of the disk with its plane nearly vertical;
the second case corresponds to the upright spinning of the disk about its
vertical diameter; and the last case represents the disk spinning with its
face nearly horizontal.

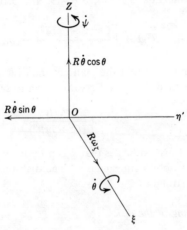

Fig. 5.16–2. Components of velocity.

Rolling of a disk with plane of the disk nearly vertical

Consider the disk rolling with the plane of the disk nearly vertical. Here
the angular velocities $\dot\theta$ and $\dot\psi$ will be small in comparison with ω_ζ. Also
$\sin\theta \cong 1$, and $\cos\theta = \alpha$, where α is the complimentary angle $\alpha = (\pi/2) -
\theta$. We can therefore replace $\dot\theta$ and $\ddot\theta$ by $-\dot\alpha$ and $-\ddot\alpha$ respectively. The
force and moment equations can then be written as,

$$f_\xi R = C\dot\omega_\zeta \tag{5.16–14}$$

$$A\ddot\psi = C\omega_\zeta\dot\alpha \tag{5.16–15}$$

$$-A\ddot\alpha - C\omega_\zeta\dot\psi = -f_Z R\alpha + f_{\eta'}R \tag{5.16–16}$$

$$-f_\xi = mR\dot\omega_\zeta \tag{5.16–17}$$

$$f_{\eta'} = mR(\ddot\alpha + \omega_\zeta\dot\psi) \tag{5.16–18}$$

$$f_Z - mg = 0 \tag{5.16–19}$$

From Eqs. 5.16–14 and 5.16–17, we have,

$$(C + mR^2)\dot{\omega}_\zeta = 0$$

Therefore,

$$\omega_\zeta = n = \text{constant} \tag{5.16–20}$$

With the initial conditions $\dot{\psi} = 0$, and $\alpha = 0$ (i.e., disk started in a straight path with its plane vertical), we integrate Eq. 5.16–15 to obtain,

$$A\dot{\psi} = Cn\alpha \tag{5.16–21}$$

We next substitute Eqs. 5.16–18, 5.16–19, and 5.16–21 into Eq. 5.16–16 and obtain,

$$(A + mR^2)\ddot{\alpha} + \left[\frac{Cn^2(C + mR^2)}{A} - mgR\right]\alpha = 0 \tag{5.16–22}$$

This equation indicates that the plane of the disk wobbles in and out of the vertical, provided the spin is great enough to satisfy the inequality,

$$n^2 > \frac{mgAR}{C(C + mR^2)} \tag{5.16–23}$$

Equation 5.16–21 also indicates that $\dot{\psi}$ is proportional to α, and hence the precession also wobbles sinusoidally. The disk then rolls in a wavy line which is nearly straight.

Upright spinning of the disk

We next consider the case where the main motion of the disk is a spin about the vertical axis. Due to some disturbance, the disk will move in a small circle, but it is evident that ω_ζ will be small and $\dot{\psi}$ large. θ will, however, remain nearly 90°, so that again $\sin\theta \simeq 1$ and $\cos\theta \simeq \alpha$, where α is a small angle.

With these approximations, the moment and force equations become,

$$f_\xi R = C\dot{\omega}_\zeta \tag{5.16–24}$$

$$A\ddot{\psi} = 0 \tag{5.16–25}$$

Therefore $\dot{\psi} = \text{constant}$.

$$-A\ddot{\alpha} - C\omega_\zeta\dot{\psi} - A\dot{\psi}^2\alpha = -f_Z R\alpha + f_{\eta'}R \tag{5.16–26}$$

$$-f_\xi = mR(\dot{\omega}_\zeta - \dot{\alpha}\dot{\psi}) \tag{5.16–27}$$

$$f_{\eta'} = mR(\omega_\zeta\dot{\psi} + \ddot{\alpha}) \tag{5.16–28}$$

$$f_Z - mg = 0 \tag{5.16–29}$$

From Eqs. 5.16–24 and 5.16–27, we have,

$$(C + mR^2)\dot{\omega}_\zeta = mR^2\dot{\alpha}\dot{\psi} \qquad (5.16\text{–}30)$$

With initial values of ω_ζ and α equal to zero, we integrate the above equation,

$$(C + mR^2)\omega_\zeta = mR^2\dot{\psi}\alpha \qquad (5.16\text{–}31)$$

Substituting Eqs. 5.16–28, 5.16–29, and 5.16–31 into Eq. 5.16–26, the final equation becomes,

$$(A + mR^2)\ddot{\alpha} + [(A + mR^2)\dot{\psi}^2 - mgR]\alpha = 0 \qquad (5.16\text{–}32)$$

Interpreting this equation, the spinning motion is stable as long as,

$$\dot{\psi}^2 > \frac{mgR}{A + mR^2} \qquad (5.16\text{–}33)$$

With this inequality satisfied, the disk oscillates sinusoidally with a small angle α about the ξ axis which is spinning around the vertical axis with speed $\dot{\psi}$.

Disk spinning nearly horizontally

Spin a coin about a vertical line and watch its final stages when the plane of the coin is nearly horizontal. You will be able to detect from the sound that the frequency increases very rapidly during the last stage of oscillation. You can also observe that the point of contact with the table spins around a circle of diameter nearly equal to that of the coin, and that ω_ζ is very small (i.e., the face on the coin is rotating very slowly). The ζ axis is nearly vertical so that θ is very small. However, the end of the ζ axis is precessing around the vertical very rapidly so that $\dot{\psi}$ is very large.

With these assumptions, the moment and force equations become,

$$f_\xi R = C\dot{\omega}_\zeta \qquad (5.16\text{–}34)$$

$$A\ddot{\psi}\theta + 2A\dot{\psi}\dot{\theta} = 0 \qquad (5.16\text{–}35)$$

$$A\ddot{\theta} - A\dot{\psi}^2\theta = -f_Z R + f_{\eta'} R\theta \qquad (5.16\text{–}36)$$

$$-f_\xi = mR\dot{\omega}_\zeta \qquad (5.16\text{–}37)$$

$$f_{\eta'} = mR\omega_\zeta\dot{\psi} \qquad (5.16\text{–}38)$$

$$f_Z - mg = mR\ddot{\theta} \qquad (5.16\text{–}39)$$

From Eqs. 5.16–34 and 5.16–37 we find that ω_ζ must be a constant.

$$(C + mR^2)\dot{\omega}_\zeta = 0$$
$$\omega_\zeta = n = \text{constant} \qquad (5.16\text{–}40)$$

From Eq. 5.16–35 we obtain,

$$\frac{\ddot{\psi}}{\dot{\psi}} = -2\frac{\dot{\theta}}{\theta} \tag{5.16–41}$$

Integrating with initial conditions $\dot{\psi}_0$ and θ_0, we obtain

$$\ln\frac{\dot{\psi}}{\dot{\psi}_0} = -2\ln\frac{\theta}{\theta_0}$$

Therefore,

$$\dot{\psi} = \dot{\psi}_0\left(\frac{\theta}{\theta_0}\right)^{-2} \tag{5.16–42}$$

Substituting Eqs. 5.16–38, 5.16–39, and 5.16–42 into Eq. 5.16–36, we arrive at the differential equation for θ.

$$(A + mR^2)\ddot{\theta} = \frac{A\dot{\psi}_0^2\theta_0^4}{\theta^3} + \left(-mgR + mR^2n\dot{\psi}_0\frac{\theta_0^2}{\theta}\right) \tag{5.16–43}$$

Due to θ and θ^3 in the denominator, the acceleration $\ddot{\theta}$ increases as θ approaches zero. Equation 5.16–42 indicates that the precession also increases to infinity as θ goes to zero.

PROBLEMS

1. Determine the least spin about a vertical axis for a 50-cent coin by making the necessary measurements.
2. Determine the equation for the least spin about the vertical diameter of a circular hoop of radius R. Compare with that of a solid disk of same radius.
3. Determine the least rolling velocity of a 50-cent coin.
4. Determine the least rolling velocity of a circular hoop of radius R.
5. With v_ξ, $v_{\eta'}$, v_Z given as in Fig. 5.16–2, verify the accelerations of Eqs. 5.16–11, 5.16–12, 5.16–13 as $\mathbf{a} = [\dot{\mathbf{v}}] + \boldsymbol{\omega} \times \mathbf{v}$, where the components of $\boldsymbol{\omega}$ are $\omega_\xi = \dot{\theta}$, $\omega_{\eta'} = 0$, and $\omega_Z = \dot{\psi}$.

Dynamics of Gyroscopic Instruments

CHAPTER 6

6.1 Small Oscillations of Gyros

If the gyro or top of Fig. 5.13–1 is given a slight disturbance from its steady state, we can show that the oscillations about the steady values will be harmonic.

We can begin with Eqs. 5.13–4, which are the moment equations about the node system coordinates shown in Fig. 6.1–1.

$$Wl \sin \theta = A\ddot{\theta} + Cn\dot{\psi} \sin \theta - A\dot{\psi}^2 \sin \theta \cos \theta$$
$$0 = A\ddot{\psi} \sin \theta + 2A\dot{\psi}\dot{\theta} \cos \theta - Cn\dot{\theta}$$
$$0 = C\dot{n} \qquad (6.1\text{–}1)$$

Letting the steady-state values of θ and $\dot{\psi}$ be θ_0 and $\dot{\psi}_0$, and designating the deviations about the steady values by $\theta_\sim$ and $\dot{\psi}_\sim$, the instantaneous values of θ and $\dot{\psi}$ are,

$$\theta = \theta_0 + \theta_\sim$$
$$\dot{\psi} = \dot{\psi}_0 + \dot{\psi}_\sim$$

For small oscillations we can make the following approximations.

$$\dot{\theta}\dot{\psi} = \dot{\theta}_\sim(\dot{\psi}_0 + \dot{\psi}_\sim) \cong \dot{\theta}_\sim \dot{\psi}_0$$
$$\sin \theta \cong \sin \theta_0 + \theta_\sim \cos \theta_0$$
$$\cos \theta \cong \cos \theta_0 - \theta_\sim \sin \theta_0$$

Substituting these expressions into the second of Eq. 6.1–1 and neglecting products and squares of the small deviations, we obtain,

$$A\ddot{\psi}_\sim \sin\theta_0 + 2A\dot{\theta}_\sim\dot{\psi}_0\cos\theta_0 - Cn\dot{\theta}_\sim = 0 \tag{6.1-2}$$

Integrating,

$$A\sin\theta_0\int_0^{\dot{\psi}_\sim}d\dot{\psi}_\sim = (Cn - 2A\dot{\psi}_0\cos\theta_0)\int_0^{\theta_\sim}d\theta_\sim$$

$$A\dot{\psi}_\sim\sin\theta_0 = (Cn - 2A\dot{\psi}_0\cos\theta_0)\theta_\sim \tag{6.1-3}$$

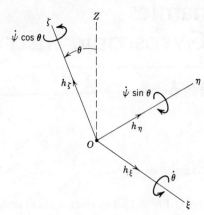

Fig. 6.1–1.   Angular momentum along node system coordinates.

Likewise, from the first of Eq. 6.1–1, we obtain,

$$A\ddot{\theta}_\sim - A(\dot{\psi}_0^2 + 2\dot{\psi}_0\dot{\psi}_\sim)(\sin\theta_0\cos\theta_0 - \theta_\sim\sin^2\theta_0 + \theta_\sim\cos^2\theta_0)$$

$$+ Cn(\dot{\psi}_0 + \dot{\psi}_\sim)(\sin\theta_0 + \theta_\sim\cos\theta_0) = Wl(\sin\theta_0 + \theta_\sim\cos\theta_0) \tag{6.1-4}$$

However, for steady precession $\ddot{\theta}$ is zero, and the steady components of Eq. 6.1–4 are,

$$-A\dot{\psi}_0^2\cos\theta_0 + Cn\dot{\psi}_0 = Wl$$

Thus by striking out these terms in Eq. 6.1–4 and introducing Eq. 6.1–3 to eliminate $\dot{\psi}_\sim$, we arrive at the second-order differential equation in $\theta_\sim$:

$$A^2\ddot{\theta}_\sim + [(Cn)^2 - 4AWl\cos\theta_0 + A^2\dot{\psi}_0^2(1 - \cos^2\theta_0)]\theta_\sim = 0 \tag{6.1-5}$$

The nodding oscillations are therefore sinusoidal with period equal to,

$$\tau_0 = \frac{2\pi}{\omega_0} = \frac{2\pi A}{\sqrt{(Cn)^2 - 4AWl\cos\theta_0 + A^2\dot{\psi}_0^2(1 - \cos^2\theta_0)}} \tag{6.1-6}$$

If n is very large, the nutation period reduces to,

$$\tau_\theta \cong \frac{2\pi A}{Cn} \qquad (6.1\text{-}7)$$

The precession period is also the same since, from Eq. 6.1–3,

$$\dot\psi_\sim = \left(\frac{Cn - 2A\dot\psi_0 \cos \theta_0}{A \sin \theta_0}\right)\theta_\sim \qquad (6.1\text{-}8)$$

and $\dot\psi_\sim$ is proportional to $\theta_\sim$.

When $\cos \theta_0 = 1$, we have the sleeping top and, in order for the denominator of Eq. 6.1–6 to be real, $(Cn)^2$ must be greater than $4AWl$. Thus, again, we verify the stability requirement for the sleeping top, which is,

$$n > \frac{2}{C}\sqrt{AWl}$$

6.2 Oscillations About Gimbal Axes

Figure 6.2–1 shows the two gimbal gyro with the mass center coinciding with the geometric center of the gimbals. We wish now to write the moment

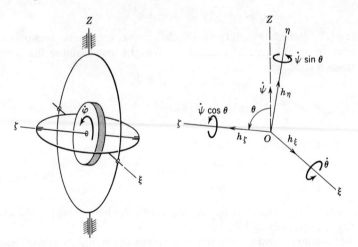

Fig. 6.2–1. Symmetric gyro with gimbal axes ξ and Z.

equations about the gimbal bearing axes. Neglecting again the mass of the gimbals, the moments of inertia about the ξ, η, ζ axes are A, A, C, of the wheel.

The angular velocities and angular momentum about the ξ, η, ζ axes are the same as in Sec. 5.13, and the moment equations about the ξ, η, ζ axes are rewritten in the following form.

$$M_\xi = A\ddot\theta + Cn\dot\psi \sin\theta - A\dot\psi^2 \sin\theta \cos\theta$$

$$M_\eta = A(\ddot\psi \sin\theta + 2\dot\psi\dot\theta \cos\theta) - Cn\dot\theta \qquad (6.2\text{--}1)$$

$$M_\zeta = C\frac{d}{dt}(\dot\varphi + \dot\psi \cos\theta)$$

The moments M_η and M_ζ can be resolved along the vertical Z axis and in the horizontal plane. The Z component is (i.e., $M_\zeta = 0$),

$$
\begin{aligned}
M_Z &= M_\eta \sin\theta + M_\zeta \cos\theta = M_\eta \sin\theta \\
&= A\ddot\psi \sin^2\theta + 2A\dot\psi\dot\theta \sin\theta \cos\theta - Cn\dot\theta \sin\theta
\end{aligned} \qquad (6.2\text{--}2)
$$

These nonlinear equations can be linearized under certain simplifying assumptions. Usually the spin $\dot\varphi$ is very large in comparison to $\dot\psi$ and $\dot\theta$. The spin $\dot\varphi$ is then approximately equal to n which is a constant of large magnitude. Neglecting the squares and products of the smaller quantities $\dot\psi$ and $\dot\theta$, the simplified equations of interest are,

$$
\begin{aligned}
M_\xi &= A\ddot\theta + Cn\dot\psi \sin\theta \\
M_Z &= A\ddot\psi \sin^2\theta - Cn\dot\theta \sin\theta
\end{aligned} \qquad (6.2\text{--}3)
$$

Several interesting solutions of the above equations are possible depending on the type of excitation. We can first examine the steady precession condition,

$$\theta = \theta_0$$
$$\dot\psi = \dot\psi_0$$
$$\dot\theta = \ddot\theta = \ddot\psi = 0$$

The above equations then become,

$$
\begin{aligned}
M_{\xi 0} &= Cn\dot\psi_0 \sin\theta_0 \\
M_{Z0} &= 0
\end{aligned} \qquad (6.2\text{--}4)
$$

which requires a constant torque $M_{\xi 0}$ about the horizontal gimbal axis, as found previously in Sec. 5.14.

We can now consider the problem where the spin axis under steady precession is given a disturbance by a moment $M_\xi(t)$.

We will assume small oscillations and introduce the equations,

$$
\begin{aligned}
\theta &= \theta_0 + \theta_\sim & \sin\theta &= \sin\theta_0 + \theta_\sim \cos\theta_0 \\
\psi &= \psi_0 + \psi_\sim & \cos\theta &= \cos\theta_0 - \theta_\sim \sin\theta_0
\end{aligned}
$$

Again neglecting products of the small oscillatory terms, Eqs. 6.2–3 become,

$$\ddot{\theta}_{\sim} + \frac{Cn}{A}(\dot{\psi}_0\theta_{\sim}\cos\theta_0 + \dot{\psi}_0\sin\theta_0 + \dot{\psi}_{\sim}\sin\theta_0) = \frac{M_\xi(t) + M_{\xi 0}}{A}$$

$$\ddot{\psi}_{\sim}\sin\theta_0 - \frac{Cn}{A}\dot{\theta}_{\sim} = 0 \tag{6.2-5}$$

Eliminating the steady-state terms, Eq. 6.2–4, from the above, and letting $Cn/A = p$, we have the final form of the differential equations for the disturbance about the steady precession:

$$\ddot{\theta}_{\sim} + p(\dot{\psi}_0\theta_{\sim}\cos\theta_0 + \dot{\psi}_{\sim}\sin\theta_0) = \frac{M_\xi(t)}{A}$$

$$\ddot{\psi}_{\sim}\sin\theta_0 - p\dot{\theta}_{\sim} = 0 \tag{6.2-6}$$

The solution of these equations is most conveniently obtained by the use of Laplace transforms with $\theta_{\sim}$ and $\psi_{\sim}$ as dependent variables.[10] Since initially $\theta = \theta_0$, $\dot{\theta} = 0$, and $\dot{\psi} = \dot{\psi}_0$, $\theta_{\sim}(0) = \dot{\theta}_{\sim}(0) = \dot{\psi}_{\sim}(0) = 0$, and the transform equations are,

$$(s^2 + p\dot{\psi}_0\cos\theta_0)\,\bar{\theta}_{\sim}(s) + (p\sin\theta_0)\,\bar{\psi}_{\sim}(s) = \frac{\bar{M}_\xi(s)}{A}$$

$$-p\,\bar{\theta}_{\sim}(s) + (\sin\theta_0)\,\bar{\psi}_{\sim}(s) = 0 \tag{6.2-7}$$

By Cramer's rule, the equations for $\bar{\theta}_{\sim}(s)$ and $\bar{\psi}_{\sim}(s)$ are,

$$\bar{\theta}_{\sim}(s) = \frac{\begin{vmatrix} \dfrac{1}{A}\bar{M}_\xi(s) & p\sin\theta_0 \\ 0 & \sin\theta_0 \end{vmatrix}}{\begin{vmatrix} (s^2 + p\dot{\psi}_0\cos\theta_0) & p\sin\theta_0 \\ -p & \sin\theta_0 \end{vmatrix}} \tag{6.2-8}$$

$$\bar{\psi}_{\sim}(s) = \frac{\begin{vmatrix} (s^2 + p\dot{\psi}_0\cos\theta_0) & \dfrac{1}{A}\bar{M}_\xi(s) \\ -p & 0 \end{vmatrix}}{\begin{vmatrix} (s^2 + p\dot{\psi}_0\cos\theta_0) & p\sin\theta_0 \\ -p & \sin\theta_0 \end{vmatrix}}$$

which may be solved for any excitation $M_\xi(t)$, and any angle θ_0.

Example 6.2–1

When the spin axis is at rest at $\theta_0 = \pi/2$, an impulse is applied to the spin axis, resulting in an impulsive moment $\mathbf{M}_\xi(t) = \hat{\mathbf{M}}\,\delta(t)$, where $\delta(t)$ is a delta function with the unit per second and $\hat{M}$ is the moment impulse in lb-in. sec. From Eqs. 6.2–8, we have

$$\bar{\theta}_{\sim}(s) = \frac{\hat{M}}{A}\frac{1}{s^2 + p^2}$$

$$\bar{\psi}_{\sim}(s) = \frac{p\hat{M}}{A}\frac{1}{s^2 + p^2}$$

and their time solutions are,

$$\dot\theta_{\sim}(t) = \frac{\hat M}{Ap} \sin pt \qquad \theta_{\sim}(t) = \frac{\hat M}{A} \cos pt$$

$$\dot\psi_{\sim}(t) = \frac{\hat M}{A} \sin pt \qquad \psi_{\sim}(t) = \frac{\hat M}{Ap} (1 - \cos pt)$$

The actual position of the spin axis at any time is then equal to

$$\theta(t) = \frac{\pi}{2} + \frac{\hat M}{Ap} \sin pt$$

$$\psi(t) = \frac{\hat M}{Ap} (1 - \cos pt)$$

These results can be interpreted as follows. Assume first that the spin axis is stationary so that $\dot\psi_0 = 0$. The moment impulse $\hat M$ suddenly shifts the angular momentum vector Cn along the equator by an angle

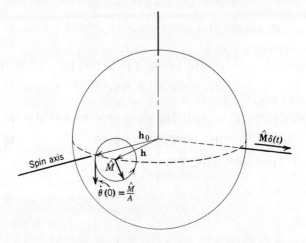

Fig. 6.2–2. Motion of spin axis due to delta function impulse (initial moment is zero).

$\hat M/Ap = \hat M/Cn$. The spin axis however cannot change instantaneously, but develops a downward velocity of $\dot\theta_{\sim}(0) = \hat M/A$ from the equatorial position. Thus the rotation of the spin axis around the new resultant angular momentum vector generates a cone of base radius $\hat M$, as shown in Fig. 6.2–2. This is, of course, consistent with the conclusions of Sec. 5.8 which indicates that, with zero moment (with a delta function moment, the moment is zero at all t except $t = 0$), the angular momentum vector is constant and stationary, and the spin axis will precess around it.

If next we consider an initial steady precession with the spin axis at $\theta_0 = \pi/2$ due to a constant moment M_{ξ_0}, the resultant angular momentum vector $\mathbf{h}$ will be above the equator by the moment component $A\dot{\psi}_0$. The moment impulse $\hat{M}$ will again suddenly shift the resultant $\mathbf{h}$ horizontally by an amount $\hat{M}$ along the latitude $\lambda = A\dot{\psi}_0/Cn$, as shown in Fig. 6.2–3. At time $t = 0+$, the spin axis will have angular velocity components $\dot{\theta}(0) = \hat{M}/A$ vertically, and $\dot{\psi}_0$ horizontally. Their resultant will however be

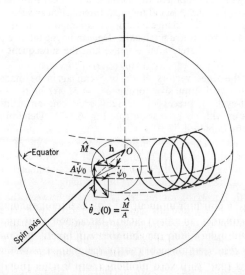

Fig. 6.2–3. Motion of spin axis dues to delta function impulse (initial moment is contant, with steady precession $\dot{\psi}_0$).

normal to the radial line from the $\mathbf{h}$ vector, as shown in Fig. 6.2–3. Thus the $\mathbf{h}$ vector of approximate length Cn will precess steadily along the latitude $\lambda = A\dot{\psi}_0/Cn$ with angular velocity $\dot{\psi}_0$ while the spin axis will rotate around $\mathbf{h}$ in a circle of radius $\sqrt{\hat{M}^2 + (A\dot{\psi}_0)^2}$. The result is a combined nutation and precession, and the curve described by the spin axis depends on the relative magnitudes of the initial velocity components $\hat{M}/A$ and $\dot{\psi}_0$. This behavior is somewhat similar to the problem of the disturbed top, the difference being that we have here imposed a constant moment about the node axis ξ, whereas in the problem of the top, the gravity moment will change with θ. For small disturbances, however, the motions are identical.

PROBLEMS

1. With the axes initially at rest with $\theta = \pi/2$, a constant moment is applied to the ξ axis. Determine its solution.
2. Repeat Prob. 1 if M is a delta function $\hat{M}\,\delta(t)$.

3. With the system initially at rest at an angle θ_0, a constant moment is applied to the OZ axis. Determine its motion, assuming Cn to be very large.

4. Consider $C\dot{\varphi}$ to be the only angular momentum of importance for $\dot{\varphi}$ large, and determine the time required for the ζ axis to rotate from the horizontal to the vertical position due to a constant moment M_Z about the Z axis.

5. A satellite with spin moment of inertia C and equal transverse moment of inertia $B = A$ has at the initial moment three collinear vectors, $\mathbf{h}$, $\boldsymbol{\omega}$, and $\mathbf{k}$, along the spin axis. To change its attitude, an impulsive torque $M_x = \hat{M}\,\delta(t)$ is applied to the transverse pitch axis. Determine the relative position of the three vectors immediately after impulse and the subsequent motion of the satellite.

6. Show that the three vectors of Prob. 5 can again be made collinear by applying an equal impulsive torque $M_x = \hat{M}\,\delta(t)$ in the same direction after a half-cycle of precession, thereby producing an attitude change of $\tan^{-1}(2\hat{M}/Cn)$ with no precession (see Fig. 6.2–2). Determine the spacing in time of the impulses.

7. It is proposed to change the attitude of the spin axis of a satellite with angular momentum $h = Cn$ by an angle of $90°$ with a series of N equal impulses, ending up with the $\mathbf{h}$, $\boldsymbol{\omega}$, and $\mathbf{k}$ vectors collinear. Determine the magnitude of the impulses and their time spacing.

8. Starting with Eq. 3.5–4 for the transformation between the body angular velocities ω_x, ω_y, ω_z, and the Euler angular rates, $\dot{\psi}$, $\dot{\varphi}$, $\dot{\theta}$, show that the general solution for the Euler angles are given by the equations,

$$\theta = \theta_0 + \int_0^t (\omega_x \cos\varphi - \omega_y \sin\varphi)\,dt$$

$$\psi = \psi_0 + \int_0^t \left(\frac{\omega_x \sin\varphi + \omega_y \cos\varphi}{\sin\theta}\right) dt$$

$$\varphi = \varphi_0 + \int_0^t \dot{\psi}\,dt$$

9. For small oscillations with $\dot{\varphi} = $ constant, discuss the solution for θ and ψ of Prob. 8 as influenced by the value of θ_0 which may be arbitrarily chosen. Why is $\theta_0 = \pi/2$ desirable for small angle solutions?

10. With $\theta_0 = \pi/2$ in the equations of Prob. 8, solve Prob. 5 for θ and ψ by first determining the angular rates about the body-fixed axes x and y.

11. Assume that the rotor of Fig. 6.2–1 is misaligned so that its normal (axis z) makes an angle α with the spin axis ζ. Let x be an axis normal to both z and ζ, y the third axis of the x, y, z set, and show that the angular velocities along x, y, z, are;

$$\omega_x = \dot{\theta}\cos\varphi + \dot{\psi}\sin\theta\sin\varphi$$

$$\omega_y = -\dot{\theta}\sin\varphi\cos\alpha + \dot{\psi}\sin\theta\cos\varphi\cos\alpha + (\dot{\varphi} + \dot{\psi}\cos\theta)\sin\alpha$$

$$\omega_z = \dot{\theta}\sin\varphi\sin\alpha - \dot{\psi}\sin\theta\cos\varphi\sin\alpha + (\dot{\varphi} + \dot{\psi}\cos\theta)\cos\alpha$$

Since the moments of inertia of the rotor are A, A, C, the angular momentum along x, y, z are; $h_x = A\omega_x$, $h_y = A\omega_y$, $h_z = C\omega_z$.

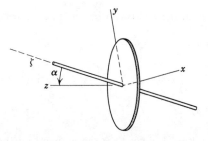

Prob. II

12. Resolve the angular momentum h_x, h_y, h_z of Prob. 11 along the gyro gimbal axes ξ, Z, and ζ, assuming α to be small (see Fig. 6.2–1).

6.3 Gimbal Masses Included (Perturbation Technique)

Referring to Fig. 6.2–1, the moment of inertia of the rotor about the ξ, η, ζ axes were A, A, C. In addition, we introduce the moment of inertia of the gimbals as follows. The moment of inertia of the inner gimbal about the ξ, η, ζ axes are A_i, B_i, C_i respectively. The moment of inertia of the outer gimbal about the OZ axis is C_0.

Noting that the ξ, η, ζ, axes rotate with angular speeds $\dot\theta$, $\dot\psi \sin\theta$, $\dot\psi \cos\theta$, and noting the masses which rotate due to these components, the moment of inertia about the three axes are,

$$I_\xi = A + A_i$$

$$I_\eta = A + B_i \qquad (6.3\text{–}1)$$

$$I_\zeta = C_i$$

The direction cosines of the OZ axis with respect to ξ, η, ζ, are $l_{Z\xi} = 0$, $l_{Z\eta} = \sin\theta$, $l_{Z\zeta} = \cos\theta$, so that the moment of inertia about the Z axis becomes,

$$I_Z = C_0 + I_\xi l_{Z\xi}^2 + I_\eta l_{Z\eta}^2 + I_\zeta l_{Z\zeta}^2$$

$$= C_0 + (A + B_i) \sin^2\theta + C_i \cos^2\theta \qquad (6.3\text{–}2)$$

We next determine the angular momenta about the ξ, η, ζ axes, which are,

$$h_\xi = (A + A_i)\dot\theta \qquad (6.3\text{–}3)$$

$$h_\eta = (A + B_i)\dot\psi \sin\theta \qquad (6.3\text{–}4)$$

$$h_\zeta = C(\dot\varphi + \dot\psi \cos\theta) + C_i\dot\psi \cos\theta \qquad (6.3\text{–}5)$$

The moment about the ζ axis can be separated into two parts, thus:

$$M_\zeta = h_\zeta - h_\xi \dot\psi \sin\theta + h_\eta \dot\theta$$

$$= C\frac{d}{dt}(\dot\varphi + \dot\psi \cos\theta) + \left[C_i \frac{d}{dt}(\dot\psi \cos\theta) + (B_i - A_i)\dot\theta\dot\psi \sin\theta \right]$$

$$= M_\zeta' + M_\zeta'' \tag{6.3-6}$$

where $M_\zeta' = C(d/dt)(\dot\varphi + \dot\psi \cos\theta)$ is the moment on the rotor axis, and M_ζ'' is the moment due to the forces exerted on the inner gimbal axis by

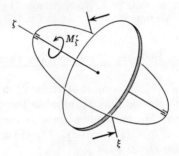

Fig. 6.3-1. Moments on inner gimbal and rotor.

the outer gimbal, as shown in Fig. 6.3-1. We will assume M_ζ' to be zero, in which case the rotor angular momentum will be a constant.

$$C(\dot\varphi + \dot\psi \cos\theta) = Cn = \text{constant} \tag{6.3-7}$$

The angular momentum components along ξ, η, ζ, shown in Fig. 6.3-2, can be resolved along the ξ', η', ζ' axes by noting the vertical and horizontal components of h_η and h_ζ. The h_ξ and $h_{\eta'}$ components now rotate about the OZ axis with angular speed $\dot\psi$, as shown in Fig. 6.3-2. The components of $\mathbf{h}$ along the ξ', η', ζ' axes are,

$$h_{\xi'} = (A + A_i)\dot\theta \tag{6.3-8}$$

$$h_{\eta'} = h_\eta \cos\theta - h_\zeta \sin\theta = (A + B_i)\dot\psi \sin\theta \cos\theta$$

$$-(Cn + C_i\dot\psi \cos\theta)\sin\theta \tag{6.3-9}$$

$$h_{\zeta'} = h_\eta \sin\theta + h_\zeta \cos\theta + C_0\dot\psi = (A + B_i)\dot\psi \sin^2\theta$$

$$+ (Cn + C_i\dot\psi \cos\theta)\cos\theta + C_0\dot\psi \tag{6.3-10}$$

We can now write the moment equations about the ζ' and ξ' axes as follows:

$$\begin{aligned}
M_{\zeta'} = M_Z = h_{\zeta'} &= (A + B_i)(\ddot{\psi} \sin^2 \theta + 2\dot{\psi}\dot{\theta} \sin \theta \cos \theta) \\
&\quad - (Cn + C_i\dot{\psi} \cos \theta)\dot{\theta} \sin \theta \\
&\quad + (C_i\ddot{\psi} \cos \theta - C_i\dot{\psi}\dot{\theta} \sin \theta) \cos \theta + C_0\ddot{\psi} \\
&= [C_0 + C_i \cos^2 \theta + (A + B_i) \sin^2 \theta]\ddot{\psi} \\
&\quad + 2(A + B_i - C_i)\dot{\psi}\dot{\theta} \sin \theta \cos \theta - Cn\dot{\theta} \sin \theta \\
&= I_Z\ddot{\psi} + \dot{\psi}\frac{dI_Z}{dt} - Cn\dot{\theta} \sin \theta \\
&= \frac{d}{dt}(I_Z\dot{\psi}) - Cn\dot{\theta} \sin \theta
\end{aligned} \tag{6.3-11}$$

$$\begin{aligned}
M_{\xi'} = h_{\xi'} &- h_{\eta'}\dot{\psi} \\
&= (A + A_i)\ddot{\theta} + Cn\dot{\psi} \sin \theta \\
&\quad + [C_i - (A + B_i)]\dot{\psi}^2 \sin \theta \cos \theta = 0
\end{aligned} \tag{6.3-12}$$

We next investigate the problem for which the axes are initially at rest, and the inner gimbal axis in position θ_0 is given an initial angular velocity

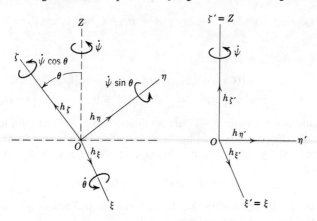

Fig. 6.3–2. Resolution of angular velocity and angular momentum.

$\dot{\theta}(0) = \alpha$, by an impulsive moment in the form of a delta function about the ξ axis. The time $t = 0$ is referenced to the instant after the impulse, in which case $M_\xi = M_Z = 0$, and the initial conditions are $\psi(0) = 0$, $\dot{\psi}(0) = 0$, $\theta(0) = \theta_0$, $\dot{\theta}(0) = \alpha$. We can safely assume that $\dot{\psi}(0) = 0$, since the initial velocity α results in a gyroscopic moment about the Z axis (through reaction of ξ bearings) which is not impulsive.

With the outer gimbal axis unrestrained, $M_Z = 0$, and Eq. 6.3–11 can be written as,

$$d(I_Z\dot\psi) = Cn \sin \theta \, d\theta \qquad (6.3\text{–}13)$$

Integrating and noting that $I_Z\dot\psi$ at $t = 0$ is zero, we obtain the equation,

$$I_Z\dot\psi = -Cn(\cos \theta - \cos \theta_0) \qquad (6.3\text{–}14)$$

We now make the small oscillation approximation,

$$\theta = \theta_0 + \theta_\sim$$
$$\sin \theta \cong \sin \theta_0 + \theta_\sim \cos \theta_0$$
$$\cos \theta \cong \cos \theta_0 - \theta_\sim \sin \theta_0$$
$$\sin \theta \cos \theta \cong \sin \theta_0 \cos \theta_0 + \theta_\sim(\cos^2 \theta_0 - \sin^2 \theta_0)$$

and rewrite Eqs. 6.3–2, 6.3–14, and 6.3–12 as Eqs. 6.3–15, 6.3–16, and 6.3–17

$$\begin{aligned}
I_Z &= [C_0 + (A + B_i) \sin^2 \theta_0 + C_i \cos^2 \theta_0] \\
&\quad + 2\theta_\sim(A + B_i - C_i) \sin \theta_0 \cos \theta_0 \qquad (6.3\text{–}15) \\
&= I_0 + 2\theta_\sim(A + B_i - C_i) \sin \theta_0 \cos \theta_0
\end{aligned}$$

$$I_0\dot\psi - (Cn \sin \theta_0)\theta_\sim + 2\theta_\sim\dot\psi(A + B_i - C_i) \sin \theta_0 \cos \theta_0 = 0 \qquad (6.3\text{–}16)$$

$$\begin{aligned}
(A + A_i)\ddot\theta_\sim &+ (Cn \sin \theta_0)\dot\psi \\
&+ \{(Cn \cos \theta_0)\theta_\sim\dot\psi - (A + B_i - C_i)[\sin \theta_0 \cos \theta_0 \\
&+ \theta_\sim(\cos^2 \theta_0 - \sin^2 \theta_0)]\dot\psi^2\} = 0 \qquad (6.3\text{–}17)
\end{aligned}$$

Equations 6.3–16 and 6.3–17 are nonlinear due to the last term in each equation. They can be solved by the perturbation technique[1,3,8] which will be illustrated by the following simple example. Consider a first-order nonlinear equation

$$\dot y + ay + by^2 = 0 \qquad (a)$$

where the coefficient b of the nonlinear term is a small quantity. We will now consider a similar equation

$$\dot y + ay + \mu by^2 = 0 \qquad (b)$$

which differs from the previous equation by an additional factor μ which may be any positive number. If the solution of Eq. b is found, then the solution of the previous equation, Eq. a, is found by letting $\mu = 1$.

We seek now a solution in the form,

$$y = y_0 + \mu y_1 + \mu^2 y_2 + \cdots \qquad (c)$$

Substituting c into b we obtain,

$$(\ddot{y}_0 + \mu\ddot{y}_1 + \mu^2\ddot{y}_2 + \cdots) + a(y_0 + \mu y_1 + \mu^2 y_2 + \cdots)$$
$$+ \mu b(y_0 + \mu y_1 + \mu^2 y_2 + \cdots)^2 = 0 \quad (d)$$

Rearranging, this equation can be written in terms of equal powers of μ as follows:

$$(\ddot{y}_0 + ay_0) + \mu(\ddot{y}_1 + ay_1 + by_0{}^2) + \mu^2(\ddot{y}_2 + ay_2 + 2by_0 y_1)$$
$$+ \mu^3(\ddot{y}_3 + \cdots) = 0 \quad (e)$$

We note now that, if $\mu = 0$, we obtain y_0 as the solution of the linear equation. The solution y_0 is called the *generating solution*, and it can be fitted to the initial conditions of the problem. If μ is allowed to increase from zero, Eq. e can be satisfied only if the coefficients of μ raised to the various powers are zero. We thus obtain the following equations,

$$\ddot{y}_1 + ay_1 + by_0{}^2 = 0 \quad (f)$$

$$\ddot{y}_2 + ay_2 + 2by_0 y_1 = 0, \text{ etc.} \quad (g)$$

which can be solved for y_1, y_2, etc.

We will now apply this technique to Eqs. 6.3–16 and 6.3–17, but will carry out the solution only to the first-order correction. Since Eq. 6.3–16 and 6.3–17 already have the symbol θ_0, we will let the solution to the linear equation (corresponding to y_0) be θ_{00} and $\dot{\psi}_{00}$. The linear equations are then,

$$I_0\dot{\psi}_{00} - (Cn \sin \theta_0)\theta_{00} = 0 \quad (6.3\text{-}16a)$$

$$(A + A_i)\ddot{\theta}_{00} + (Cn \sin \theta_0)\dot{\psi}_{00} = 0 \quad (6.3\text{-}17a)$$

Eliminating $\dot{\psi}_{00}$, we obtain the equation

$$\ddot{\theta}_{00} + \frac{(Cn \sin \theta_0)^2}{I_0(A + A_i)}\theta_{00} = 0 \quad (6.3\text{-}18)$$

Letting

$$\omega = \frac{Cn \sin \theta_0}{\sqrt{I_0(A + A_i)}} \quad (6.3\text{-}19)$$

the generating solution fitting the initial conditions is,

$$\theta_{00} = \frac{\dot{\theta}(0)}{\omega} \sin \omega t \quad (6.3\text{-}20)$$

$$\dot{\psi}_{00} = \frac{Cn \sin \theta_0}{I_0}\left[\frac{\dot{\theta}(0)}{\omega} \sin \omega t\right] \quad (6.3\text{-}21)$$

We next consider the first-order correction corresponding to y_1 in Eq. f. From the generating solution we determine the nonlinear terms

$$\theta_{00}\dot{\psi}_{00} = \frac{\dot{\theta}(0)^2 Cn \sin\theta_0}{\omega^2 I_0} \sin^2\omega t$$

$$\dot{\psi}_{00}^2 = \left(\frac{Cn\sin\theta_0}{I_0}\right)^2 \left(\frac{\dot{\theta}(0)}{\omega}\right)^2 \sin^2\omega t$$

which, substituted into Eqs. 6.3–16 and 6.3–17, results in a new set of differential equations,

$$I_0\dot{\psi}_1 - (Cn\sin\theta_0)\theta_1$$

$$= -\frac{2(A + B_i - C_i)\sin^2\theta_0 \cos\theta_0}{I_0} \frac{Cn\dot{\theta}(0)^2}{\omega^2} \sin^2\omega t \quad (6.3\text{–}16b)$$

$$(A + A_i)\ddot{\theta}_1 + (Cn\sin\theta_0)\dot{\psi}_1 = -(Cn\cos\theta_0)\frac{\dot{\theta}(0)^2}{\omega^2}\frac{Cn\sin\theta_0}{I_0}\sin^2\omega t$$

$$+ (A + B_i - C_i)\sin\theta_0\cos\theta_0\left(\frac{\dot{\theta}(0)Cn\sin\theta_0}{\omega I_0}\right)^2 \sin^2\omega t \quad (6.3\text{–}17b)$$

We will now eliminate θ_1. From Eq. 6.3–16b we obtain

$$\ddot{\theta}_1 = \left(\frac{I_0}{Cn\sin\theta_0}\right)\dddot{\psi}_1$$

$$+ \left[\frac{4\dot{\theta}(0)^2(A + B_i - C_i)\cos\theta_0\sin\theta_0}{I_0}\right](\cos^2\omega t - \sin^2\omega t)$$

Substituting into Eq. 6.3–17b, we obtain the differential equation for ψ_1.

$$\left[\frac{I_0(A + A_i)}{Cn\sin\theta_0}\right]\dddot{\psi}_1 + (Cn\sin\theta_0)\dot{\psi}_1 = \left[\frac{\dot{\theta}(0)Cn}{\omega}\right]^2$$

$$\times \left[\frac{(A + B_i - C_i)\sin^3\theta_0\cos\theta_0}{I_0^2} - \frac{\sin\theta_0\cos\theta_0}{I_0}\right]\frac{1}{2}(1 - \cos 2\omega t)$$

$$- \frac{A + A_i}{I_0}[4\dot{\theta}(0)^2(A + B_i - C_i)\sin\theta_0\cos\theta_0]\cos 2\omega t \quad (6.3\text{–}17c)$$

In examining this equation, the solution of the homogeneous equation for $\dot{\psi}_1$ is again harmonic of frequency ω as given by Eq. 6.3–19. The particular solution will have harmonic terms of frequency 2ω and, in addition, a constant term equal to the constant term on the right side of the equation divided by the coefficient of $\dot{\psi}_1$ on the left side. We are interested in the constant term since it results in a steady drift which rotates the outer gimbal according to the equation $\psi_s = \dot{\psi}_s t$.

The constant term of the solution is

$$\dot{\psi}_s = -\frac{\dot{\theta}(0)^2 Cn}{2\omega^2 I_0^2 \sin \theta_0} [(A + B_i - C_i) \sin^3 \theta_0 \cos \theta_0 - I_0 \sin \theta_0 \cos \theta_0]$$

Substituting for I_0 from Eq. 6.3–15, it reduces to

$$\dot{\psi}_s = -\frac{\dot{\theta}(0)^2 Cn(C_0 + C_i) \cos \theta_0}{2\omega^2 I_0^2} \qquad (6.3\text{–}22)$$

It is evident, then, that the outer gimbal oscillates and drifts in a negative direction, a phenomenon referred to as "gimbal walk."[6] It should be noted that gimbal walk cannot take place at $\theta_0 = 90°$ or if the moment of inertia $(C_0 + C_i)$ is zero.

PROBLEMS

1. The periodic solution for Eq. 6.3–17c is

$$\dot{\psi}_1 = \alpha + \beta \cos 2\omega t$$

Evaluate the coefficient β.

2. Discuss the solution of a nonhomogeneous equation,

$$\dot{y} + ay + by^n = f(t)$$

by the procedure of Sec. 6.3.

3. Solve the nonlinear differential equation,

$$m\ddot{y} + ky - by^3 = F \sin pt$$

by the perturbation method outlined in Sec. 6.3. The solution with the use of only two terms, $y = y_0 + \mu y_1$, is

$$y = \left[a + \frac{3ba^3}{4m(\omega^2 - p^2)}\right] \sin pt - \frac{ba^3}{4m(\omega^2 - 9p^2)} \sin 3pt$$

where

$$a = \frac{F}{m(\omega^2 - p^2)} \quad \text{and} \quad \omega^2 = \frac{k}{m}$$

4. Show that if only the first term $y = a \sin pt$ of Prob. 3 is substituted into the differential equation, the amplitude relationship

$$\frac{3b}{4m\omega^2}a^3 = \left(1 - \frac{p^2}{\omega^2}\right) a - \frac{F}{m\omega^2}$$

is obtained. Letting the ordinate $v = (3b/4m\omega^2)a^3$ be plotted against a, discuss the solution for the amplitude a versus p/ω, where $\omega^2 = k/m$.

5. For the problem of Sec. 6.3, investigate the equation for θ and establish whether there is a unidirectional motion about the node axis.

6.4 The Gyrocompass

The requirement of a gyrocompass is to point north at any latitude at any time. The high-speed, two-gimballed gyro, with a pendulous weight w on the $-\eta$ axis to give it moment $wl \cos \theta$ about the ξ axis when the axis

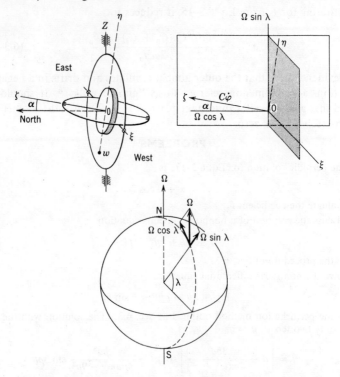

Fig. 6.4–1. Gyrocompass and angular velocity components.

is tilted above the horizon, as shown in Fig. 6.4–1, will satisfy this requirement.

In Fig. 6.4–1, the rotation of the earth from west to east is indicated by the angular rotation vector Ω pointing in the northerly direction. Its numerical value is $\Omega = 2\pi/(24 \times 3600) = 7.27 \times 10^{-5}$ rad/sec. At any latitude λ, Ω will have components in the meridian plane, equal to $\Omega \cos \lambda$ horizontally, and $\Omega \sin \lambda$ vertically.

With the Z axis of the gyrocompass in the local vertical direction, in order for the ζ axis to remain in the meridian plane, and hence point north, the outer gimbal must precess steadily by an amount $\dot{\psi} = \Omega \sin \lambda$ and in addition have an angular velocity $\Omega \cos \lambda$ about the $-\eta$ axis perpendicular to the outer gimbal plane. We assume that the gyro is constrained to move in this manner, and investigate the moment requirement satisfying the motion.

Letting $\theta = 90° - \alpha_0$, where α_0 is a small angle above the horizontal plane at latitude λ, the angular velocities of the ξ, η, ζ axes are,

$$\omega_\xi = 0$$
$$\omega_\eta = \Omega(\sin \lambda \cos \alpha_0 - \cos \lambda \sin \alpha_0) \tag{6.4-1}$$
$$\omega_\zeta = \Omega(\sin \lambda \sin \alpha_0 + \cos \lambda \cos \alpha_0)$$

Assuming the spin to be very large, we can neglect all other components of $\mathbf{h}$. With $M_Z = 0$, $\dot{\varphi} = $ constant. The required moment about the ξ axis, is

$$M_\xi = (C\dot{\varphi})\omega_\eta$$
$$wl \sin \alpha_0 = C\dot{\varphi}\Omega(\sin \lambda \cos \alpha_0 - \cos \lambda \sin \alpha_0) \tag{6.4-2}$$

Dividing by $\sin \alpha_0$, the required inclination of the spin axis above the horizon is,

$$\tan \alpha_0 = \frac{C\dot{\varphi}\Omega \sin \lambda}{wl + C\dot{\varphi}\Omega \cos \lambda} \tag{6.4-3}$$

which depends on the latitude λ.

The moment required for the angular velocity $\Omega \cos \lambda$ about the $-\eta'$ axis is supplied by the reaction of the bearings on the outer gimbal axis Z.

6.5   Oscillation of the Gyrocompass

If the axis of the gyrocompass is disturbed from the meridian plane, as shown in Fig. 6.5–1, the oscillation which takes place will have two components, one perpendicular to the meridian plane and the other in the meridian plane. Both oscillations will have the same frequency, and so the end of the axis of the gyrocompass will describe an ellipse.

Letting ψ be the angular deviation of the spin axis out of the meridian plane, and α its inclination above the horizontal, we will assume both these angles to be small, in which case the angular velocities about the ξ, η, ζ axes will be,

$$\omega_\xi = -\dot{\alpha} - \Omega\psi \cos \lambda$$
$$\omega_\eta = \dot{\psi} + \Omega \sin \lambda - \Omega\alpha \cos \lambda \tag{6.5-1}$$
$$\omega_\zeta = (\dot{\psi} + \Omega \sin \lambda)\alpha + \Omega \cos \lambda$$

We will also assume the spin to be large and the angular momentum about the ξ and η axes to be negligible in comparison.

$$h_\zeta = C\dot{\varphi} = \text{constant for } M_\zeta = 0$$
$$h_\xi = h_\eta = 0$$

The moment equations of interest are then,

$$M_\eta = -h_\zeta \omega_\xi = C\dot{\varphi}(\dot{\alpha} + \Omega\psi \cos \lambda) = 0$$
$$M_\xi = h_\zeta \omega_\eta = C\dot{\varphi}(\psi + \Omega \sin \lambda - \Omega\alpha \cos \lambda) = wl\alpha \qquad (6.5\text{--}2)$$

where wl is the mass unbalance of the pendulous weight on the $-\eta$ axis. Rearranging these equations, we have

$$\dot{\alpha} + (\Omega \cos \lambda)\psi = 0$$
$$(C\dot{\varphi})\psi - (wl + C\dot{\varphi}\Omega \cos \lambda)\alpha = -C\dot{\varphi}\Omega \sin \lambda \qquad (6.5\text{--}3)$$

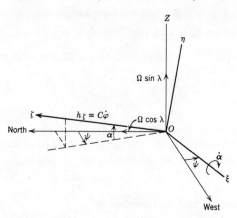

Fig. 6.5–1. Vector components for the gyrocompass.

Eliminating ψ between these equations, the differential equation for α becomes

$$\ddot{\alpha} + \frac{(wl + C\dot{\varphi}\Omega \cos \lambda)(\Omega \cos \lambda)}{C\dot{\varphi}}\alpha = \Omega^2 \sin \lambda \cos \lambda \qquad (6.5\text{--}4)$$

with the general solution,

$$\alpha = C_1 \sin pt + C_2 \cos pt + \frac{C\dot{\varphi}\Omega \sin \lambda}{wl + C\dot{\varphi}\Omega \cos \lambda} \qquad (6.5\text{--}5)$$

$$p = \sqrt{\frac{wl\Omega \cos \lambda + C\dot{\varphi}\Omega^2 \cos^2 \lambda}{C\dot{\varphi}}} \cong \sqrt{\frac{wl\Omega \cos \lambda}{C\dot{\varphi}}} \qquad (6.5\text{--}6)$$

From Eq. 6.5–3, the equation for ψ is,

$$\psi = \frac{-p}{\Omega \cos \lambda}\left(C_1 \cos pt - C_2 \sin pt\right) \tag{6.5–7}$$

These equations indicate that the spin axis oscillates horizontally about the meridian plane and vertically about the stationary angle α_0, given by Eq. 6.4–3. The frequency of oscillation, Eq. 6.5–6 is a function of the latitude λ, and is very small due to $C\dot\varphi$ in the denominator. The frequency p approaches zero, as the gyrocompass nears the north polar axis, where the reliability of the instrument diminishes.

High-frequency oscillation

In addition to the slow oscillation given by the foregoing equations, there is a high-frequency oscillation which was not revealed because the angular momentum about ξ and η were assumed to be zero. By adding $I_\eta\ddot\psi$ and $-I_\xi\ddot\alpha$ to Eqs. 6.5–2, we have,

$$\begin{aligned}
M_\eta &= C\dot\varphi(\dot\alpha + \Omega\psi \cos \lambda) + I_\eta\ddot\psi = 0 \\
M_\xi &= C\dot\varphi(\psi + \Omega \sin \lambda - \Omega\alpha \cos \lambda) - I_\xi\ddot\alpha = wl\alpha
\end{aligned} \tag{6.5–8}$$

Rearranging, and letting,

$$a = C\dot\varphi\Omega \cos \lambda$$

$$b = wl + C\dot\varphi\Omega \cos \lambda$$

the above equations become,

$$\begin{aligned}
\ddot\psi + \frac{C\dot\varphi}{I_\eta}\dot\alpha + \frac{a}{I_\eta}\psi &= 0 \\
\ddot\alpha - \frac{C\dot\varphi}{I_\xi}\dot\psi + \frac{b}{I_\xi}\alpha &= \frac{C\dot\varphi\Omega \sin \lambda}{I_\xi}
\end{aligned} \tag{6.5–9}$$

Assuming harmonic oscillations, e^{ipt}, the natural frequencies are given by the determinant,

$$\begin{vmatrix} -p^2 + \dfrac{b}{I_\xi} & -\dfrac{C\dot\varphi}{I_\xi}ip \\[2ex] \dfrac{C\dot\varphi}{I_\eta}ip & -p^2 + \dfrac{a}{I_\eta} \end{vmatrix} = 0$$

or

$$p^4 - \left[\frac{(C\dot{\varphi})^2 + aI_\xi + bI_\eta}{I_\xi I_\eta}\right]p^2 + \frac{ab}{I_\xi I_\eta} = 0 \qquad (6.5\text{--}10)$$

Since $(C\dot{\varphi})^2$ is very much larger than aI_ξ or bI_η, the natural frequency equation simplifies to

$$p^2 = \frac{(C\dot{\varphi})^2}{2I_\xi I_\eta}\left[1 \pm \sqrt{1 - \frac{4abI_\xi I_\eta}{(C\dot{\varphi})^4}}\right] = \frac{(C\dot{\varphi})^2}{2I_\xi I_\eta}\left\{1 \pm \left[1 - \frac{2abI_\xi I_\eta}{(C\dot{\varphi})^4} + \cdots\right]\right\}$$

$$(6.5\text{--}11)$$

The two frequencies are, therefore,

$$p_1{}^2 = \frac{ab}{(C\dot{\varphi})^2} = \frac{(wl + C\dot{\varphi}\Omega\cos\lambda)\Omega\cos\lambda}{C\dot{\varphi}} \qquad p_2{}^2 = \frac{(C\dot{\varphi})^2}{I_\xi I_\eta} \qquad (6.5\text{--}12)$$

We find then that p_1 corresponds to Eq. 6.5–6, and an additional high-frequency oscillation of frequency p_2 is introduced. With $h_\xi = h_\eta = 0$, $p_2 = \infty$ did not enter into the previous solution. The amplitude of the high-frequency oscillation, however, is extremely small, and hence, the slow oscillation at frequency p_1 is generally the only one detectable.

Effect of damping

Damping for the slow oscillation of the gyrocompass can be provided by introducing a moment about the η axis as follows. We move the pendulous weight w a distance e to the east of the center line so that its coordinate $(\xi, \eta, \zeta) = (-e, -l, 0)$. The equation for the moment about the axis is then modified as follows:

$$M_\eta = C\dot{\varphi}(\dot{\alpha} + \Omega\psi\cos\lambda) = -we\alpha \qquad (6.5\text{--}13)$$

or,

$$\dot{\alpha} = -\Omega\psi\cos\lambda - \frac{we\alpha}{C\dot{\varphi}} \qquad (6.5\text{--}14)$$

Differentiating the equation for M_ξ (second of Eq. 6.5–8 without $I_\xi\dot{\alpha}$ term), and substituting for $\dot{\alpha}$ from above and α from M_ξ, we arrive at the following differential equation for ψ:

$$\ddot{\psi} + \left(\frac{we}{C\dot{\varphi}}\right)\dot{\psi} + \frac{(wl + C\dot{\varphi}\Omega\cos\lambda)\Omega\cos\lambda}{C\dot{\varphi}}\,\psi = -\frac{we\Omega\sin\lambda}{C\dot{\varphi}} \qquad (6.5\text{--}15)$$

The effect of the offset e is then to damp the ψ oscillations and shift the equilibrium position to the east by the angle,

$$\psi_0 = -\frac{we\tan\lambda}{wl + C\dot{\varphi}\Omega\cos\lambda} \qquad (6.5\text{--}16)$$

Compass heading error due to vehicle motion

When a vehicle carrying a gyrocompass moves in a northerly direction along a meridian with velocity v, an angular velocity v/R pointing west,

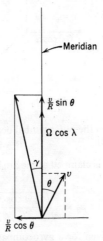

Fig. 6.5–2. Angular velocities in horizontal plane due to vehicle motion v.

where R is the radius of the earth, is introduced. By combining this vector with the horizontal component of the earth's rotation $\Omega\cos\lambda$, the resultant angular velocity in the horizontal plane deviates to the west by an angle

$$\gamma \cong \frac{v/R}{\Omega\cos\lambda}$$

and the gyrocompass will now point in the direction of the resultant, introducing a heading error of γ.

If the vehicle is traveling in a direction making an angle θ with the meridian plane, v can be replaced by $v\cos\theta$ to give,

$$\gamma \cong \frac{v\cos\theta}{R\Omega\cos\lambda + v\sin\theta}$$

where the effect of the component $v\sin\theta$ is neglected due to the fact that it is small in comparison to $\Omega\cos\lambda$.

PROBLEMS

1. Resolve $h = C\dot\varphi$ along the vertical and horizontal directions, so that we have the vector diagram shown. Derive Eq. 6.4–3 from this configuration.

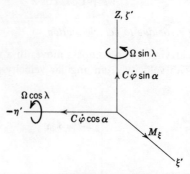

Prob. I

2. Determine the reactions on the outer gimbal bearings ξ for the gyrocompass of Sec. 6.4 (see Fig. 6.4–1).

3. If the gyrocompass axis ξ is clamped at the angle α_0, show that the frequency of oscillation becomes,

$$p_3 = \sqrt{\frac{C\dot\varphi\Omega \cos \lambda}{I_\eta}}$$

4. The following data are given for a gyrocompass.

$$C = 3.0 \text{ in. lb sec}^2$$
$$A = 1.80 \text{ in. lb sec}^2$$
$$\dot\varphi = 1000 \text{ rad/sec}$$
$$wl = 75 \text{ lb in.}$$

Determine p_1, p_2, and α_0 for any latitude λ.

5. Determine the gyrocompass heading error for a ship traveling at a constant speed of 15 knots in a direction N 20° E at latitude 48° N. Would the heading error be different if the ships direction were N 20° W?

6. Derive the equation for the heading error of a gyrocompass, taking into account the latitude component $v \sin \theta$ of the carrier vehicle.

7. A gyropendulum is a spherical pendulum with a spinning disk with angular momentum Cn along its pendulum length l, as shown in the sketch. Letting A be the moment of inertia through O perpendicular to l, show that the moment equations for small angles are,

$$Cn\dot\varphi + A\ddot\theta = -Wl\theta$$
$$Cn\dot\theta - A\ddot\varphi = Wl\varphi$$

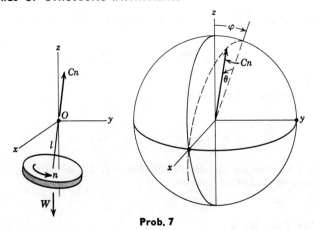

Prob. 7

8. Show that the gyropendulum of Prob. 7 has natural frequencies given by the equation,

$$p_{1,2}^2 = \frac{(C^2n^2 + 2AWl) \pm Cn\sqrt{C^2n^2 + 4AWl}}{2A^2}$$

Approximate equations for the lower and higher natural frequencies, neglecting the term AWl/C^2n^2, are,

$$p_1 = \frac{Wl}{Cn} \quad \text{and} \quad p_2 = \frac{Cn}{A}$$

9. The gyropendulum of Prob. 7 is mounted on a vehicle traveling in the x direction (a great circle) with velocity v. Show that the pendulum must tilt through a small angle φ about the x axis according to the equation,

$$\varphi = \frac{Cnv}{WlR}$$

where R is the radius of the earth.

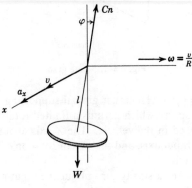

Prob. 9

10. If the vehicle of Prob. 9 accelerates along the x axis so that $dv/dt = a_x$, show that

$$\frac{d\varphi}{dt} = \frac{Cna_x}{WlR} \quad \text{and} \quad \frac{Cn}{Wl} = \sqrt{\frac{R}{g}}$$

The lower natural period of the pendulum, according to Prob. 8, now becomes,

$$\tau_1 = 2\pi \frac{Cn}{Wl} = 2\pi \sqrt{\frac{R}{g}} = 84.4 \text{ min}$$

which is called the Schuler period, after Max Schuler of Germany, who extensively studied the problem.

6.6 The Rate Gyro

High-speed gyros serve as basic elements in many instruments for guidance and control of moving vehicles. Figure 6.6–1 shows the essential

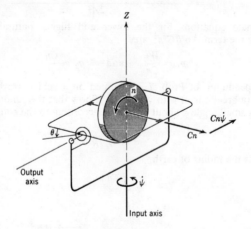

Fig. 6.6–I.   Rate gyro.

elements of a rate gyro. The inner gimbal supporting the spinning wheel is restrained by a spring which permits a limited rotation about the outer gimbal which is fixed in the vehicle. The Z axis about which the vehicle turns is called the input axis, and the axis of rotation of the inner gimbal is called the output axis.

If the vehicle makes a steady turn about the input axis at a rate $\dot\psi$, the rate of change of the angular momentum vector Cn is $Cn\dot\psi$, which requires a moment equal to it about the output axis. This moment is supplied by the torsional spring of stiffness K, as the output axis tilts by a small angle θ, as shown in Fig. 6.6–1. Equating the two moments, we have,

$$Cn\dot{\psi} = K\theta$$

or

$$\theta = \frac{Cn}{K}\dot{\psi} \qquad (6.6\text{–}1)$$

and the output angle θ is proportional to the rate of turn $\dot{\psi}$ of the input axis or the vehicle itself.

The output angle θ is in general read electrically by a pickoff device. One such device is the E-pickoff shown in Fig. 6.6–2. The middle leg of

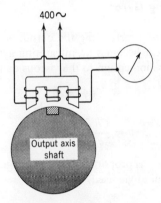

400 ∿

Output axis
shaft

Fig. 6.6–2. E-pickoff on shaft.

the E is supplied by an alternating current, generally of 400 cps. The two outer legs of the E are wound in opposition, so that when the armature, attached to the output axis, is centered about the middle leg, no voltage appears across the opposed outer coils connected in series. When the armature is displaced due to θ, the magnetic flux path is unbalanced, resulting in a voltage reading in the outer coils.

In the undamped instrument, the output axis will overshoot the steady angle θ and oscillate about it. To prevent this undesirable condition, damping is generally provided, and its behavior can be established from the following differential equation,

$$A\ddot{\theta} + c\dot{\theta} + K\theta = Cn\dot{\psi} \qquad (6.6\text{–}2)$$

where A is the moment of inertia of the wheel and inner gimbal about the output axis, and c the coefficient of viscous damping. Thus the transient characteristics of the instrument can be obtained from the homogeneous equation,

$$\ddot{\theta} + 2\zeta\omega\dot{\theta} + \omega^2\theta = 0 \qquad (6.6\text{--}3)$$

where

$$\omega = \sqrt{\frac{K}{A}} = \text{undamped natural frequency}$$

$$\zeta = \frac{c}{c_{\text{cr}}} = \text{damping factor}$$

$$c_{\text{cr}} = 2\sqrt{KA} = \text{critical damping}$$

6.7 The Integrating Gyros

If the torsional spring restraining the output is replaced by a viscous damper, the instrument becomes an integrating gyro. Equating the rate of change of angular momentum to the viscous damping torque,

$$Cn\dot{\psi} = c\dot{\theta}$$

or

$$\theta = \frac{Cn}{c}\int \dot{\psi}\, dt = \frac{Cn}{c}\psi \qquad (6.7\text{--}1)$$

Thus the output angle θ is proportional to the integral of the input angular rate which is the input angle itself.

6.8 The Stable Platform

The principal function of the stable platform is to maintain a space-fixed angular reference. It is an essential part of an inertial guidance system. The platform makes use of the property of the gyroscope, that a torque about an input axis (excluding the spin axis) produces an angular velocity about the orthogonal (output) axis. In general, three single-degree-of-freedom gyros oriented in mutually perpendicular directions are mounted on the platform, as shown in Fig. 6.8–1. The platform, in turn, is mounted on two gimbals which allow it three degrees of angular freedom. If the platform is perfectly balanced and the bearings are frictionless, no torque will be experienced by the platform, and its orientation will be maintained regardless of the motion of the carrier. However, due to unbalance and friction which cannot be eliminated entirely, disturbing torques will be felt by the platform. It is the function of the gyros to sense this disturbance

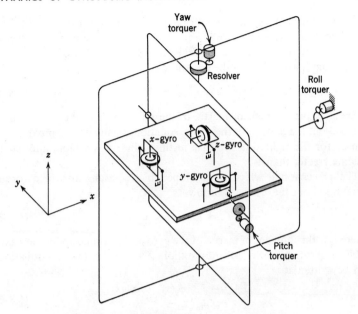

Fig. 6.8–1. Stable platform for inertial guidance.

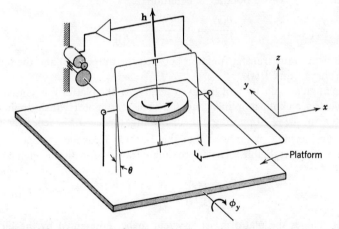

Fig. 6.8–2. Single-axis platform to maintain angular orientation about y axis.

and, through a servo system, counteract the disturbing torque to produce essentially a torque-free system.

The understanding of the dynamics of the stable platform can be obtained by a discussion of the single-axis platform shown in Fig. 6.8–2,[4] where the y axis is the input axis and the x axis (rotation of the spin axis) is the output axis. A disturbing torque T_y about the y axis will rotate the spin axis, and, therefore, **h** through the angle θ, and the applied torque minus the inertia torque about the y axis must equal the rate of change of the angular momentum **h** according to the equation,

$$T_y - J_y\ddot{\phi}_y = h\dot{\theta} \qquad (6.8\text{--}1)$$

where J_y is the moment of inertia of the platform, and the gyro with its frame about the y axis. In the above equation, small angle approximation is used for the right side which is justified since θ is seldom allowed to become greater than $1°$.

The precession $\dot{\theta}$ which is developed by T_y results also in a torque equation about the x axis as follows:

$$- T_x = I_x\ddot{\theta} - h\dot{\phi}_y = 0 \qquad (6.8\text{--}2)$$

where I_x is the moment of inertia of the gyro and its frame about the x axis.

Using the Laplace transform notation $\mathscr{L}\dot{\theta} = s\,\bar{\theta}(s)$, the two equations can be written as

$$\begin{aligned}
\bar{T}_y(s) - J_y s^2\,\bar{\phi}_y(s) &= hs\,\bar{\theta}(s) \\
I_x s^2\,\bar{\theta}(s) &= hs\,\bar{\phi}_y(s)
\end{aligned} \qquad (6.8\text{--}3)$$

Eliminating $\bar{\phi}_y(s)$, we obtain the equation

$$\frac{\bar{\theta}(s)}{\bar{T}_y(s)} = \frac{h/J_y I_x}{s[s^2 + (h^2/J_y I_x)]} \qquad (6.8\text{--}4)$$

which defines the transfer function between the output $\bar{\theta}(s)$ and the input disturbing torque $\bar{T}_y(s)$.

The angular velocity $\dot{\theta}$ of the gyro relative to the platform is sensed by the electric pickoff, amplified and fed to a servomotor which applies a counter torque T_s opposing the disturbing torque T_y. Generally the platform inertia J_y is large so that the nutation frequency $\sqrt{h^2/J_y I_x}$ (see Eq. 6.3–19) is negligible. The approximate transfer function is then equal to,

$$\frac{\bar{\theta}(s)}{\bar{T}_y(s)} = \frac{h}{J_y I_x s^3} \qquad (6.8\text{--}5)$$

which enables the platform servosystem to be represented by the block diagram of Fig. 6.8–3, where $A(s)$ is the transfer function of the electric pickoff, amplifier, and the servo motor. With $A(s)$ known, the stable platform's dynamical behavior can be studied for stability and drift characteristics.

The three-axis stable platform can be considered to be an assembly of three single-axis platforms similar to those of the previous section, but mounted in a single stable unit, as shown in Fig. 6.8–1.[4] The analysis is more complex due to coupling between the three rotations. For the discussion of the inertial guidance system, it is sufficient to assume that

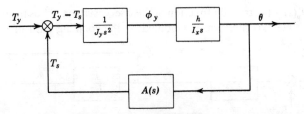

Fig. 6.8–3. Block diagram for the single-axis platform.

there is a platform which will successfully maintain a given orientation in space.

6.9 Three-Axis Platform (Resolution of Motion)

The analysis of a three-axis platform is more complex owing to coupling between the three rotations and the necessity of resolution of the platform pickoff signals because of nonalignment of the gimbal and platform axes.

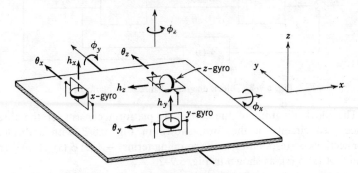

Fig. 6.9–1. Coupling in three-axis platform.

Figure 6.9–1 shows the angular momentum vectors of the x, y, z gyros. Letting θ_x, θ_y, θ_z be the outputs of the x, y, z gyros due to rotations ϕ_x, ϕ_y, ϕ_z of the input axes, the pickoff of each gyro must be,

$$\sigma_x = \theta_x - \phi_y$$
$$\sigma_y = \theta_y + \phi_x \qquad (6.9\text{--}1)$$
$$\sigma_z = \theta_z - \phi_y$$

With the gimbal axes lined up with the platform axes, the counteracting torques called for by the pickoff signals are of the form,

$$T_{sy}(s) = A_y(s) \; \sigma_y(s) = A_y(s)[\theta_y(s) + \phi_x(s)] \qquad (6.9\text{--}2)$$

where $A_y(s)$ is the transfer function of the y servosystem. Thus the behavior of the single axis platform is modified by the coupling term of the form $A_y(s) \; \phi_x(s)$.

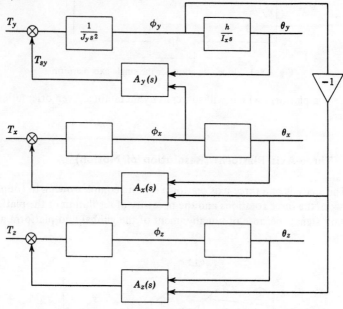

Fig. 6.9–2. Block diagram of three-axis platform

The block diagram of the three-axis platform consists of the three uncoupled circuits of the form shown in Fig. 6.8–3 with additional connections corresponding to the coupling terms, $-A_x(s) \; \phi_y(s)$, $A_y(s) \; \phi_x(s)$, and $-A_z(s) \; \phi_y(s)$, as shown in Fig. 6.9–2.

Assuming that the outer gimbal axis, originally parallel to the platform x axis, is attached to the vehicle, as shown in Fig. 6.9–3, and assuming that the vehicle is roll-stabilized, the motion of the vehicle in pitch and yaw will cause the gimbal axes to deviate from the platform axes. It is evident then that the platform torques T_x, T_y, T_z must now be resolved along the displaced gimbal axes where the torque servomotors act. Since the countertorques are proportional to the platform pickoff signals, the proper torques about the new gimbal axes are found by resolution of the platform pickoff signals along the gimbal axes.

With the vehicle roll-stabilized, we first allow the vehicle to nose down through a pitch angle Φ_y. Letting the new gimbal axes be indicated by

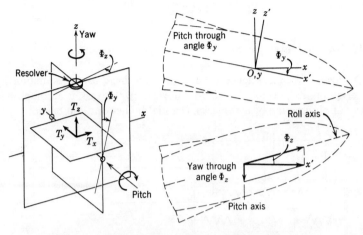

Fig. 6.9–3. Gimbal rotation requires resolution of torque.

primes, the components of the platform pickoff signals along the gimbal axes are

$$\sigma_{x'} = \sigma_x \cos \Phi_y - \sigma_z \sin \Phi_y$$
$$\sigma_{y'} = \sigma_y$$
$$\sigma_{z'} = \sigma_x \sin \Phi_y + \sigma_z \cos \Phi_y$$

Next allow a rotation Φ_z in yaw about the z' axis, and resolve $\sigma_{x'}$ along the pitch gimbal axis and the new roll axis,

$$\sigma_{\text{roll}} = \sigma_{x'} \sec \Phi_z$$
$$\sigma_{\text{pitch}} = -\sigma_{x'} \tan \Phi_z$$

The resulting signal about the new gimbal axes of roll, pitch, and yaw due to both Φ_y and Φ_z are, then,

$$\sigma_{\text{roll}} = (\sigma_x \cos \Phi_y - \sigma_z \sin \Phi_y) \sec \Phi_z$$
$$\sigma_{\text{pitch}} = -(\sigma_x \cos \Phi_y - \sigma_z \sin \Phi_y) \tan \Phi_z + \sigma_y$$
$$\sigma_{\text{yaw}} = (\sigma_x \sin \Phi_y + \sigma_z \cos \Phi_y) = \sigma_{z'}$$

which can be expressed in the following matrix notation.

$$
\begin{bmatrix} \sigma_{\text{roll}} \\ \sigma_{\text{pitch}} \\ \sigma_{\text{yaw}} \end{bmatrix} = \begin{bmatrix} \cos \Phi_y \sec \Phi_z & 0 & -\sin \Phi_y \sec \Phi_z \\ -\cos \Phi_y \tan \Phi_z & 1 & \sin \Phi_y \tan \Phi_z \\ \sin \Phi_y & 0 & \cos \Phi_y \end{bmatrix} \begin{bmatrix} \sigma_x \\ \sigma_y \\ \sigma_z \end{bmatrix}
$$

The function of the resolver is then to resolve the platform pickoff signals σ_x, σ_y, σ_z to the components σ_{roll}, σ_{pitch}, σ_{yaw} along the displaced gimbal axes housing the roll, pitch, and yaw servomotors.

6.10   Inertial Navigation

Navigation is the science of directing a vehicle to a destination by determining its position. In inertial navigation this task is accomplished without observation of landmarks, celestial bodies or radio beams.

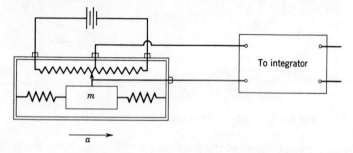

Fig. 6.10–1.   Accelerometer and integrator.

A vehicle moving in space possesses six degrees of freedom, three translational and three rotational. Consequently, six sensors are needed. The stable platform discussed in Sec. 6.8 offers a reference for the rotational motion, whereas the accelerometer is an instrument capable of detecting translational motion. In fact, the three gyros of the stable platform and the three accelerometers oriented in mutually perpendicular directions can supply all the information for establishing the motion of a rigid body, and the high degree of accuracy with which this is being done has made inertial navigation a practical reality.[2,5,7]

Figure 6.10–1 shows a schematic of an accelerometer and integrator. Acceleration along its axis displaces the mass against the spring according to Newton's equation $F = ma$, where F is the spring force. The displacement of the mass which is proportional to the acceleration is picked off by a potentiometer and integrated to velocity and displacement of the vehicle.

The accelerometers are mounted on a table which is always maintained normal to the local radius of the earth. This is accomplished by means of a computer and a clock which rotates the table relative to the stable platform, as indicated in Fig. 6.10–2. In some cases the accelerometers are mounted directly on the stable platform which is torqued to the normal position.

To obtain an understanding of how the inertial navigator works, we assume that the vehicle starts at the equator and that the plane of the stable platform is horizontal with the arrow pointing in the N polar direction. If the vehicle moves towards the north along a longitude, and the accelerometer table is always kept normal to the geocentric radius r, the N-S accelerometer will measure the acceleration a_x (see Fig. 6.10-2). The

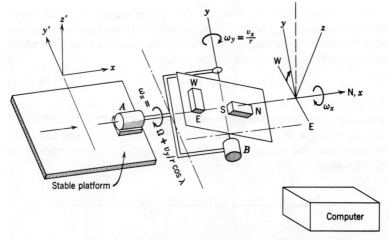

Fig. 6.10–2. Components of an inertial navigator.

proper rate of rotation of the table about the y axis is then $\omega_y = v_x/r$, where v_x is determined from the first integral of a_x. The latitude motor B then rotates the table at a rate ω_y to keep the N-S line on the table normal to r.

Due to rotation of the earth towards the east, the E-W line of the table must be rotated by the longitude motor A to unwind the earth's rotation. Since during the motion of the vehicle the orientation of the stable platform remains fixed in inertial space (towards the N star) the required rotation of the accelerometer table about the x axis of the stable platform at any latitude is Ω or $15°$/hr.

To this rotation must be added the rotation about the platform x axis due to the E-W motion of the vehicle relative to the original longitude. By integrating the output of the E-W accelerometer and dividing by $r \cos \lambda$, the additional rotation to maintain the E-W line of the table normal to r is $\omega_x = v_y/r \cos \lambda$.

These computations are performed by a computer which must be an integral part of the inertial system. Thus the inertial navigator must consist of the stable platform, accelerometers with integrators, a computer to compute the proper angular rates of the table due to vehicle motion, a clock to unwind the earth's rotation, and the servomotors to actually carry out these functions.

6.11 Oscillation of Navigational Errors

The accelerometers mounted on the vehicle measure only the non-gravitational force $\mathbf{F}_{ng}$ acting on the vehicle, and therefore one must add to it the gravitational force $\mathbf{F}_g$ in order to determine the total force which determines the acceleration $\mathbf{a}_v$ of the vehicle.

$$\mathbf{F}_{ng} + \mathbf{F}_g = m\mathbf{a}_v \tag{6.11-1}$$

For example if the vehicle is resting on the surface of the earth, the verticle accelerometer will indicate the upward force (thrust) of the ground on the vehicle, or $\mathbf{F}_{ng} = \mathbf{W}$. To this must be added the gravitational attraction of the earth on the vehicle $\mathbf{F}_g = -\mathbf{W}$ which results in the zero acceleration of the vehicle.

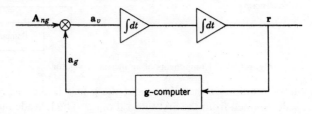

Fig. 6.11–1. Simplified block diagram of gravity and position computation.

Equation 6.11-1 can be expressed entirely in terms of acceleration by dividing by m

$$\mathbf{a}_v = \mathbf{A}_{ng} + \mathbf{a}_g \tag{6.11-2}$$

where $\mathbf{A}_{ng}$ is the nongravitational (thrust) acceleration indicated by the accelerometers. The vehicle position is then found by a double integration of the vehicle acceleration $\mathbf{a}_v$ as shown by the simplified block diagram of Fig. 6.11-1. The gravitational acceleration $\mathbf{a}_g$ which depends only on the position $\mathbf{r}$ is computed and added to the output of the accelerometers to give the vehicle acceleration $\mathbf{a}_v$.

It is evident that an accelerometer error would result in incorrect rotation rates of the accelerometer table which would result in a position error, an incorrect value of $\mathbf{a}_g$, and a deviation of the accelerometer table from the normal to the true geocentric radius $\mathbf{r}$. These errors are oscillatory for subsatellite speeds, and we will now investigate their nature.

We will define the correct position of the vehicle by a vector **r** referenced to inertial coordinates with origin at the center of earth. We will also define another set of coordinates x, y, z with origin coinciding with the correct position of the vehicle and with the z axis parallel to **r** as shown in Fig. 6.11–2. Thus the angular velocity of the vehicle is specified

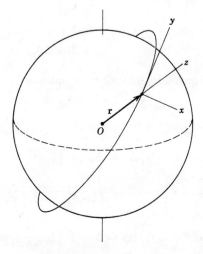

Fig. 6.11–2. Vehicle position indicated by **r**. Origin of x, y, z coinciding with vehicle.

by ω_x, ω_y, ω_z respectively and the xy plane is always normal to the local geocentric radius **r**.

We now assume that the position of the vehicle is in error by

$$\Delta \mathbf{r} = \Delta x \mathbf{i} + \Delta y \mathbf{j} + \Delta z \mathbf{k} \qquad (6.11\text{–}3)$$

and examine first the error in $\mathbf{a}_g = -\mathbf{g}$. Since **g** is inversely proportional to the square of the distance from the center of the earth, the incorrect components of **g** computed from $\mathbf{r} + \Delta \mathbf{r}$ are

$$g_z = -g\left(\frac{r}{r + \Delta z}\right)^2 = -g\left(1 - \frac{2\,\Delta z}{r}\right) = -g + 2\omega_0^2\,\Delta z$$

$$g_x = -g\,\frac{\Delta x}{r} = -\omega_0^2\,\Delta x \qquad (6.11\text{–}4)$$

$$g_y = -g\,\frac{\Delta y}{r} = -\omega_0^2\,\Delta y$$

where $-g$ is the correct value and $\omega_0^2 = g/r$.

The acceleration error of the vehicle can be determined from the general equation for acceleration by replacing $\mathbf{r}$ by $\Delta\mathbf{r}$.

$$
\begin{aligned}
\Delta\mathbf{a}_v = & [\Delta\ddot{x} + \omega_x\omega_y\,\Delta y + \omega_x\omega_z\,\Delta z - (\omega_y{}^2 + \omega_z{}^2)\,\Delta x + \dot{\omega}_y\,\Delta z - \dot{\omega}_z\,\Delta y \\
& \qquad\qquad + 2(\omega_y\,\Delta\dot{z} - \omega_z\,\Delta\dot{y})]\mathbf{i} \\
& + [\Delta\ddot{y} + \omega_x\omega_y\,\Delta x + \omega_y\omega_z\,\Delta z - (\omega_x{}^2 + \omega_z{}^2)\,\Delta y + \dot{\omega}_z\,\Delta x - \dot{\omega}_x\,\Delta z \\
& \qquad\qquad + 2(\omega_z\,\Delta\dot{x} - \omega_x\,\Delta\dot{z})]\mathbf{j} \\
& + [\Delta\ddot{z} + \omega_x\omega_z\,\Delta x + \omega_y\omega_z\,\Delta y - (\omega_x{}^2 + \omega_y{}^2)\,\Delta z + \dot{\omega}_x\,\Delta y - \dot{\omega}_y\,\Delta x \\
& \qquad\qquad + 2(\omega_x\,\Delta\dot{y} - \omega_y\,\Delta\dot{x})]\mathbf{k} \qquad (6.11\text{--}5)
\end{aligned}
$$

Substituting these quantities into the error equation,

$$
\Delta\mathbf{a}_v = \Delta\mathbf{A} + \Delta\mathbf{g}
$$

its component equations can be written as

$$
\left[\frac{d^2}{dt^2} + \omega_0{}^2 - (\omega_y{}^2 + \omega_z{}^2)\right]\Delta x = \Delta A_x + \left(2\omega_z\frac{d}{dt} + \dot{\omega}_z - \omega_x\omega_y\right)\Delta y
$$
$$
- \left(2\omega_y\frac{d}{dt} + \dot{\omega}_y + \omega_x\omega_y\right)\Delta z \qquad (6.11\text{--}6)
$$

$$
\left[\frac{d^2}{dt^2} + \omega_0{}^2 - (\omega_x{}^2 + \omega_z{}^2)\right]\Delta y = \Delta A_y + \left(2\omega_x\frac{d}{dt} + \dot{\omega}_x - \omega_y\omega_z\right)\Delta z
$$
$$
- \left(2\omega_z\frac{d}{dt} + \dot{\omega}_z + \omega_x\omega_y\right)\Delta x \qquad (6.11\text{--}7)
$$

$$
\left[\frac{d^2}{dt^2} - 2\omega_0{}^2 - (\omega_x{}^2 + \omega_y{}^2)\right]\Delta z = \Delta A_z + \left(2\omega_y\frac{d}{dt} + \dot{\omega}_y - \omega_x\omega_z\right)\Delta x
$$
$$
- \left(2\omega_x\frac{d}{dt} + \dot{\omega}_x + \omega_y\omega_z\right)\Delta y \qquad (6.11\text{--}8)
$$

To interpret these equations, assume the vehicle to be traveling with velocity v in the y direction along a great circle at constant altitude. Then $\omega_y = \omega_z = 0$, and $\omega_x = -v/r$. The above equations reduce to,

$$
\left(\frac{d^2}{dt^2} + \omega_0{}^2\right)\Delta x = \Delta A_x \qquad (6.11\text{--}9)
$$

$$
\left[\frac{d^2}{dt^2} + \left(\omega_0{}^2 - \frac{v^2}{r^2}\right)\right]\Delta y = \Delta A_y + \left(2\omega_x\frac{d}{dt} + \dot{\omega}_x\right)\Delta z \qquad (6.11\text{--}10)
$$

$$
\left[\frac{d^2}{dt^2} - \left(2\omega_0{}^2 + \frac{v^2}{r^2}\right)\right]\Delta z = \Delta A_z - \left(2\omega_x\frac{d}{dt} + \dot{\omega}_x\right)\Delta y \qquad (6.11\text{--}11)
$$

The first two equations have solutions which are harmonic oscillations of frequency $\omega_0 = \sqrt{g/r}$ and $\sqrt{\omega_0^2 - (v^2/r^2)}$. The solution of the third equation is hyperbolic and Δz must diverge.

For ordinary altitudes, the period as computed from ω_0 is,

$$\tau = 2\pi\sqrt{\frac{r}{g}} = 2\pi\sqrt{\frac{3960 \times 5280}{32.2 \times 60^2}} = 84 \text{ min} \qquad (6.11\text{--}12)$$

and the inertial system is often referred to as the 84-min Schuler pendulum. As v approaches orbital speeds for satellites, $\omega_0^2 - (v^2/r^2)$ will approach zero and the desirable oscillatory nature of the position error disappears.

In addition, we might mention briefly the error introduced by the deviation from normal of the acceleration table. If the table tilts by a small angle $\boldsymbol{\phi} = \phi_x\mathbf{i} + \phi_y\mathbf{j} + \phi_z\mathbf{k}$, the error in the accelerometer output will be,

$$\boldsymbol{\phi} \times \mathbf{A} = (\phi_y A_z - \phi_z A_y)\mathbf{i} + (\phi_z A_x - \phi_x A_z)\mathbf{j} + (\phi_x A_y - \phi_y A_x)\mathbf{k}$$
$$(6.11\text{--}13)$$

PROBLEMS

1. An aircraft directional gyro has a spin angular momentum of $h = 3.0$ lb-in./sec. If the drift rate is specified as $0.01°$/hr, determine the torque producing the drift.

2. An inertial system is to guide an airplane traveling at a speed of 600 mph to a destination of 1000 miles with an accuracy of $\frac{1}{2}$ mile. Determine the allowable drift rate.

3. Assume that for the single-axis stable platform of Fig. 6.8–2 there is damping and spring stiffness restraining the rotation of the output axis. Write the equation for the torque about the output axis.

4. Write the subsidiary equation for the single-axis platform including damping and spring stiffness, and draw the new block diagram.

5. For the single-axis platform of Prob. 3, determine the transfer function $\theta(s)/\phi(s)$ and discuss the special cases when: (a) damping $= 0$; (b) spring stiffness $= 0$; (c) damping and spring stiffness $= 0$. Indicate the type of gyro obtained in each case.

6. For case (c) of Prob. 5, obtain the transfer function T_s/T_d, where T_d and T_s are the disturbing torque and the servo countertorque, and discuss the influence of $A(s)$ on the system.

7. Obtain the stiffness characteristics of the single-axis platform by examining the transfer functions $T_d(s)/\theta(s)$ and $T_d(s)/\phi(s)$.

8. A three-axis stable platform has gyros mounted as shown in the sketch. Identify which are the x, y, and z gyros and determine the equations for the pickoff signals, using ϕ_i for input and θ_i for output.

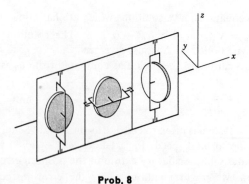

Prob. 8

9. If the three-axis platform has gyros arranged as shown, identify the x, y, and z gyros and determine the equations for the pickoff signals.

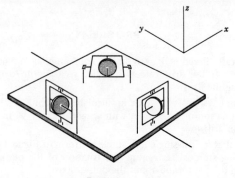

Prob. 9

10. Draw a block diagram for the dynamics of the platform of Prob. 9.

11. If the platform of Prob. 9 is mounted on a missile with the roll-stabilized axis along the x axis, determine the equations for the resolver signal along the servo roll, pitch, and yaw axes due to angular rotations Φ_y and Φ_z.

12. An airplane with an inertial navigator is headed in the direction N 60° W at latitude 32° N, at speed 600 mph. Determine the angular rates about the meridian and latitude axes due to the motion and the required angular rates of the accelerometer table.

13. A rocket ship guided by an inertial navigator is traveling along a great circle route at constant altitude of 100 miles and at a speed of 12,000 mph. With the horizontal axis y oriented along the flight path, discuss the nature of the navigational errors and calculate the frequency of the oscillatory error.

REFERENCES

1. Cunningham, W. J., *Nonlinear Analysis*, McGraw-Hill Book Co., New York (1958), p. 125.
2. Frye, E., "Fundamentals of Inertial Guidance and Navigation," *J. Astronaut. Sci.*, V, No. 1 (1958), 1.
3. Malkin, I. G., "Some Problems in the Theory of Nonlinear Oscillations," Translation Series, *Atomic Energy Commission, AEC-tr-3766* (Book 1), Chapter 1.
4. Mitsutomi, T., "Characteristics and Stabilization of an Inertial Platform," *Trans. IRE, Aeronautical and Navigational Electronics* (June 1958), 95–105.
5. O'Donnell, C. F., "Inertial Navigation," *J. Franklin Institute*, **266**, Nos. 4 and 5 (Oct. and Nov. 1958).
6. Plymale, B. T., and Goodstein, R., "Nutation of a Free Gyro Subject to an Impulse," *J. Appl. Mech.*, **22**, No. 3 (Sept. 1955), 365–366.
7. Slater, J. M., and Duncan, D. B., "Inertial Navigation," *Aeronaut. Eng. Review*, **15**, No. 1 (Jan. 1956).
8. Stoker, J. J., *Nonlinear Vibrations*, Interscience Publishers, New York (1950), p. 98.
9. Streeter, J. R., "Error Analysis of an Inertial System in Vehicles of High Speed," Master of Science Thesis in Engineering, University of California, Los Angeles (June 1960).
10. Thomson, W. T., *Laplace Transformation* (2nd ed.), Prentice-Hall, Englewood Cliffs, N.J. (1960).
11. Wrigley, W., Woodbury, R. B., and Hovorka, J., "Application of Inertial Guidance Principles" Automatic Control, 7, No. 1, 22 (July 1957).
12. Pitman, G. R., "Inertial Guidance" John Wiley & Sons, Inc. New York (1962).

Space Vehicle Motion

CHAPTER 7

7.1 General Equations in Body Coordinates

For rockets and space vehicles it is often necessary to consider the general problem of the spinning body under thrust. The concern here is the body attitude and the motion of the center of mass. We will first consider problems where the rate of mass variation is small enough to be negligible.

To outline the problem at hand, we will consider a rigid body and define a set of body-fixed axes x, y, z rotating with angular velocity $\boldsymbol{\omega}$, and with the origin coinciding with the center of mass. Although it is always desirable to let the body axes coincide with the principal axes, this is often not possible, so that, in the general case, the moments and products of inertia will be defined as

$$I_x = A \qquad I_{xy} = D$$
$$I_y = B \qquad I_{xz} = E$$
$$I_z = C \qquad I_{yz} = F$$

The angular momentum (Eq. 5.2–7) in the above notation becomes

$$\mathbf{h} = (A\omega_x - D\omega_y - E\omega_z)\mathbf{i} + (B\omega_y - F\omega_z - D\omega_x)\mathbf{j}$$
$$+ (C\omega_z - E\omega_x - F\omega_y)\mathbf{k} \qquad (7.1\text{–}1)$$

and the moment equation about the body axes

$$\mathbf{M} = [\dot{\mathbf{h}}] + \boldsymbol{\omega} \times \mathbf{h}$$

can be written out in terms of the components given by Eq. 5.6–3 as

$$M_x = (A\dot{\omega}_x - D\dot{\omega}_y - E\dot{\omega}_z) + (C\omega_z - E\omega_x - F\omega_y)\omega_y$$
$$- (B\omega_y - F\omega_z - D\omega_x)\omega_z$$
$$M_y = (B\dot{\omega}_y - F\dot{\omega}_z - D\dot{\omega}_x) + (A\omega_x - D\omega_y - E\omega_z)\omega_z$$
$$- (C\omega_z - E\omega_x - F\omega_y)\omega_x \qquad (7.1\text{–}2)$$
$$M_z = (C\dot{\omega}_z - E\dot{\omega}_x - F\dot{\omega}_y) + (B\omega_y - F\omega_z - D\omega_x)\omega_x$$
$$- (A\omega_x - D\omega_y - E\omega_z)\omega_y$$

We next let the velocity of the center of mass be expressed by the equation

$$\mathbf{v} = v_x\mathbf{i} + v_y\mathbf{j} + v_z\mathbf{k} \qquad (7.1\text{–}3)$$

and the force as

$$\mathbf{F} = F_x\mathbf{i} + F_y\mathbf{j} + F_z\mathbf{k} \qquad (7.1\text{–}4)$$

Since the x, y, z coordinates are rotating with the body, the force components in the x, y, z directions are determined from the equation

$$\mathbf{F} = m\left[\frac{d\mathbf{v}}{dt}\right] + \boldsymbol{\omega} \times m\mathbf{v} \qquad (7.1\text{–}5)$$

to be

$$F_x = m(\dot{v}_x + v_z\omega_y - v_y\omega_z)$$
$$F_y = m(\dot{v}_y + v_x\omega_z - v_z\omega_x) \qquad (7.1\text{–}6)$$
$$F_z = m(\dot{v}_z + v_y\omega_x - v_x\omega_y)$$

If the resultant of the above forces does not pass through the center of mass coinciding with the origin of the x, y, z, axes, Eqs. (7.1–2) and (7.1–6) become coupled owing to the moment of the force. Also these equations define the motion of the body only in terms of the linear and angular velocities referred to body axes, and their solution and transformation to displacements and angles relative to inertial coordinates are problems of considerable difficulty which can be accomplished only under simplifying assumptions.

7.2 Thrust Misalignment

We will consider first a simple problem of a spinning missile with a misalignment of the thrust line. We will assume that the missile is symmetric so that the x, y, z, axes coincide with the principal axes 1, 2, 3 with $I_1 = I_2 = A$ and $I_3 = C$. With $A = B$, we can rotate the 1, 2 axes so that one of these axes, say 1, is perpendicular to the plane containing the thrust and axis 3, as shown in Fig. 7.2–1.

Euler's equation for the missile is then

$$A\dot{\omega}_1 + (C - A)\omega_2\omega_3 = M_1 \qquad (7.2\text{--}1)$$

$$A\dot{\omega}_2 - (C - A)\omega_1\omega_3 = 0 \qquad (7.2\text{--}2)$$

$$C\dot{\omega}_3 = 0 \qquad (7.2\text{--}3)$$

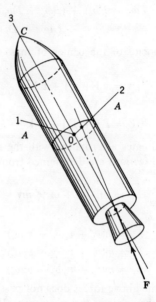

Fig. 7.2-1. Thrust misalignment resulting in moment M_1.

The third equation tells us that $\omega_3 = n$, a constant. Although C is generally less than A for missiles, we let

$$\lambda = n\frac{C - A}{A} \qquad (7.2\text{--}4)$$

as in Sec. 5.8–3 and rewrite the first two equations as

$$\dot{\omega}_1 + \lambda\omega_2 = \frac{M_1}{A} \qquad (7.2\text{--}5)$$

$$\dot{\omega}_2 - \lambda\omega_1 = 0 \qquad (7.2\text{--}6)$$

For the solution of these equations, we will use the technique used in Sec. 5.8 of adding ω_1 and ω_2 in quadrature. Multiplying Eq. 7.2–6 by

$i = \sqrt{-1}$, adding it to Eq. 7.2–5, and letting $\omega_{12} = \omega_1 + i\omega_2$, the two equations for ω_1 and ω_2 are replaced by one

$$\dot{\omega}_{12} - i\lambda\omega_{12} = \frac{M_1(t)}{A} \qquad (7.2\text{–}7)$$

Using the method of Laplace transformation, its subsidiary equation becomes

$$\bar{\omega}_{12}(s) = \frac{\omega_{12}(0)}{s - i\lambda} + \frac{\bar{M}_1(s)}{A(s - i\lambda)} \qquad (7.2\text{–}8)$$

and its solution given by its inverse is

$$\omega_{12}(t) = \omega_{12}(0)e^{i\lambda t} + \frac{1}{A} \int_0^t M_1(\tau)e^{i\lambda(t-\tau)} \, d\tau \qquad (7.2\text{–}9)$$

The separation of this equation to ω_1 and ω_2 is easily accomplished by noting its real and imaginary parts.

PROBLEMS

1. From Eq. 7.2–6, $\omega_1 = (1/\lambda)\dot{\omega}_2$. Substitution into Eq. 7.2–5 results in

$$\ddot{\omega}_2 + \lambda^2\omega_2 = \frac{\lambda}{A} M_1(t)$$

Solve this equation for ω_2 and show that

$$\omega_2(t) = \omega_2(0) \cos \lambda t + \frac{\dot{\omega}_2(0)}{\lambda} \sin \lambda t + \frac{1}{A} \int_0^t M_1(\tau) \sin \lambda(t - \tau) \, d\tau$$

2. From $\omega_1 = (1/\lambda)\dot{\omega}_2$ and the solution of Prob. 1, determine the solution for $\omega_1(t)$.

3. Note that the procedure of Prob. 2 encounters differentiation of an integral. Let $\phi(t) = \int_a^b F(\tau, t) \, d\tau = \int_0^t M_1(\tau) \sin \lambda(t - \tau) \, d\tau$ and use

$$\frac{d\phi(t)}{dt} = \int_a^b \frac{\partial F}{\partial t} \, d\tau + F(b, t) \frac{db}{dt} - F(a, t) \frac{da}{dt}$$

The result is

$$\frac{d}{dt} \int_0^t M_1(\tau) \sin \lambda(t - \tau) \, d\tau = \lambda \int_0^t M_1(\tau) \cos \lambda(t - \tau) \, d\tau$$

4. Separate the Eq. 7.2–9, $\omega_{12} = \omega_1 + i\omega_2$, into its real and imaginary parts and verify the solutions for ω_1 and ω_2 of Probs. 1, 2, and 3.

5. A symmetric body, A, A, C, is damped with moments about the body-fixed axes as follows: $M_1 = -k\omega_1$, $M_2 = -k\omega_2$, $M_3 = -k\omega_3$. Show that the angle between the spin axis, (axis 3) and the angular momentum vector is

$$\tan \theta = \frac{|\omega_{12}(0)|}{\omega_3(0)} \frac{A}{C} \exp\left[-\left(\frac{C}{A} - 1\right)\frac{kt}{C} \right]$$

6. A symmetrical satellite with moments of inertia A, A, C is spinning with angular velocity ω about the axis of C. If a constant torque M_1 is applied to the transverse body-fixed pitch axis, show that the angular velocity of the satellite is

$$\omega_z = n$$

$$\omega_{12}(t) = \omega_{12}(0)e^{i\lambda t} - \frac{M_1}{i\lambda A}(1 - e^{i\lambda t})$$

7.3 Rotations Referred to Inertial Coordinates

The solution of the previous problem is in terms of body-fixed coordinates which are rotating. In order to transform from the body-fixed coordinates to the inertial coordinates X, Y, Z it is necessary to introduce Euler's angles. From Eq. 3.5–1 the transformation is

$$\omega_1 = \dot{\psi} \sin \theta \sin \varphi + \dot{\theta} \cos \varphi$$

$$\omega_2 = \dot{\psi} \sin \theta \cos \varphi - \dot{\theta} \sin \varphi \qquad (7.3\text{–}1)$$

$$\omega_3 = \dot{\psi} \cos \theta + \dot{\varphi} = n \quad \text{a constant for } M_3 = 0$$

Adding ω_1 and ω_2 in quadrature,

$$\omega_{12} = \omega_1 + i\omega_2 = (\dot{\theta} + i\dot{\psi} \sin \theta)e^{-i\varphi} \qquad (7.3\text{–}2)$$

From ω_3

$$\dot{\psi} = \frac{n - \dot{\varphi}}{\cos \theta} \qquad (7.3\text{–}3)$$

which, substituted into Eq. 7.3–2, results in

$$\omega_{12} = [\dot{\theta} + i(n - \dot{\varphi}) \tan \theta]e^{-i\varphi} \qquad (7.3\text{–}4)$$

Although this equation relates the angular velocity ω_{12} about the body-fixed coordinates in terms of Euler's angles referenced to inertial axes, further simplification generally requires a small angle approximation for θ. Such an approximation is often justified when dealing with rockets and missiles whose spin axis must not deviate greatly from a fixed direction of flight.

When θ is small $\tan \theta$ can be replaced by θ

$$\omega_{12} = [\dot{\theta} + i(n - \dot{\varphi})\theta]e^{-i\varphi} \qquad (7.3\text{–}5)$$

At this point we introduce a complex angle of attack, proposed by H. Leon[4,5], which uncouples Eq. 7.3–5.

$$\theta_{12} = \theta e^{-i\varphi} \qquad (7.3\text{–}6)$$

Differentiating

$$\dot{\theta}_{12} = (\dot{\theta} - i\dot{\varphi}\theta)e^{-i\varphi} \tag{7.3-7}$$

Equation 7.3-5 can then be written as

$$\dot{\theta}_{12} + in\theta_{12} = \omega_{12} \tag{7.3-8}$$

so that when ω_{12} is a known function of time we have a first-order ordinary differential equation in θ_{12} to solve. It must be remembered however that the above procedure is limited to problems where θ is small.

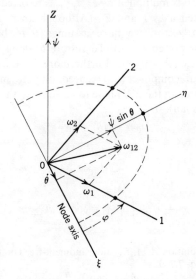

Fig. 7.3–1. Velocity components in transverse plane tilted by angle θ.

At this point the significance of the term $e^{-i\varphi}$ appearing in the various equations should be pointed out. For example, consider Eq. 7.3-2, which is

$$\omega_{12} = \omega_1 + i\omega_2 = (\dot{\theta} + i\dot{\psi}\sin\theta)e^{-i\varphi}$$

All of the components in this equation lie in the tilted transverse plane which are shown in Fig. 7.3–1.

Writing $e^{-i\varphi} = (\cos\varphi - i\sin\varphi)$, ω_{12} becomes

$$\omega_{12} = (\dot{\theta}\cos\varphi + \dot{\psi}\sin\theta\sin\varphi) + i(\dot{\psi}\sin\theta\cos\varphi - \dot{\theta}\sin\varphi) \tag{7.3-9}$$

The real and imaginary parts of this equation are, however, equal to the components of $\dot{\theta}$ and $\dot{\psi}\sin\theta$ along axes 1 and 2. Thus the multiplication of the components $(\dot{\theta} + i\dot{\psi} \sin \theta)$ along the node coordinate system by $e^{-i\varphi}$ results in $\omega_1 + i\omega_2$, the components along the body-fixed axes rotated through an angle φ from the node axis. It follows logically then that, if we multiply the components $\omega_1 + i\omega_2$ along the body axes by $e^{i\varphi}$, we should obtain the vector ω_{12} in terms of the node axis components as follows:

$$(\omega_1 + i\omega_2)e^{i\varphi} = \omega_{12}e^{i\varphi} = (\dot{\theta} + i\dot{\psi} \sin \theta) = \omega_{\xi\eta} \qquad (7.3\text{--}10)$$

We can now attach physical significance to the complex angle of attack $\theta_{12} = \theta e^{-i\varphi}$. Since θ is multiplied by $e^{-i\varphi}$, θ_{12} is resolved into components along the body-fixed axes 1 and 2. (Although an angle is strictly not representable as a vector, we have assumed θ to be small, thereby justifying its vector presentation.) To restore θ along its node axis we multiply θ_{12} by $e^{i\varphi}$, i.e., $\theta = \theta_{12}e^{i\varphi}$. Furthermore, if we wish to examine θ_{12} in terms of inertial components, we need only to multiply θ by $e^{i\psi}$ (assuming ψ measured in the XY plane to be equal to that measured in the tilted plane for θ small), or

$$\theta e^{i\psi} = (\theta_{12}e^{i\varphi})e^{i\psi}$$
$$= \theta_{12}e^{i(\varphi+\psi)} \qquad (7.3\text{--}11)$$
$$\simeq \theta_{12}e^{int}$$

where

$$\varphi + \psi \simeq \int (\dot{\varphi} + \dot{\psi} \cos \theta)\, dt \simeq nt$$

Example 7.3–I

For a body of revolution A, A, C, under moment-free condition, the complex angular rate from Eq. 7.2–9 is

$$\omega_{12} = \omega_{12}(0)e^{i\lambda t} \qquad (a)$$

where $\lambda = n[(C - A)/A]$. Substitution into Eq. 7.3–8 results in the differential equation for the complex angle of attack

$$\dot{\theta}_{12} + in\theta_{12} = \omega_{12}(0)e^{i\lambda t} \qquad (b)$$

Letting $\bar{\theta}_{12}(s)$ be the Laplace transform of $\theta_{12}(t)$, the subsidiary equation becomes

$$\bar{\theta}_{12}(s) = \frac{\theta_{12}(0)}{s + in} + \frac{\omega_{12}(0)}{(s + in)(s - i\lambda)} \qquad (c)$$

From its inverse, the solution for $\theta_{12}(t)$ is

$$\theta_{12}(t) = \theta_{12}(0)e^{-int} + \omega_{12}(0)e^{-int}\int_0^t e^{i(\lambda+n)\tau}\, d\tau$$

$$= \theta_{12}(0)e^{-int} + i\omega_{12}(0)\frac{e^{-int}}{n + \lambda}[1 - e^{i(n+\lambda)t}] \qquad (d)$$

which is referred to body-fixed axes.

To examine the angle of attack in terms of the inertial axes, we multiply by e^{int} to obtain

$$\theta_{XY}(t) = \theta_{12}(0) + i\frac{\omega_{12}(0)}{n + \lambda}[1 - e^{i(n+\lambda)t}]$$

$$= \left[\theta_{12}(0) + \frac{\omega_{12}(0)}{n + \lambda}\sin(n + \lambda)t\right] + i\frac{\omega_{12}(0)}{n + \lambda}[1 - \cos(n + \lambda)t] \quad (e)$$

The real and imaginary components of $\theta_{12}(t)$ are along the X and Y axes, the end of the vector $\theta_{XY}(t)$ prescribing a circle of radius $[\omega_{12}(0)]/(n + \lambda)$ about the center $\theta_{12}(0) + i[\omega_{12}(0)]/(n + \lambda)$, as shown in Fig. 7.3–2.

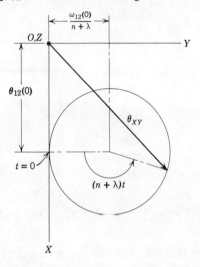

Fig. 7.3–2. Inertial components of angle of attack θ.

7.4 Near Symmetric Body of Revolution with Zero Moment

When the geometric axes x, y, z corresponding to yaw, pitch, and spin of a missile are not principal axes, the solution in terms of such body coordinates will require the solving of the general Eq. 7.1–2. These equations do not lead to a simple solution, even for small products of inertia, and it is desirable to take a different approach.

We recognize first that every body has a set of principal axes 1, 2, 3. For the near symmetric body, the principal axis 3 deviates only by a small angle β from the spin axis z, as shown in Fig. 7.4–1. Without loss of generality, the transverse axis x can be chosen normal to the plane $z03$, and the other two axes 1 and 2 are defined by the angle Φ between axis 1 and the transverse axis x which intersects the planes 1, 2 and xy. Since both the 1, 2, 3 and x, y, z axes are body-fixed axes, β and Φ are constants.

If we assume that the principal inertias $I_1 \cong I_2$, then elementary solutions are available in terms of principal axes 1, 2, 3. The motion of the geometric axes x, y, z can then be obtained by a transformation of coordinates with β and Φ known.

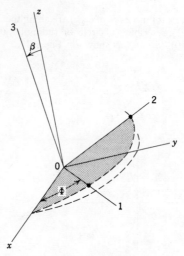

Fig. 7.4–1. Principal axes 1, 2, 3 displaced from missile axes x, y, z. Axis x is normal to plane $z03$.

We will assume that the moments and products of inertia about the missile axes x, y, z are

$$
\begin{aligned}
I_x &= A & I_{xy} &= D \\
I_y &= B & I_{xz} &= E \\
I_z &= C & I_{yz} &= F
\end{aligned}
$$

They are related to the principal moments of inertia I_1, I_2, I_3 by the equations of Sec. 5.4 as follows:

$$
A = l_{x1}^2 I_1 + l_{x2}^2 I_2 + l_{x3}^2 I_3 = I_1 \cos^2 \Phi + I_2 \sin^2 \Phi
$$

$$
B = I_1 \sin^2 \Phi + I_2 \cos^2 \Phi + \beta^2 I_3
$$

$$
C = I_1 \beta^2 \sin^2 \Phi + I_2 \beta^2 \cos^2 \Phi + I_3
$$

$$
-D = l_{x1} l_{y1} I_1 + l_{x2} l_{y2} I_2 + l_{x3} l_{y3} I_3 = (I_1 - I_2) \sin \Phi \cos \Phi
$$

$$
-E = (I_1 - I_2)\beta \sin \Phi \cos \Phi
$$

$$
-F = (I_1 \sin^2 \Phi + I_2 \cos^2 \Phi)\beta - \beta I_3
$$

$$(7.4\text{–}1)$$

The direction cosines used in the above equations are obtained from the matrix transformation between coordinates x, y, z and 1, 2, 3 with lengths x', y', z' along it (see Sec. 5.4). Since β is small, the approximation $\sin \beta = \beta$ and $\cos \beta = 1$ is used:

$$\begin{bmatrix} x \\ y \\ z \end{bmatrix} = \begin{bmatrix} \cos \Phi & -\sin \Phi & 0 \\ \sin \Phi \cos \beta & \cos \Phi \cos \beta & -\sin \beta \\ \sin \Phi \sin \beta & \cos \Phi \sin \beta & \cos \beta \end{bmatrix} \begin{bmatrix} x' \\ y' \\ z' \end{bmatrix} \qquad (7.4\text{--}2)$$

If we assume $I_1 = I_2$, Eqs. 7.4–1 reduce to the following

$$\begin{aligned} A &= I_1 & D &= 0 \\ B &= I_1 + \beta^2 I_3 & E &= 0 \\ C &= \beta^2 I_1 + I_3 & F &= -\beta(I_1 - I_3) \end{aligned} \qquad (7.4\text{--}3)$$

and the angle β becomes

$$\beta = \frac{F}{I_3 - I_1} \qquad (7.4\text{--}4)$$

To solve for the angular velocities, we first write down the transformation from the missile axes to the principal axes, assuming β to be small:

$$\begin{bmatrix} \omega_1 \\ \omega_2 \\ \omega_3 \end{bmatrix} = \begin{bmatrix} \cos \Phi & \sin \Phi & \beta \sin \Phi \\ -\sin \Phi & \cos \Phi & \beta \cos \Phi \\ 0 & -\beta & 1 \end{bmatrix} \begin{bmatrix} \omega_x \\ \omega_y \\ \omega_z \end{bmatrix} \qquad (7.4\text{--}5)$$

Adding ω_1 and $i\omega_2$, where $i = \sqrt{-1}$, we obtain the complex angular velocity

$$\omega_{12} = \omega_1 + i\omega_2 = [(\omega_x + i\omega_y) + i\beta\omega_z]e^{-i\Phi}$$
$$\cong (\omega_{xy} + i\beta n)e^{-i\Phi} \qquad (7.4\text{--}6)$$

where $\omega_z \cong n$. By multiplying Eq. 7.4–6 by $e^{i\Phi}$, this equation may also be written in the inverse form

$$\omega_{xy} = \omega_{12}e^{i\Phi} - i\beta n \qquad (7.4\text{--}7)$$

Figure 7.4–2 shows the relationship between the missile axes x, y, z, the principal axes 1, 2, 3, the inertial axes X, Y, Z, and the line of nodes ξ. The missile axis x is normal to axes 3 and z, whereas the line of nodes ξ is normal to axes z and Z. The position of the missile axes x, y, z is obtained by starting with the missile spin axis z coinciding with Z and performing three rotations as follows: (1) rotation of ψ about Z; (2) rotation of θ about ξ; and (3) rotation of φ about the spin axis z. The principal axes 1, 2, 3 are then referenced to the x, y, z axes by the fixed angles β and Φ.

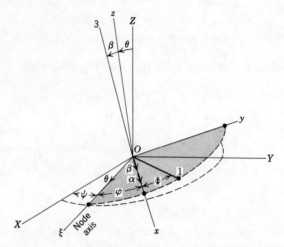

Fig. 7.4–2. Principal axes 1, 2, 3 referred to missile axes x, y, z, which, in turn, are referred to node axis ξ and inertial axes X, Y, Z.

From Ex. 7.3–1 we have the solution for the moment-free body in terms of the principal axes with $I_1 = I_2$,

$$\omega_{12} = \omega_{12}(0)e^{i\lambda t} \tag{7.4–8}$$

$$\theta_{12} = \theta_{12}(0)e^{-int} + i\frac{\omega_{12}(0)}{n+\lambda}e^{-int}[1 - e^{i(n+\lambda)t}] \tag{7.4–9}$$

where $\lambda = n[(I_3 - I_1)/I_1]$. From Eq. 7.4–6, the initial value $\omega_{12}(0)$ is found to be

$$\omega_{12}(0) = [\omega_{xy}(0) + i\beta n]e^{-i\Phi} \tag{7.4–10}$$

Substituting Eq. 7.4–10 into Eq. 7.4–8, and Eq. 7.4–8 into Eq. 7.4–7, we have

$$\omega_{xy} = \omega_{xy}(0)e^{i\lambda t} - i\beta n(1 - e^{i\lambda t}) \tag{7.4–11}$$

which transforms the complex angular rate solution to the missile axes x, y, z.

As discussed in Sec. 7.3, $\theta_{xy} = \theta e^{-i\varphi}$ resolves θ into components along the body-fixed axes x, y. It is also noted (see Fig. 7.4–2) that the angle β is a vector along axis x so that the components of θ and β along the x and y axes are

$$(\theta \cos \varphi + \beta) - i(\theta \sin \varphi) = \theta e^{-i\varphi} + \beta$$
$$= \theta_{xy} + \beta \qquad (7.4\text{–}12)$$

To reference $\theta_{xy} + \beta$ to the principal axes 1, 2, we multiply by $e^{-i\Phi}$ and designate it as θ_{12}:

$$\theta_{12} = (\theta_{xy} + \beta)e^{-i\Phi} \qquad (7.4\text{–}13)$$

Multiplying Eq. 7.4–13 by $e^{i\Phi}$, we obtain its inverse

$$\theta_{xy} = \theta_{12}e^{i\Phi} - \beta \qquad (7.4\text{–}14)$$

Substituting for θ_{12} from Eq. 7.4–9,

$$\theta_{xy} = \left\{ \theta_{12}(0)e^{-int} + i\frac{\omega_{12}(0)e^{-int}}{n + \lambda}[1 - e^{i(n+\lambda)t}] \right\}e^{i\Phi} - \beta \quad (7.4\text{–}15)$$

However, from Eq. 7.4–13,

$$\theta_{12}(0) = [\theta_{xy}(0) + \beta]e^{-i\Phi} \qquad (7.4\text{–}16)$$

so by substituting from Eqs. 7.4–10 and 7.4–16 into Eq. 7.4–15,

$$\theta_{xy} = \theta_{xy}(0)e^{-int} + i\frac{\omega_{xy}(0)e^{-int}}{n + \lambda}[1 - e^{i(n+\lambda)t}]$$
$$+ \beta\left\{ e^{-int} - \frac{n}{n + \lambda}e^{-int}[1 - e^{i(n+\lambda)t}] - 1 \right\} \quad (7.4\text{–}17)$$

Equation 7.4–17 expresses the angle θ between the missile spin axis and the inertial Z axis as a vector in the transverse plane with components along the rotating x and y axes, with x real and y imaginary. To examine this vector in the transverse inertial plane XY, we multiply θ_{xy} by e^{int} to obtain

$$\theta_{XY} = \theta_{xy}(0) + i\frac{\omega_{xy}(0)}{n + \lambda}[1 - e^{i(n+\lambda)t}] + \beta\left[\frac{\lambda}{n + \lambda} - e^{int} + \frac{n}{n + \lambda}e^{i(n+\lambda)t} \right]$$
$$(7.4\text{–}18)$$

Comparing Eq. 7.4–18 with Eq. 7.3e, we find that the vector θ in the inertial plane has an added term due to β. In addition to the precession speed $(n + \lambda)$, the effect of the product of inertia has introduced a component βe^{int} which rotates at the spin speed n. The result is a motion of the spin axis indicated by a curve, as shown in Fig. 7.4–3.

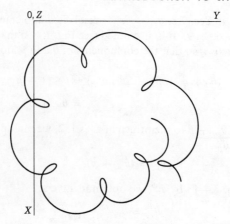

Fig. 7.4–3. Possible motion of spin axis projected on inertial plane X, Y.

PROBLEMS

1. Assume that the fuel consumption rate of a given missile with $A = 10C$ is small enough to justify a constant mass analysis. If the thrust misalignment in the body coordinates is a constant and equal to Te, where T is the thrust parallel to the longitudinal axis, and e is its offset, determine the equation for the complex angular velocity ω_{12}.

2. When θ, the angle between the principal axis 3 and an inertial axis Z is small, it can be represented as a vector. Along what axis do we represent this vector, and how do we resolve it into components along the rotating body axes with spin velocity $\dot{\varphi}$. How do we resolve it into components along the inertial XY axes.

3. If we assume the velocity vector $\mathbf{V}$ of the center of mass of a missile to be fixed in space, determine for the moment-free missile the angle between $\mathbf{V}$ and the angular momentum vector $\mathbf{h}$.

4. Defining the angle of attack of a missile as the angle between the longitudinal axis and the velocity vector $\mathbf{V}$, show how this varies for the moment-free missile.

5. For a body of revolution (A, A, C), the motion under moment-free conditions is described by a constant precession angle θ, the plane containing the angular momentum $\mathbf{h}$, the angular velocity $\boldsymbol{\omega}$, and the spin axis 3 rotating about the fixed $\mathbf{h}$ vector at a rate $\dot{\psi} = C\dot{\varphi}/[(A - C)\cos\theta] = Cn/(A\cos\theta)$. Show that the results of Example 7.3–1 are consistent with this requirement provided θ_{12}, the angle of attack, is small.

6. If the missile of Prob. 1 is spinning at a rate $n = 2\pi$ rad/sec, determine the complex angle of attack relative to inertial space, and plot the results in the XY plane.

7. If the geometric axes x, y, z of a near symmetric missile deviates from the principal axes 1, 2, 3 by the angles β and Φ, where β is a small angle between

the z axis and the principal axis 3, show that the assumption of $I_1 \cong I_2$ results in $\Phi = \pi$. Physically what does this mean?

8. Show that in the general case for small β, where Φ may not be π, Eq. 7.4–1 can be solved for β as

$$ \beta = \sqrt{\frac{FD + E(B - C)}{FD}} $$

9. The principal moment of inertia ratio I_3/I_1 for a near symmetrical satellite is given as 1.20, and the principal axis 3 is tilted from the geometric axis z by the angle $\beta = 0.05$ rad. If the spin rate $n = 2\pi$ rad/sec and $[\omega_{xy}(0)]/n = \frac{1}{10}$, determine the complex angle of attack relative to inertial space and show how the geometric axis z is moving in a plane normal to the velocity vector.

10. Using the equation for $\omega_{12}(t)$ of Prob. 6, p. 198, in Eq. 7.3–8 shows that the angle θ_{XY} referred to inertial axes is given by the equation

$$ \theta_{XY} = \theta_{12}(0) + \frac{i\omega_{12}(0)}{n + \lambda}[1 - e^{i(n+\lambda)t}] - \frac{M_1}{n\lambda A}\left[\frac{\lambda}{n + \lambda} - e^{int} + \frac{n}{n + \lambda}e^{i(n+\lambda)t}\right] $$

11. Body axes x, y, and z initially coinciding with the inertial axes X, Y, and Z are given the following sequence of rotations. Rotation θ_3 about z followed by rotation θ_2 about the displaced y axis and a rotation θ_1 about the final position of the x axis. Derive the transfer matrix expressing the body axes in terms of the inertial axes, and its inverse.

12. Assume angular velocities $\dot\theta_3$, $\dot\theta_2$ and $\dot\theta_1$ about axes z, y, and x in the sequence given by Prob. 11, and write the equations for the angular velocities ω_1, ω_2, and ω_3 about the final position of the body axes x, y, z.

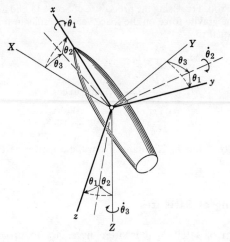

Prob. 12

13. Referring to the figure of Prob. 12, assume the missile to be symmetric so that $I_y = I_z$, and determine the equation for the attitude deviation $\theta_2 + i\theta_3$ of the longitudinal axis due to a constant yawing torque M_z.

14. A space vehicle of moment of inertia I_1, I_2, and I_3 is in a circular orbit with constant angular velocity ω_0 about the axis 2 to maintain the direction of axis 1 always tangent to the orbit as shown in the sketch. Assuming small disturbances θ_1, θ_2, θ_3, derive the differential equation of motion for the torques about the body axes 1, 2, and 3.

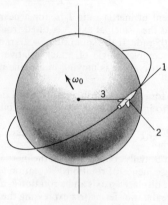

Prob. 14

15. Assume the body axes 1, 2, and 3 of the space vehicle of Prob. 14 to deviate from the orbit axes $1'$, $2'$ and $3'$ by angles θ_3, θ_2, and θ_1 in the sequence specified in Prob. 11. Using the procedure of Sec. 4.18 and a spherical earth, show that the gravity force on the space vehicle results in torques about the body axes equal to

$$M_1 = \frac{3K}{2R_0{}^3}(I_3 - I_2)\sin 2\theta_1 \cos^2 \theta_2$$

$$M_2 = \frac{3K}{2R_0{}^3}(I_3 - I_1)\sin 2\theta_2 \cos \theta_1$$

$$M_3 = \frac{3K}{2R_0{}^3}(I_1 - I_2)\sin 2\theta_2 \sin \theta_1$$

where $K = Gm_e$ and R_0 is distance from the center of earth to the vehicle center of mass.

7.5 Despinning of Satellites

In the design of satellites it is often necessary to provide means for reducing the spin of a spin-stabilized satellite while in orbit to allow proper functioning of instruments. Figure 7.5–1* shows a simple device,[7] used in the Pioneer III lunar probe, which is capable of reducing the spin to zero. It consists of a small mass m on the end of a light cord wrapped around

* To maintain symmetry two such masses are released.

the symmetrical spinning body. With the satellite spinning with speed ω_0 about its axis of symmetry, the mass m is released. The cord will now unwind and the angular speed of the satellite will gradually decrease.

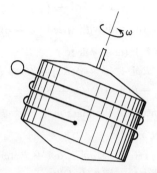

Fig. 7.5–1. Despinning device for a satellite.

When the cord is completely unwound, it is released and allowed to fly away. By choosing the length of the cord properly, the spin of the satellite can be reduced to any value less than the initial value.

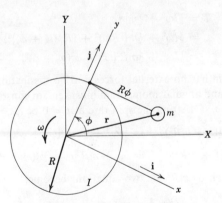

Fig. 7.5–2. Unwinding of mass m.

The device may be analyzed as follows. Since m is small, the body may be assumed to spin about the geometric axis of symmetry of the body 0, with moment of inertia I and angular velocity ω. We attach the X, Y coordinate axes to the body and allow a second set of axes x, y to rotate relative to the body so that the y axis always passes through the tangent point of the cord, as shown in Fig. 7.5–2.

We will assume that initially m was in contact with the cylinder at the X axis, in which case the length of cord extending beyond the tangent point is equal to the arc length $R\phi$. The position of m is then,

$$\mathbf{r} = R\phi\mathbf{i} + R\mathbf{j} \qquad (7.5\text{-}1)$$

Since the axes x, y are rotating with speed $(\omega + \dot{\phi})\mathbf{k}$, the velocity of m is

$$\begin{aligned}
\mathbf{v} &= \dot{\mathbf{r}} + (\omega + \dot{\phi})\mathbf{k} \times \mathbf{r} \\
&= R\dot{\phi}\mathbf{i} + (\omega + \dot{\phi})R\phi\mathbf{j} - (\omega + \dot{\phi})R\mathbf{i} \\
&= -R\omega\mathbf{i} + R\phi(\omega + \dot{\phi})\mathbf{j} \qquad (7.5\text{-}2)
\end{aligned}$$

The angular momentum of the mass m is,

$$\begin{aligned}
\mathbf{h} &= \mathbf{r} \times m\mathbf{v} \\
&= (R\phi\mathbf{i} + R\mathbf{j}) \times m[-R\omega\mathbf{i} + R\phi(\omega + \dot{\phi})\mathbf{j}] \\
&= mR^2[\omega + \phi^2(\omega + \dot{\phi})]\mathbf{k} \qquad (7.5\text{-}3)
\end{aligned}$$

and the total angular momentum is

$$\mathbf{H} = \{I\omega + mR^2[\omega + \phi^2(\omega + \dot{\phi})]\}\mathbf{k} \qquad (7.5\text{-}4)$$

The system kinetic energy T is the sum of the kinetic energy of the satellite and m.

$$\begin{aligned}
T &= \tfrac{1}{2}I\omega^2 + \tfrac{1}{2}mv^2 \\
&= \tfrac{1}{2}I\omega^2 + \tfrac{1}{2}m\{(R\omega)^2 + [R\phi(\omega + \dot{\phi})]^2\} \\
&= \tfrac{1}{2}I\omega^2 + \tfrac{1}{2}mR^2[\omega^2 + \phi^2(\omega + \dot{\phi})^2] \qquad (7.5\text{-}5)
\end{aligned}$$

Since the system has no external forces and no dissipation of energy, the kinetic energy and angular momentum must remain constant and equal to their initial values. Letting the spin rate at $t = 0$ be ω_0,

$$T = \tfrac{1}{2}(I + mR^2)\omega_0{}^2 = \tfrac{1}{2}I\omega^2 + \tfrac{1}{2}mR^2[\omega^2 + \phi^2(\omega + \dot{\phi})^2] \quad (7.5\text{-}6)$$

$$H = (I + mR^2)\omega_0 = I\omega + mR^2[\omega + \phi^2(\omega + \dot{\phi})] \qquad (7.5\text{-}7)$$

Dividing through by mR^2 the two equations become,

$$C(\omega_0{}^2 - \omega^2) = \phi^2(\omega + \dot{\phi})^2 \qquad (7.5\text{-}8)$$

$$C(\omega_0 - \omega) = \phi^2(\omega + \dot{\phi}) \qquad (7.5\text{-}9)$$

where

$$C = \frac{I}{mR^2} + 1 \qquad (7.5\text{-}10)$$

Dividing the first equation by the second, we find,

$$\omega + \omega_0 = \omega + \dot{\phi}$$

Therefore,

$$\omega_0 = \dot{\phi}$$
$$\omega_0 t = \phi$$

(7.5–11)

which tells us that the mass m unwinds at a constant rate. Substituting for ϕ and $\dot{\phi}$ in Eq. 7.5-9, the spin rate at any time becomes

$$\omega = \omega_0\left(\frac{C - \phi^2}{C + \phi^2}\right) = \omega_0\left(\frac{C - \omega_0^2 t^2}{C + \omega_0^2 t^2}\right)$$

(7.5–12)

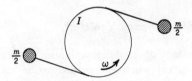

Fig. 7.5–3. Despinning of satellites.

The spin may be reduced to any desired value ω_f by choosing the proper length of cord, and releasing it when completely unwound. If l_f is the length selected, the terminal value of ϕ is,

$$\phi_f = \frac{l_f}{R}$$

(7.5–13)

and from Eq. 7.5-12,

$$\omega_f = \omega_0\left(\frac{C - \phi_f^2}{C + \phi_f^2}\right) = \omega_0\left(\frac{CR^2 - l_f^2}{CR^2 + l_f^2}\right)$$

(7.5–14)

Solving for l_f, the required length of cord is,

$$l_f = R\sqrt{C\frac{\omega_0 - \omega_f}{\omega_0 + \omega_f}}$$

(7.5–15)

If the terminal angular velocity is to be zero, l_f becomes,

$$l_f = R\sqrt{C}$$
$$= \sqrt{R^2 + \frac{I}{m}}$$

(7.5–16)

For symmetry, two cords with masses $\frac{1}{2}m$ can be used as shown in Fig. 7.5-3, the result being the same as that for one mass of value m.

Example 7.5–I
The Pioneer III lunar probe was launched with an initial spin rate of 400 rpm. It was desired to reduce this to 5.5 rpm by using a yo-yo with two weights of 0.2 oz each.

For Pioneer III

$$gI = 92 \text{ lb in.}^2$$
$$R = 5 \text{ in.}$$
$$C = \frac{92}{(0.4/16)5^2} + 1 = 148$$

From Eq. 7.5–15 the proper length of cord is,

$$l = 5\sqrt{148\frac{400-5.5}{400+5.5}} = 60 \text{ in.}$$

7.6 Attitude Drift of Space Vehicles[8]

The attitude of a body of revolution spinning in the absence of external forces is not a constant when energy dissipation takes place. Elastic vibrations, induced by gyroscopic action, result in a dissipation of energy and a change in the precession cone angle θ. In this section we examine the effect of energy dissipation on the spinning body and evaluate the time required for a body of given configuration to undergo a specified change in attitude.[8]

The moment-free motion of an unsymmetric body with principal moments of inertia A, B, C is an unsteady periodic precession and nutation about the resultant angular momentum vector **h** fixed in space. Steady rotation is possible only about the principal axis of maximum or minimum moment of inertia, the principal axis of intermediate moment of inertia being unstable.

For a body of revolution A, A, C, the moment-free motion is a steady precession of the spin axis at a constant angle θ about the resultant angular momentum vector **h** fixed in space. Steady rotation is again possible about the axis of maximum or minimum moment of inertia, and the axis of intermediate moment of inertia does not exist.

In either case, the axis of maximum or minimum moment of inertia is considered to be stable in that, if the spin axis deviates slightly from the resultant angular momentum vector, there is no tendency for this deviation to grow. This statement is true only for a perfectly rigid body in the absence of external moment.

In an elastic body, deformation between particles will always take place, resulting in some dissipation of energy. When the dissipation of energy is taken into account, we must revise our statement of stability in that it is possible for a small deviation of the spin axis to grow into a large one and, eventually, to result in a complete change in attitude of the body. For such bodies, only the principal axis of maximum moment of inertia is stable, and the axis of minimum moment of inertia is one of unstable equilibrium.

These facts were actually observed in the Explorer I satellite,[6] which was spin-stabilized about the longitudinal axis of minimum moment of inertia. The flexible antennas of the satellite provided an excellent source for energy dissipation, and in one revolution around its orbit (approximately 90 min) the Explorer I was observed to be tumbling at an attitude of $\theta = 60°$ instead of spinning about its longitudinal axis at $\theta = 0$. The remedy for this behavior is obviously to shorten the longitudinal dimensions

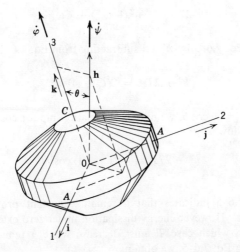

Fig. 7.6-1. Coordinate system of body axes 1, 2, 3.

of the satellite so that the moment of inertia about the longitudinal spin axis is greater than that about the transverse pitch or yaw axis. However, the problem still exists for missiles which are long, slender bodies and inherently unstable about the spin axis. Here the important question is how long can the spinning missile coast in a moment-free condition without an appreciable change in its attitude.

Energy considerations of stability

We will now examine the basis for stability from an energy point of view. For a body of revolution with principal moments of inertia A, A, C, as shown in Fig. 7.6-1, the moment-free motion is that of steady precession described by the equations

$$\dot{\psi} = \frac{C\dot{\varphi}}{(A - C)\cos\theta} = \frac{C}{A}\frac{\omega_3}{\cos\theta} \qquad (7.6-1)$$

$$\omega_3 = \dot{\varphi} + \dot{\psi}\cos\theta$$

Since the moment is zero, the angular momentum vector **h** is a constant and we can write for the square of its magnitude the equation

$$h^2 = A^2(\omega_1{}^2 + \omega_2{}^2) + C^2\omega_3{}^2 \qquad (7.6\text{–}2)$$

We next examine the kinetic energy of rotation, T, which is given by the equation

$$2T = A(\omega_1{}^2 + \omega_2{}^2) + C\omega_3{}^2 \qquad (7.6\text{–}3)$$

Multiplying Eq. 7.6–3 by A and subtracting from Eq. 7.6–2,

$$h^2 - 2TA = C(C - A)\omega_3{}^2 \qquad (7.6\text{–}4)$$

and since $C\omega_3 = h \cos \theta$, we obtain the relationship for $\cos \theta$ in terms of h and T as follows

$$h^2 - 2TA = \frac{h^2}{C}(C - A)\cos^2 \theta \qquad (7.6\text{–}5)$$

Equation 7.6–5 indicates that θ remains constant provided T and h are constant. However energy dissipation under zero external moment is possible, in which case T must decrease while h remains constant. Differentiating Eq. 7.6–5, we obtain

$$\dot{T} = \frac{h^2}{C}\left(\frac{C}{A} - 1\right)(\sin \theta \cos \theta)\dot{\theta} \qquad (7.6\text{–}6)$$

and with $\dot{T}$ a negative quantity, $\dot{\theta}$ is negative for $C/A > 1$ and positive for $C/A < 1$. Thus the principal axis of minimum moment of inertia is one of unstable equilibrium, and a small deviation of the spin axis will increase due to energy dissipation when $C/A < 1$.

Dissipation of energy

Assuming an elastic body, the energy dissipated per unit volume per cycle of stress can be assumed to be

$$\frac{\gamma\sigma^2}{2E} \qquad (7.6\text{–}7)$$

where γ is a hysteretic damping factor establishing the fraction of the elastic energy which is dissipated as shown by the shaded area in Fig. 7.6–2. Dividing by the time t_0 per cycle of stress, and integrating over the entire structure, the rate of energy dissipation can be found. Thus the equation to be solved is of the general form

$$\int \frac{\gamma\sigma^2}{2Et_0}\,dV = \frac{h^2}{C}\left(\frac{C}{A} - 1\right)(\sin\theta\cos\theta)\dot\theta \tag{7.6–8}$$

In examining the source of cyclic stressing, free vibrations can be discarded since they will soon damp out. Steady cycling of stress is

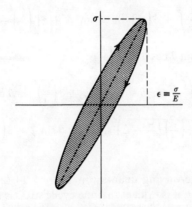

Fig. 7.6–2. Energy dissipated by hysteresis damping.

however induced by the gyroscopic precession, and these stresses are repeated at the rate $\dot\varphi$ and $2\dot\varphi$, as we will presently show.

The excitation for the cyclic stress is the acceleration. Choosing an arbitrary point on the structure and orienting the plane 1, 0, 3 through it, the position vector for the specified point is

$$\mathbf{r} = \xi\mathbf{i} + z\mathbf{k} \tag{7.6–9}$$

With $\dot\theta$ small, the angular velocity and acceleration of the coordinate axes 1, 2, 3 are

$$\boldsymbol{\omega} = (\dot\psi\sin\theta\sin\varphi)\mathbf{i} + (\dot\psi\sin\theta\cos\varphi)\mathbf{j} + (\dot\varphi + \dot\psi\cos\theta)\mathbf{k} \tag{7.6–10}$$

$$\dot{\boldsymbol{\omega}} = \dot\varphi\dot\psi\sin\theta\,(\cos\varphi\mathbf{i} - \sin\varphi\mathbf{j}) \tag{7.6–11}$$

Substituting into the general vector equation for the acceleration

$$\mathbf{a} = \mathbf{a}_0 + [\mathbf{a}] + \boldsymbol{\omega}\times(\boldsymbol{\omega}\times\mathbf{r}) + \dot{\boldsymbol{\omega}}\times\mathbf{r} + 2\boldsymbol{\omega}\times[\mathbf{v}] \tag{7.6–12}$$

and noting that the following quantities are zero

$$\mathbf{a}_0 = [\mathbf{a}] = [\mathbf{v}] = 0$$

the result after some algebraic reduction is

$$\mathbf{a} = [-\xi(\dot{\varphi}^2 + \dot{\psi}^2) + \xi\dot{\psi}^2 \sin^2 \theta \sin^2 \varphi - 2\xi\dot{\varphi}\dot{\psi} \cos \theta + z\dot{\psi}^2 \sin \theta \cos \theta \sin \varphi]\mathbf{i}$$
$$+ (\xi\dot{\psi}^2 \sin^2 \theta \sin \varphi \cos \varphi + z\dot{\psi}^2 \sin \theta \cos \theta \cos \varphi)\mathbf{j}$$
$$+ (2\xi\dot{\varphi}\dot{\psi} \sin \theta \sin \varphi + \xi\dot{\psi}^2 \sin \theta \cos \theta \sin \varphi - z\dot{\psi}^2 \sin^2 \theta)\mathbf{k} \qquad (7.6\text{--}13)$$

A somewhat more convenient form of Eq. 7.6–13 results by eliminating $\dot{\varphi}$ and $\dot{\psi}$.

$$\mathbf{a} = \omega_0{}^2 \left\{ -\xi\left(\frac{C}{A}\right)^2 + \xi\left(\frac{C}{A}\right)^2 \sin^2 \theta \sin^2 \varphi + \xi\left[\left(\frac{C}{A}\right)^2 - 1\right] \cos^2 \theta \right.$$
$$\left. + z\left(\frac{C}{A}\right)^2 \sin \theta \cos \theta \sin \varphi \right\} \mathbf{i}$$
$$+ \omega_0{}^2 \left[\xi\left(\frac{C}{A}\right)^2 \sin^2 \theta \sin \varphi \cos \varphi + z\left(\frac{C}{A}\right)^2 \sin \theta \cos \theta \cos \varphi \right]\mathbf{j}$$
$$+ \omega_0{}^2 \left[-\xi\left(\frac{C}{A}\right)\left(\frac{C}{A} - 2\right) \sin \theta \cos \theta \sin \varphi - z\left(\frac{C}{A}\right)^2 \sin^2 \theta \right]\mathbf{k}$$
$$(7.6\text{--}14)$$

Since the only time-varying quantity in Eq. 7.6–14 (assuming $\dot{\theta}$ to be negligible) is $\varphi = \dot{\varphi}t$, it is evident that the cyclic stress is repeated at a rate $\dot{\varphi}$ and $2\dot{\varphi}$. It should be pointed out that, for slender bodies like missiles, C/A is small compared to unity, and the predominant variable acceleration term is

$$\mathbf{a} = 2\omega_0{}^2 \xi\left(\frac{C}{A}\right) \sin \theta \cos \theta \sin \varphi \mathbf{k} \qquad (7.6\text{--}14a)$$

which is repeated in the time

$$t_0 = \frac{2\pi}{\dot{\varphi}} = \frac{2\pi}{[1 - (C/A)]\omega_0 \cos \theta} \qquad (7.6\text{--}15)$$

Example 7.6–I

As an example of the simplest kind, we will consider two solid disks connected by a flexible tube, as shown in Fig. 7.6–3. We will let C_1 and A_1 be the moments of inertia of each disk about its own polar and diametric axes. The gyroscopic moment required by each disk is

$$M_g = C_1(\dot{\varphi} + \dot{\psi} \cos \theta)\dot{\psi} \sin \theta - A_1\dot{\psi}^2 \sin \theta \cos \theta \qquad (a)$$

Since the moments of inertia about the center of mass of the body are

$$C = 2C_1$$
$$A \cong 2(A_1 + m_1 l^2)$$

Equation a can be rewritten as

$$M_g = \tfrac{1}{2}[C(\dot\varphi + \dot\psi \cos\theta)\dot\psi \sin\theta - A\dot\psi^2 \sin\theta \cos\theta] + m_1 l^2 \dot\psi^2 \sin\theta \cos\theta$$

The first term, however, is the moment about the center of mass which is zero and from which Eq. 7.6–1 is obtained. We are thus left with

$$M_g = m_1 l^2 \dot\psi^2 \sin\theta \cos\theta$$
$$= Fl \cos\theta \tag{b}$$

where $F = m_1 l \dot\psi^2 \sin\theta$ is the centripital force of the precessing disk.

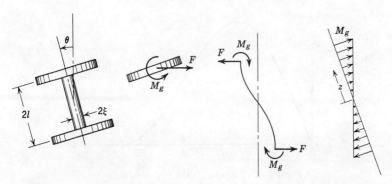

Fig. 7.6–3. Satellite configuration, displacement, and moment distribution.

The effect of M_g and F on the flexible tube is shown in Fig. 7.6–3. At point z along the tube, measured from the center of mass, the bending moment is

$$M_z = M_g \frac{z}{l} \tag{c}$$

and the expression for the maximum stress becomes

$$\sigma = \frac{M_z \xi}{I} = m_1 l^2 \dot\psi^2 \frac{z}{l} \frac{\xi}{I} \sin\theta \cos\theta$$
$$= \frac{1}{2} ml^2 \left(\frac{C}{A}\right)^2 \omega_0^2 \frac{z}{l} \frac{\xi}{I} \sin\theta \cos\theta \tag{d}$$

which is repeated at the rate given by Eq. 7.6–15. The rate of energy dissipation as given by the left side of Eq. 7.6–8 is then

$$\frac{\gamma}{48\pi E}\left(\frac{ml^2 \xi}{I}\right)^2 V \left(\frac{C}{A}\right)^4 \left(\frac{C}{A} - 1\right) \omega_0^5 \sin^2\theta \cos^3\theta \tag{e}$$

and the rate of change of the attitude angle θ becomes

$$\dot\theta = \frac{\gamma}{48\pi E}\left(\frac{ml^2 \xi}{I}\right)^2 \frac{V}{C}\left(\frac{C}{A}\right)^4 \omega_0^3 \sin\theta \cos^2\theta$$
$$= K \sin\theta \cos^2\theta \tag{f}$$

A plot of Eq. *f* is shown in Fig. 7.6–4. Since $\dot\theta$ is zero for $\theta = 0$, tumbling cannot be initiated unless the initial value of θ is finite. However $\theta = 0$ is never attainable in practice for many reasons, and $\dot\theta$ will build up when C/A is less than unity. By differentiating Eq. *f*, $\dot\theta$ can be shown to have a maximum at $\theta = \tan^{-1} 1/\sqrt{2} = 35°\,20'$. Due to $\cos^2\theta$, $\dot\theta$ will diminish to a small value near $\theta = 90°$, and an infinite time will be required to reach this angle.

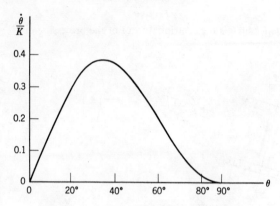

Fig. 7.6–4. Variation in the rate of tumbling.

For small values of θ, Eq. *f* is approximately equal to

$$\dot\theta = K\theta \tag{g}$$

and the time required for the attitude angle to change from θ_0 to θ_1 is

$$t = \frac{1}{K}\ln\frac{\theta_1}{\theta_0} \tag{h}$$

Numerical example

Let the two solid disks be aluminum, ½ in. thick and 24 in. diameter, and the flexible tube be 0.032 in. stainless steel, 6 in. in diameter and 24 in. long. The quantities required for the computation of K are:

$$C = 8.16 \text{ lb-in.-sec}^2$$
$$A = 20.44 \text{ lb-in.-sec}^2$$
$$m = 0.1136 \text{ lb-in.}^{-1}\text{-sec}^2$$
$$V = 14.5 \text{ in.}^3 = \text{volume of flexed tube}$$
$$\xi = 3.0 \text{ in.}$$
$$I = 2.71 \text{ in.}^4$$
$$l = 12.0 \text{ in.}$$
$$E = 29 \times 10^6 \text{ lb-in.}^{-2}$$

Assuming $\gamma = 0.05$ and $\omega_0 = 50\pi/\text{sec}$, the value of K is 662×10^6. Thus for the body to undergo an attitude change from $1°$ to $10°$, the time required, as calculated from Eq. h, is

$$t = \frac{2.303}{662} \times 10^6 = 3480 \text{ sec}$$

$$= 58.0 \text{ min}$$

PROBLEMS

1. A satellite has a moment of inertia of $I = 1.20$ lb-in.-sec^2 about its spin axis. It is desired to reduce the initial spin of 200 rpm to zero by two weights of $\frac{1}{20}$ lb each wrapped around a section having a radius of 10 in. Determine the proper length of cord.

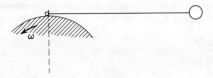

Prob. 2

2. In Prob. 1, determine the speed of the $\frac{1}{20}$-lb weights as they fly off. As shown in the sketch, the pin holding the string will slide out when the string goes beyond the tangent to the circle.

3. It is proposed to despin a satellite by four weights of mass m each, hinged by stiff arms as shown in the sketch. Show that the spin is given by

$$\omega = \frac{\omega_0(C_0 + 4mr_0^2)}{C_0 + 4m(r_0 + l \sin \theta)^2}$$

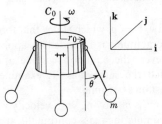

Prob. 3

4. Determine the $\mathbf{i}$, $\mathbf{j}$, $\mathbf{k}$ components of the acceleration of m in Prob. 3.
5. Verify the relations given by Eq. 7.6–8.
6. Verify Eq. 7.6–14.

7. Equation 7.6–5 can be written as

$$T = \frac{h^2}{2C}\left[1 + \left(\frac{C}{A} - 1\right) \sin^2 \theta\right]$$

Plot T versus θ for $-\frac{\pi}{2} < \theta < \frac{\pi}{2}$ when (a) $C/A > 1$; (b) $C/A < 1$, and discuss its stability.

8. For a symmetric body A, A, C acted upon by moments M_1, M_2, M_3 about body axes 1, 2, 3, show that the dissipation rate of energy is equal to

$$\omega_1 M_1 + \omega_2 M_2 + \omega_3 M_3 = \frac{1}{2} A \frac{d}{dt}(\omega_1{}^2 + \omega_2{}^2) + \frac{1}{2} C \frac{d}{dt} \omega_3{}^2$$

9. Show that the hysteretic damping factor γ of Eq. 7.6–7 is related to the structural damping factor α by the relationship $\alpha = \gamma/2\pi$.
Hint: The work done per cycle by a damping force F_d for harmonic oscillations is $W = \pi F_d X$, where X is the amplitude leading the damping force by 90°. The structural damping force can be taken as $i\alpha kx$, where k is the stiffness.

7.7 Variable Mass

In the previous sections we have limited our discussion to a constant-mass system. In many cases the mass variation rate is large, which requires us to consider the problem of variable mass.

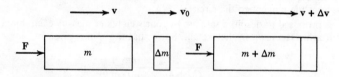

Fig. 7.7–1.   Momentum of the system at times t and $t + \Delta t$.

Newton's second law, $\mathbf{F} = \dot{\mathbf{p}}$, which states that force is equal to the time rate of change of momentum, is intended to apply only to a system of definite mass. However, the equation can be applied to a system of varying mass provided the same mass is examined for the change in momentum at two instances of time.

We will consider a mass m moving with velocity $\mathbf{v}$ at time t, and assume that our system is accumulating mass continually at a rate $\dot{m}$ (if the system is losing mass as in a rocket, $\dot{m}$ is negative). We will define our system to be the mass $m + \Delta m$ at time t as shown in Fig. 7.7–1. Its initial momentum at time t is

$$\mathbf{p} = m\mathbf{v} + \mathbf{v}_0 \Delta m \tag{7.7–1}$$

where $\mathbf{v}_0$ is the initial velocity of Δm before it is acquired by m. The momentum at time $t + \Delta t$ is

$$
\begin{aligned}
\mathbf{p} + \Delta\mathbf{p} &= (m + \Delta m)(\mathbf{v} + \Delta v) \\
&= m\mathbf{v} + m\,\Delta\mathbf{v} + \mathbf{v}\,\Delta m
\end{aligned}
\tag{7.7-2}
$$

where the negligible second-order term $(\Delta m)(\Delta\mathbf{v})$ has been omitted. Subtracting to determine the change in momentum and dividing by Δt, the equation for the variable mass system becomes,

$$
\mathbf{F} = m\frac{d\mathbf{v}}{dt} + (\mathbf{v} - \mathbf{v}_0)\frac{dm}{dt}
\tag{7.7-3}
$$

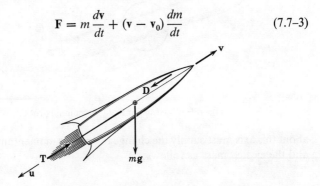

Fig. 7.7-2. Forces on a rocket.

Equation 7.7-3 indicates that the force $\mathbf{F}$ is expended in accelerating the mass m and changing the momentum of the acquired mass from $\mathbf{v}_0\,dm$ to $\mathbf{v}\,dm$.

For rockets $\mathbf{u}\dfrac{dm}{dt} = \mathbf{T}$ is the thrust exerted by the jet, where $\mathbf{u} = -(\mathbf{v} - \mathbf{v}_0)$ is the velocity of the gas jet relative to the engine.* Thus the equation for the rocket in rectilinear motion, Fig. 7.7-2, can be written as

$$
\mathbf{F} + \mathbf{T} = m\frac{d\mathbf{v}}{dt}
\tag{7.7-4}
$$

where the external forces of gravity and aerodynamic drag can be included in $\mathbf{F}$. If the rocket is not spinning or turning and $\mathbf{T}$ acts through the center of mass, the moment on the rocket is zero, and we are concerned only with its translational motion.

7.8   Jet Damping (Nonspinning Variable Mass Rocket)

When a nonspinning rocket rotates about a transverse axis, as shown in Fig. 7.8-1, the ejected gas acquires a momentum component $-\dot{m}l\omega$

---

* $\mathbf{u}$ is the velocity of the jet relative to the nozzle. When it is positive we have a retro-rocket. $\dot{m}$ is negative for any rocket.

perpendicular to the longitudinal axis, where m is the mass of the rocket at any time, $\dot{m} = dm/dt$ is its rate of change ($\dot{m}$ is negative), and $l\omega$ is the transverse velocity of the nozzle exit due to rotational velocity ω. Letting the transverse rotational axis coincide at all times with the center of mass and letting the moment of inertia of the rocket about this axis be $I = mk^2$, where k is the radius of gyration about the transverse axis, the moment M

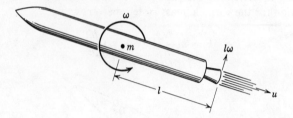

Fig. 7.8–I.   Jet damping of nonspinning rocket.

about this axis must supply the change in angular momentum of the rocket and the ejected mass as follows;

$$M = \frac{d}{dt} I\omega - \dot{m}l^2\omega \qquad (7.8\text{--}1)$$

Substituting for I and carrying out the differentiation,

$$M = I\frac{d\omega}{dt} + \omega\left(m\frac{d}{dt}k^2 + k^2\dot{m}\right) - \dot{m}l^2\omega$$
$$= I\dot{\omega} - \omega\left[\dot{m}(l^2 - k^2) - m\frac{d}{dt}k^2\right] \qquad (7.8\text{--}2)$$

Assuming the applied moment M to be zero and $\dfrac{d}{dt}k^2$ to be negligible (i.e., burning proceeds radially), this equation can be solved in the following manner:

$$\frac{d\omega}{\omega} = \frac{(l^2 - k^2)}{I}\, dm = \left(\frac{l^2}{k^2} - 1\right)\frac{dm}{m} \qquad (7.8\text{--}3)$$

$$\ln\left(\frac{\omega}{\omega_0}\right) = \left(\frac{l^2}{k^2} - 1\right)\ln\frac{m}{m_0}$$

$$\frac{\omega}{\omega_0} = \left(\frac{m}{m_0}\right)^{\frac{l^2}{k^2} - 1} \qquad (7.8\text{--}4)$$

Thus with an initial angular rate of ω_0, the angular speed ω decreases if l/k is greater than unity and increases if l/k is less than unity. Since in

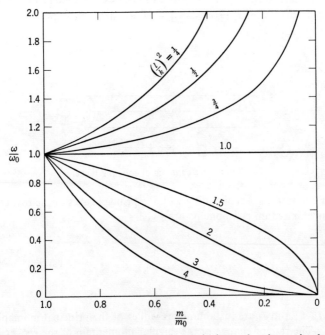

Fig. 7.8–2. Change in pitch angular rate of nonspinning rocket due to jet damping.

most configurations $l/k > 1$, the angular speed decreases and the action of the jet is that of a damper. Figure 7.8–2 shows how the angular speed changes with mass ratio for various values of l/k.

7.9 Euler's Dynamical Equations for Spinning Rockets[2]

The statement that the force is equal to the time rate of change of momentum can be applied to problems of varying mass provided the momentum of a definite mass is examined for its rate of change. The result is the rate of change of the momentum of the varying mass plus the rate of momentum transfer from the varying mass. The moment of the force about its center of mass is then the time rate of change of the moment of momentum which, can be written as

$$\mathbf{M} = [\dot{\mathbf{h}}] + \boldsymbol{\omega} \times \mathbf{h} + \text{rate of angular momentum transfer}$$
$$\text{from the variable mass system} \qquad (7.9\text{–}1)$$

Consider a general motion of a symmetric rocket with body axes x, y, z fixed in the rocket with the origin at the center of mass. The jet is considered to be ejected through a cluster of nozzles, the center of each being defined by the vector $\mathbf{r}_i = x_i\mathbf{i} + y_i\mathbf{j} - l\mathbf{k}$, as shown in Fig. 7.9–2.

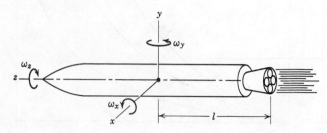

Fig. 7.9–1. Jet damping of spinning rocket.

With the x, y, z axes coinciding with the principal axes of the rocket, the angular momentum components are

$$h_x = I_x\omega_x$$
$$h_y = I_y\omega_y \tag{7.9–2}$$
$$h_z = I_z\omega_z$$

where I_x, I_y, and I_z are instantaneous values of the principal moments of inertia. The rate of change of the angular momentum of the rocket is

$$[\dot{\mathbf{h}}] + \boldsymbol{\omega} \times \mathbf{h}$$

which results in the components

$$\dot{I}_x\omega_x + I_x\dot{\omega}_x + (I_z - I_y)\omega_y\omega_z$$
$$\dot{I}_y\omega_y + I_y\dot{\omega}_y + (I_x - I_z)\omega_x\omega_z \tag{7.9–3}$$
$$\dot{I}_z\omega_z + I_z\dot{\omega}_z + (I_y - I_x)\omega_x\omega_y$$

In considering the angular momentum imparted to the jet, the cluster of nozzles is assumed to be symmetrically located relative to the z axis. If all the angular velocities of the missile are zero, the angular momentum imparted to the jet is zero. The velocity of the nozzle exit due to ω_x, ω_y, and ω_z is shown in Fig. 7.9–2. Multiplying these velocities by their mass rate of flow $\dot{m}$, we obtain the linear momentum rates in the direction of the velocities, from which the angular momentum rates can be determined by multiplying with proper distances from the coordinate axes.

For the ith nozzle, the linear momentum rate due to ω_z is $-\dot{m}_i r_i \omega_z$, and its angular momentum rate about the z axis is $-\dot{m}_i r_i^2 \omega_z$, where $\dot{m}_i = dm_i/dt$ (a negative quantity). Summing over all nozzles, the total rate of angular momentum transfer to the jet, about the z axis is

$$-\omega_z \sum_i \dot{m}_i r_i^2 = -\dot{m}\rho^2 \omega_z$$

where   (7.9–4)

$$\rho^2 = \frac{\sum_i \dot{m}_i r_i^2}{\dot{m}}$$

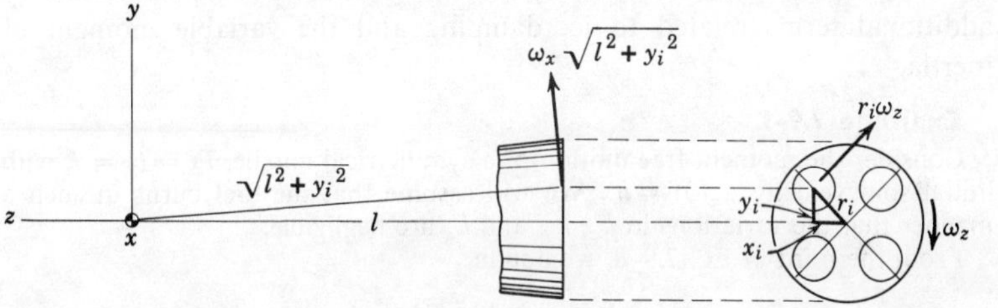

Fig. 7.9-2. Velocity of nozzle due to pitch and spin.

Due to ω_x the rate of change of the momentum of the ith jet is

$$-\dot{m}_i \sqrt{l^2 + y_i^2}\, \omega_x,$$

and its moment is $-\dot{m}_i(l^2 + y_i^2)\omega_x$. Summing over all nozzles, the total rate of change of the angular momentum of the jet about the x axis is

$$-\sum_i \dot{m}_i(l^2 + y_i^2)\omega_x = -\dot{m}(l^2 + \tfrac{1}{2}\rho^2)\omega_x \qquad (7.9\text{–}5)$$

In a similar manner, the rate of change of the angular momentum of the jet about the y axis is

$$-\sum_i \dot{m}_i(l^2 + x_i^2)\omega_y = -\dot{m}(l^2 + \tfrac{1}{2}\rho^2)\omega_y \qquad (7.9\text{–}6)$$

Putting together all these terms, the moment equations become

$$M_x = I_x \dot{\omega}_x + (I_z - I_y)\omega_y \omega_z - \dot{m}(l^2 + \tfrac{1}{2}\rho^2)\omega_x + \dot{I}_x \omega_x$$

$$M_y = I_y \dot{\omega}_y + (I_x - I_z)\omega_x \omega_z - \dot{m}(l^2 + \tfrac{1}{2}\rho^2)\omega_y + \dot{I}_y \omega_y \qquad (7.9\text{–}7)$$

$$M_z = I_z \dot{\omega}_z + (I_y - I_x)\omega_x \omega_y - \dot{m}\rho^2 \omega_z + \dot{I}_z \omega_z$$

Substituting $I_x = mk_x{}^2$, we have $\dot{I}_x = +\dot{m}k_x{}^2 + m\dfrac{d}{dt}k_x{}^2$, and the above equations can be rewritten as

$$M_x = I_x\dot{\omega}_x + (I_z - I_y)\omega_y\omega_z - \left[\dot{m}\left(l^2 + \frac{1}{2}\rho^2 - k_x{}^2\right) - m\frac{d}{dt}k_x{}^2\right]\omega_x$$

$$M_y = I_y\dot{\omega}_y + (I_x - I_z)\omega_x\omega_z - \left[\dot{m}\left(l^2 + \frac{1}{2}\rho^2 - k_y{}^2\right) - m\frac{d}{dt}k_y{}^2\right]\omega_y$$

$$M_z = I_z\dot{\omega}_z + (I_y - I_x)\omega_x\omega_y - \left[\dot{m}(\rho^2 - k_z{}^2) - m\frac{d}{dt}k_z{}^2\right]\omega_z \qquad (7.9\text{-}8)$$

We find, therefore, that the usual Euler's equations are supplemented by additional terms related to jet damping and the variable moment of inertia.

Example 7.9–I

Consider the moment-free motion of a symmetrical missile, $I_x = I_y = I$, with initial spin velocity $\omega_z(0) = n$. We will assume that the fuel burns in such a manner that the variations in k_x, k_y, and k_z are negligible.

From the third of Eq. 7.9-8, we obtain

$$\int \frac{d\omega_z}{\omega_z} = \left(\frac{\rho^2}{k_z{}^2} - 1\right)\int \frac{dm}{m} \qquad (a)$$

which leads to the solution,

$$\frac{\omega_z}{n} = \left(\frac{m}{m_0}\right)^{\frac{\rho^2}{k_z{}^2}-1} \qquad (b)$$

We now multiply the second of Eq. 7.9-8 by $i = \sqrt{-1}$ and add it to the first equation letting

$$\omega_x + i\omega_y = \omega_{xy} \qquad (c)$$

The first two equations of Eq. 7.9-8 then reduce to the following, where ω_z from above has been substituted.

$$\int \frac{d\omega_{xy}}{\omega_{xy}} = \left(\frac{l^2 + \frac{1}{2}\rho^2 - k^2}{k^2}\right)\int \frac{dm}{m} - i\left(1 - \frac{k_z{}^2}{k^2}\right)n\int \left(\frac{m}{m_0}\right)^{\frac{\rho^2}{k_z{}^2}-1} dt \qquad (d)$$

If we assume m to vary linearly with time so that $m = m_0 - m't$, this equation reduces to

$$\ln \frac{\omega_{xy}}{\omega_{xy}(0)} = \frac{l^2 + \frac{1}{2}\rho^2 - k^2}{k^2}\ln \frac{m}{m_0} - i\left(1 - \frac{k_z{}^2}{k^2}\right)n\int_0^t \left(1 - \frac{m'}{m_0}t\right)^{\frac{\rho^2}{k_z{}^2}-1} dt \qquad (e)$$

By letting $1 - (m'/m_0)t = \xi$ and $[(\rho^2/k_z{}^2) - 1] = K$, the last integral is

$$-\frac{m_0}{m'}\int_1^{1-\frac{m'}{m_0}t} \xi^K d\xi = -\frac{m_0}{m'}\frac{\xi^{K+1}}{K+1}\bigg|_1^{1-\frac{m'}{m_0}t}$$

$$= -\frac{m_0}{m'}\frac{k_z{}^2}{\rho^2}\left[\left(1 - \frac{m'}{m_0}t\right)^{\frac{\rho^2}{k_z{}^2}} - 1\right]$$

and the solution becomes

$$\ln \frac{\omega_{xy}}{\omega_{xy}(0)} \left(\frac{m_0}{m_0 - m't}\right)^{\frac{l^2 + \frac{1}{2}\rho^2 - k^2}{k^2}} = in\left(\frac{m_0}{m'}\right)\left(\frac{k_z^2}{\rho^2}\right)\left(1 - \frac{k_z^2}{k^2}\right)\left[\left(1 - \frac{m'}{m_0}t\right)^{\frac{\rho^2}{k_z^2}} - 1\right]$$

or

$$\frac{\omega_{xy}}{\omega_{xy}(0)} = \left(\frac{m_0 - m't}{m_0}\right)^{\frac{l^2 + \frac{1}{2}\rho^2 - k^2}{k^2}}$$

$$\times \exp\left\{-in\frac{m_0}{m'}\frac{k_z^2}{\rho^2}\left(1 - \frac{k_z^2}{k^2}\right)\left[1 - \left(1 - \frac{m'}{m_0}t\right)^{\frac{\rho^2}{k_z^2}}\right]\right\} \quad (f)$$

Figure 7.9–3 shows how the various terms of Eq. f vary with time. The oscillatory amplitude of ω_{xy} diminishes with time due to jet damping, and the frequency of oscillation increases.[1,3]

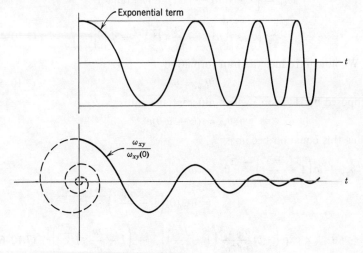

Fig. 7.9–3. Terms of Eq. f.

7.10 Angle of Attack of the Missile

The angular velocity ω_{xy} is referred to coordinates x, y, which are rotating with the missile. To establish the angle of attack of the missile, it is necessary to determine the Euler angle θ measured from a fixed inertial axis. For this determination we start with the angular velocities ω_x, ω_y, ω_z expressed in terms of Euler's angles

$$\omega_x = \dot{\psi}\sin\theta\sin\varphi + \dot{\theta}\cos\varphi$$
$$\omega_y = \dot{\psi}\sin\theta\cos\varphi - \dot{\theta}\sin\varphi \qquad (7.10\text{–}1)$$
$$\omega_z = \dot{\psi}\cos\theta + \dot{\varphi}$$

Adding the first two in quadrature, we have

$$\omega_{xy} = \omega_x + i\omega_y = (\dot\theta + i\dot\psi \sin\theta)e^{-i\varphi} \tag{7.10-2}$$

From Eq. 7.9-b and the third of the above equations, we obtain

$$n\left(\frac{m}{m_0}\right)^{(\rho^2/k_z{}^2)-1} = \dot\psi \cos\theta + \dot\varphi$$

or

$$\dot\psi = \frac{1}{\cos\theta}\left[n\left(\frac{m}{m_0}\right)^{(\rho^2/k_z{}^2)-1} - \dot\varphi\right] \tag{7.10-3}$$

By substituting into the equation for ω_{xy}, Eq. 7.10-2,

$$\omega_{xy} = \left\{\dot\theta + i\left[n\left(\frac{m}{m_0}\right)^{\frac{\rho^2}{k_z{}^2}-1} - \dot\varphi\right]\tan\theta\right\}e^{-i\varphi}$$

$$\simeq \left\{\dot\theta + i\left[n\left(1 - \frac{m'}{m_0}t\right)^{\frac{\rho^2}{k_z{}^2}-1} - \dot\varphi\right]\theta\right\}e^{-i\varphi} \tag{7.10-4}$$

We now introduce the transformation

$$\theta_{xy} = \theta e^{-i\varphi} \tag{7.10-5}$$

proposed by H. Leon,[4] which differentiates into

$$\dot\theta_{xy} = (\dot\theta - i\dot\varphi\theta)e^{-i\varphi}$$

Thus this equation becomes

$$\dot\theta_{xy} + in\left(1 - \frac{m'}{m_0}t\right)^{\frac{\rho^2}{k_z{}^2}-1}\theta_{xy} = \omega_{xy}$$

$$= \omega_{xy}(0)\left(1 - \frac{m'}{m_0}t\right)^{\frac{l^2 + \frac{1}{2}\rho^2 - k^2}{k^2}}$$

$$\times \exp\left\{-in\frac{m_0}{m'}\frac{k_z{}^2}{\rho^2}\left(1 - \frac{k_z{}^2}{k^2}\right)\left[1 - \left(1 - \frac{m'}{m_0}t\right)^{\frac{\rho^2}{k_z{}^2}}\right]\right\} \tag{7.10-6}$$

where the previous solution for ω_{xy}, Eq. f has been substituted.

This equation differs from that of the constant mass missile, first, by the fact that the coefficient of θ_{xy} which is in for the constant mass missile, is now a time function,

$$in\left(1 - \frac{m'}{m_0}t\right)^{\frac{\rho^2}{k_z{}^2}-1}$$

and, secondly, by the right-hand forcing term which is also different due to ω_z slowing down by jet damping and variable mass. The equation is a time-variable linear differential equation which can be solved for θ_{xy}.*

* The equation is in the form

$$\dot\theta_{xy} + P(t)\theta_{xy} = Q(t)$$

with solution

$$\theta_{xy} = e^{-\int P\,dt}\int Qe^{\int P\,dt}\,dt + Ce^{-\int P\,dt}$$

The angle θ_{xy} is referenced to the rotating body axes x, y and must be multiplied by $e^{i\omega_z t} = \exp\left\{in[1 - (m'/m_0)t]^{\frac{\rho^2}{k_z^2}-1}\right\}t$ in order to reference with respect to the inertial axes. Thus the complete solution for the angle of attack as a function of time and the variation of mass is possible by the foregoing procedure.

PROBLEMS

1. Water is flowing out relative to the nozzle shown in the sketch at a speed of 30 ft/sec, and at a rate of 0.10 ft³/sec for each nozzle. If $R = 1.5$ ft and the nozzles are rotated at 60 rpm, determine the torque necessary.

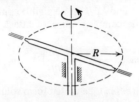

Prob. 1

2. The ends of the nozzle of Prob. 1 are bent back 30° so that the sprinkler will rotate by itself. If the resisting torque due to friction is 1.72 ft. lb, determine the speed in rpm with which the sprinkler will rotate.

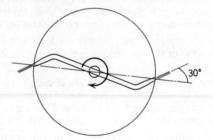

Prob. 2

3. A jet engine takes in air at a rate $\dot{m}_a$, compresses it, mixes it with kerosene at a rate $\dot{m}_k$, and ejects the ignited mixture at a speed u relative to the nozzle. If the jet plane is traveling at a speed of v, show that the thrust of the engine is

$$T = \dot{m}_k u + \dot{m}_a(u - v)$$

4. A nonspinning rocket of total mass m_0, half of which is fuel, is rotating about the pitch axis with an initial angular velocity of 0.5 rad/sec. If l/k for the rocket remains constant at 2, determine the pitch rate at burnout.

5. If in Prob. 4 the radius of gyration about the pitch axis decreased with time, would the pitch rate at burnout be larger or smaller?

6. The ratio of the fuel to the total mass for a particular missile is 0.70, and burning takes place with negligible change in the value of $l/k = \sqrt{3}$. If the rate of fuel consumption is $m'/m_0 = \frac{1}{100}$ sec^{-1}, and the missile is rotating about a transverse axis without spin, plot the variation in its rotation speed against time.

7. If for the spinning rocket with variable mass, the ratio $\rho/k_z = 1$ and the quantity ρ/k is negligible compared to l/k, show that the equation for the complex angular velocity is

$$\frac{\omega_{xy}}{\omega_{xy}(0)} = \left(\frac{m}{m_0}\right)^{\frac{l^2}{k^2}-1} e^{-i\left(1-\frac{k_z^2}{k^2}\right)nt}$$

How does this equation differ from that of the nonrotating rocket?

8. For the case $\rho^2/k_z^2 = 1$ and k_z^2/k^2 is negligible compared to unity, the differential equation for the complex angle of attack, (Eq. 7.10–6) reduces to (see also Prob. 7),

$$\dot{\theta}_{xy} + in\theta_{xy} = \omega_{xy} = \left(\frac{m}{m_0}\right)^{\frac{l^2}{k^2}-1} \omega_{xy}(0)e^{-int}$$

Letting $m/m_0 = 1 - [(m'/m_0)t]$, a closed form solution is possible when l^2/k^2 is an integer. Letting $l^2/k^2 = 4$, carry out this solution and show that the motion of the missile longitudinal axis is a converging spiral.

9. Assuming the angle of attack θ of a spinning missile to be small (angle of attack is measured from the velocity vector $\mathbf{V}$ which can be considered fixed in space) draw the inertial coordinates X, Y, Z, the node axis ξ, and the rotating body axes x, y, z, where z is the longitudinal axis of the missile at an angle θ from the vector $\mathbf{V}$ placed along the Z axis. On this diagram show the complex angular velocity ω_{xy}, ω_z, and the resultant angular velocity $\boldsymbol{\omega}$.

10. Assuming small angle of attack, determine the inertial components of the angular velocity $\boldsymbol{\omega}$ of Prob. 9, by resolving it along the Z axis and in the XY plane.

11. If the moment of inertia of the missile about the transverse and longitudinal axes are A and C, determine the position of the angular momentum vector for Prob. 9, and find the angle between it and the velocity vector. How does the angular momentum vector vary in the inertial space?

12. Compare the solution for the complex angle of attack θ_{XY} of a missile with constant thrust misalignment M_1 (Prob. 10, Sec. 7.4) with that of the near symmetrical missile with principal axis misalignment of β. Determine the product of inertia F in terms of the misalignment moment M_1, which will give the same motion.

7.11 General Motion of Spinning Bodies with Varying Configuration and Mass

In the previous sections the origin of the body axes always coincided with the mass center. In the most general case, a body under translation and rotation may have relative motion between particles leading to a varying configuration, and may be undergoing a change in mass with time. The origin of the body coordinates attached to the system will then not coincide with the center of mass at all times. Relative motion between particles could take place when motors and other moving parts are present or when the body contains liquids such as fuel. Vibration due to flexibility

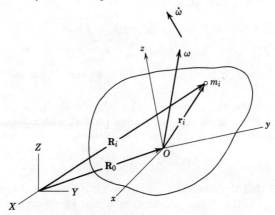

Fig. 7.11-1.   Body of varying mass and configuration.

is another contributing factor. Mass variation would take place owing to jets ejected from the body.

To examine the motion of such a general system, it is advisable to view the problem as a system of particles with the origin of the body axes not coinciding with the center of mass.[9]* Such a procedure will account for every conceivable configuration of the system and eliminate the possibility of omitting terms. In spite of this generality, the terms of the equation can be regrouped to more familiar forms of rigid body, jet ejection, center of mass shift, and relative motion.

We define the system by a group of particles within a specified boundary with body coordinates x, y, z moving with the system as shown in Fig. 7.11-1. Variation in mass is allowed by particles leaving the system through the boundary. The angular momentum of the system about the moving origin 0 at time t is

$$\mathbf{h}_0 = \sum \mathbf{r}_i \times m_i \dot{\mathbf{R}}_i \qquad (7.11\text{-}1)$$

where $\mathbf{r}_i$ is drawn from the moving origin 0, and $\dot{\mathbf{R}}_i$ is the absolute velocity

* In Ref. 9 the origin for the moment is chosen to coincide with the center of mass and the various subbodies are considered to be rigid. Also no provision is made for the variation of the total system mass.

of m_i referenced to the inertial axes X, Y, Z. If we differentiate this equation, we obtain,

$$\dot{\mathbf{h}}_0 = \sum \mathbf{r}_i \times \frac{d}{dt}(m_i \dot{\mathbf{R}}_i) + \sum \dot{\mathbf{r}}_i \times m_i \dot{\mathbf{R}}_i \qquad (7.11\text{--}2)$$

From Fig. 7.11–1, $\mathbf{R}_i = \mathbf{R}_0 + \mathbf{r}_i$, and the last term of Eq. 7.11–2 can be reduced by the following steps,

$$\sum \dot{\mathbf{r}}_i \times m_i \dot{\mathbf{R}}_i = \sum \dot{\mathbf{r}}_i \times m_i(\dot{\mathbf{R}}_0 + \dot{\mathbf{r}}_i)$$
$$= -\dot{\mathbf{R}}_0 \times \sum m_i \dot{\mathbf{r}}_i = -\dot{\mathbf{R}}_0 \times m\dot{\bar{\mathbf{r}}}$$

where m is the total mass at time t, and $\bar{\mathbf{r}}$ its center of mass relative to the body axes.

Referring to the first term of Eq. 7.11–2, $(d/dt)(m_i \dot{\mathbf{R}}_i)$ is equal to the force applied to the mass m_i, and its cross product with $\mathbf{r}_i$ is the moment about 0. Equation 7.11–2 can then be rewritten as

$$\mathbf{M}_0 = \dot{\mathbf{h}}_0 + \dot{\mathbf{R}}_0 \times m\dot{\bar{\mathbf{r}}} \qquad (7.11\text{--}3)$$

which states that the moment about an arbitrary point 0 is equal to the rate of change of the angular momentum $\mathbf{h}_0$ plus a term depending on the velocity of the origin and the velocity of the center of mass with respect to the origin. It is evident, then, that the moment is equal to the rate of change of the angular momentum only under the following conditions: (1) When 0 is stationary; (2) when the velocity of the center of mass relative to the origin is zero; or (3) when the two velocities $\dot{\mathbf{R}}_0$ and $\dot{\bar{\mathbf{r}}}$ are parallel.

The moment equation for the general system can be found directly from the equation

$$\mathbf{M}_0 = \sum \mathbf{r}_i \times \frac{d}{dt}(m_i \dot{\mathbf{R}}_i) \qquad (7.11\text{--}4)$$

However, to clarify certain concepts, we will examine the angular momentum at two instances of time and determine $\dot{\mathbf{h}}_0$ to be substituted into Eq. 7.11–3.

Figure 7.11–2 shows a mass m_i at time t, which at a later time $t = \Delta t$ occupies a different position $(\mathbf{r}_i + \Delta \mathbf{r}_i)$ and has separated into two parts, $m_i + \dot{m}_i \Delta t$ and $(-\dot{m}_i \Delta t)$, with relative velocity $\mathbf{u}_i$ between them. In separating into two parts, m is decreasing and $\dot{m} = dm/dt$ is a negative quantity. The angular momentum at $t + \Delta t$ is

$$\mathbf{h}_0 + \Delta \mathbf{h}_0 = \sum (\mathbf{r}_i + \Delta \mathbf{r}_i) \times (m_i + \dot{m}_i \Delta t)(\dot{\mathbf{R}}_i + \Delta \dot{\mathbf{R}}_i)$$
$$+ \sum (\mathbf{r}_i + \Delta \mathbf{r}_i) \times (-\dot{m}_i \Delta t)(\dot{\mathbf{R}}_i + \mathbf{u}_i) \qquad (7.11\text{--}5)^*$$

_____

* $\mathbf{u}_i$ is negative when mass is ejected in the opposite sense to $\dot{\mathbf{R}}$ (see p. 221).

and, by neglecting higher order infinitesimals and approaching the limit $\Delta \mathbf{h}_0 / \Delta t$, as $(\Delta t \to 0)$ the rate of change of the angular momentum becomes

$$\dot{\mathbf{h}}_0 = \sum \mathbf{r}_i \times m_i \ddot{\mathbf{R}}_i + \sum \dot{\mathbf{r}}_i \times m_i \dot{\mathbf{R}}_i - \sum \mathbf{r}_i \times \dot{m}_i \mathbf{u}_i$$

$$= \sum \mathbf{r}_i \times m_i \ddot{\mathbf{R}}_i - \dot{\mathbf{R}}_0 \times m \dot{\bar{\mathbf{r}}} - \sum \mathbf{r}_i \times \dot{m}_i \mathbf{u}_i \qquad (7.11\text{-}6)$$

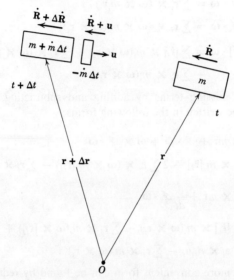

Fig. 7.11–2. Angular momentum of element m at times t and $t + \Delta t$.

Substituting Eq. 7.11–6 into 7.11–3, the moment equation becomes

$$\mathbf{M}_0 = \sum \mathbf{r}_i \times m_i \ddot{\mathbf{R}}_i - \sum \mathbf{r}_i \times \dot{m}_i \mathbf{u}_i \qquad (7.11\text{-}7)$$

which could have been obtained directly from Eq. 7.11–4 by recognizing that $(d/dt)(m_i \dot{\mathbf{R}}_i) = m_i \ddot{\mathbf{R}}_i - \dot{m}_i \mathbf{u}_i$ (see Sec. 7.7).*

We now replace $\ddot{\mathbf{R}}_i$ by the general expression for acceleration,

$$\ddot{\mathbf{R}}_i = \ddot{\mathbf{R}}_0 + \dot{\boldsymbol{\omega}} \times \mathbf{r}_i + \boldsymbol{\omega} \times (\boldsymbol{\omega} \times \mathbf{r}_i) + [\ddot{\mathbf{r}}_i] + 2\boldsymbol{\omega} \times [\dot{\mathbf{r}}_i] \qquad (7.11\text{-}8)$$

where $[\mathbf{r}]$ and $[\ddot{\mathbf{r}}_i]$ are velocity and acceleration relative to the moving coordinate system,

$$\mathbf{M}_0 = -\ddot{\mathbf{R}}_0 \times m\bar{\mathbf{r}} + \sum \mathbf{r}_i \times (\dot{\boldsymbol{\omega}} \times m_i \mathbf{r}_i) + \sum \mathbf{r}_i \times \{\boldsymbol{\omega} \times m_i(\boldsymbol{\omega} \times \mathbf{r}_i)\}$$

$$+ \sum \mathbf{r}_i \times m_i[\ddot{\mathbf{r}}_i] + 2 \sum \mathbf{r}_i \times (\boldsymbol{\omega} \times m_i[\dot{\mathbf{r}}_i]) - \sum \mathbf{r}_i \times \dot{m}_i \mathbf{u}_i \qquad (7.11\text{-}9)$$

To recognize the moment equation in terms of familiar expressions, we

---

* $\mathbf{F} = m \dfrac{d\mathbf{v}}{dt} - \mathbf{u} \dfrac{dm}{dt} =$ applied external force, therefore M_0 is the moment of the externally applied force.

introduce the moment of inertia diadic of Sec. 5.2 and identify the following:

$$\frac{d}{dt}(\mathscr{I} \cdot \boldsymbol{\omega}) = \mathscr{I} \cdot \dot{\boldsymbol{\omega}} + \boldsymbol{\omega} \times \mathscr{I} \cdot \boldsymbol{\omega} + [\dot{\mathscr{I}}] \cdot \boldsymbol{\omega}$$

$$\mathscr{I} \cdot \dot{\boldsymbol{\omega}} = \sum \mathbf{r}_i \times (\dot{\boldsymbol{\omega}} \times m_i \mathbf{r}_i)$$

$$\boldsymbol{\omega} \times \mathscr{I} \cdot \boldsymbol{\omega} = \sum \mathbf{r}_i \times \{\boldsymbol{\omega} \times m_i(\boldsymbol{\omega} \times \mathbf{r}_i)\}$$

$$[\dot{\mathscr{I}}] \cdot \boldsymbol{\omega} = \sum [\dot{\mathbf{r}}_i] \times m_i(\boldsymbol{\omega} \times \mathbf{r}_i) + \sum \mathbf{r}_i \times (\boldsymbol{\omega} \times m_i[\dot{\mathbf{r}}_i])$$

$$+ \sum \mathbf{r}_i \times \dot{m}_i(\boldsymbol{\omega} \times \mathbf{r}_i) \qquad (7.11\text{--}10)\dagger$$

Supplying the missing terms by adding and subtracting, the moment equation can be written in the following forms:

$$\mathbf{M}_0 = -\ddot{\mathbf{R}}_0 \times m\bar{\mathbf{r}} + \mathscr{I} \cdot \dot{\boldsymbol{\omega}} + \boldsymbol{\omega} \times \mathscr{I} \cdot \boldsymbol{\omega}$$

$$+ \sum \mathbf{r}_i \times m_i[\ddot{\mathbf{r}}_i] + 2 \sum \mathbf{r}_i \times (\boldsymbol{\omega} \times m_i[\dot{\mathbf{r}}_i]) - \sum \mathbf{r}_i \times \dot{m}_i \mathbf{u}_i \quad (7.11\text{--}11)$$

$$\mathbf{M}_0 = -\ddot{\mathbf{R}}_0 \times m\bar{\mathbf{r}} + \frac{d}{dt}\mathscr{I} \cdot \boldsymbol{\omega}$$

$$- \sum [\dot{\mathbf{r}}_i] \times m_i(\boldsymbol{\omega} \times \mathbf{r}_i) + \sum \mathbf{r}_i \times m_i(\boldsymbol{\omega} \times [\dot{\mathbf{r}}_i]) + \sum \mathbf{r}_i \times m_i[\ddot{\mathbf{r}}_i]$$

$$- \sum \mathbf{r}_i \times \dot{m}_i \mathbf{u}_i - \sum \mathbf{r}_i \times \dot{m}_i(\boldsymbol{\omega} \times \mathbf{r}_i) \qquad (7.11\text{--}12)$$

A third and a more convenient form can be found by reducing the first two terms of the second line in Eq. 7.11–12 into a single term by the following equation.*

$$\mathbf{a} \times (\mathbf{b} \times \mathbf{c}) + (\mathbf{b} \times \mathbf{a}) \times \mathbf{c} = \mathbf{b} \times (\mathbf{a} \times \mathbf{c})$$

$$\sum \mathbf{r}_i \times m_i(\boldsymbol{\omega} \times [\dot{\mathbf{r}}_i]) + \sum m_i(\boldsymbol{\omega} \times \mathbf{r}_i) \times [\dot{\mathbf{r}}_i] = \boldsymbol{\omega} \times \sum (\mathbf{r}_i \times m_i[\dot{\mathbf{r}}_i])$$

The third form of the moment equation then becomes

$$\mathbf{M}_0 = -\ddot{\mathbf{R}}_0 \times m\bar{\mathbf{r}} + \frac{d}{dt}(\mathscr{I} \cdot \boldsymbol{\omega})$$

$$+ \boldsymbol{\omega} \times \sum (\mathbf{r}_i \times m_i[\dot{\mathbf{r}}_i]) + \sum \mathbf{r}_i \times m_i[\ddot{\mathbf{r}}_i]$$

$$- \sum \mathbf{r}_i \times \dot{m}_i \mathbf{u}_i - \sum \mathbf{r}_i \times \dot{m}_i(\boldsymbol{\omega} \times \mathbf{r}_i) \qquad (7.11\text{--}13)$$

The various terms of these equations can now be identified. We have in the first term the effect of the origin of the body coordinates not coinciding with the mass center. The terms $\mathscr{I} \cdot \dot{\boldsymbol{\omega}} + \boldsymbol{\omega} \times \mathscr{I} \cdot \boldsymbol{\omega}$ correspond to the

---

* This equation results from the application of the relationship

$$\mathbf{a} \times (\mathbf{b} \times \mathbf{c}) = \mathbf{b}(\mathbf{a} \cdot \mathbf{c}) - \mathbf{c}(\mathbf{a} \cdot \mathbf{b}).$$

† See Probs. 18 and 19 p. 111.

usual Euler equation, whereas $(d/dt)(\mathscr{I} \cdot \boldsymbol{\omega})$ includes the additional term $[\dot{\mathscr{I}}] \cdot \boldsymbol{\omega}$ which accounts for the rate of change of the inertia diadic resulting from the position change of the particles in relative motion and the variation of mass. The term $\sum \mathbf{r}_i \times \dot{m}_i \mathbf{u}_i$ is the thrust misalignment moment,* while the term $-\sum \mathbf{r}_i \times \dot{m}_i (\boldsymbol{\omega} \times \mathbf{r}_i)$ is the jet damping due to rotation of the body. All other terms are due to relative motion of particles. The three forms of the moment equation, Eqs. 7.11–11, 7.11–12, and 7.11–13 are presented here to show the origin of the various terms, some of which were inserted due to $[\dot{\mathscr{I}}] \cdot \boldsymbol{\omega}$.

To complete the discussion, it must be recognized that the external moment may result from the forces not directed through the origin of the body coordinates. The external forces accelerate the instantaneous center of mass and change the linear momentum of the ejected particles according to the equation

$$\mathbf{F} = m[\ddot{\mathbf{R}}_0 + \boldsymbol{\omega} \times (\boldsymbol{\omega} \times \bar{\mathbf{r}}) + \dot{\boldsymbol{\omega}} \times \bar{\mathbf{r}} + 2\boldsymbol{\omega} \times [\dot{\bar{\mathbf{r}}}] + [\ddot{\bar{\mathbf{r}}}]] - \sum \dot{m}_i \mathbf{u}_i$$
(7.11–14)

Thus, in the general case, the force equations are coupled to the moment equations.

Example 7.11–1
A space vehicle is moving under a force-free condition. If a motor located at $\mathbf{r}_1$ is started, determine its perturbation torque.

The perturbation torque is the contribution from the relative motion terms of Eqs. 7.11–11, 7.11–12, or 7.11–13. We will use the form given by Eq. 7.11–13, which is (with total mass a constant, $\dot{m} = 0$)

$$M_p = [\dot{\mathscr{I}}] \cdot \boldsymbol{\omega} + \boldsymbol{\omega} \times \sum (\mathbf{r}_i \times m_i[\dot{\mathbf{r}}_i]) + \sum \mathbf{r}_i \times m_i[\ddot{\mathbf{r}}_i]$$

Owing to the symmetry of the motor rotor, the change in the inertia diadic relative to the body coordinates resulting from the spin of the rotor is zero, which eliminates the first term $[\dot{\mathscr{I}}] \cdot \boldsymbol{\omega} = 0$.

From Fig. 7.11–3, we have,

$$\mathbf{r}_i = \mathbf{r}_1 + \boldsymbol{\rho}_i$$
$$[\dot{\mathbf{r}}_i] = \boldsymbol{\omega}_1 \times \boldsymbol{\rho}_i$$
$$[\ddot{\mathbf{r}}_i] = \dot{\boldsymbol{\omega}}_1 \times \boldsymbol{\rho}_i$$

and noting that $\sum m_i \boldsymbol{\rho}_i = 0$ for a symmetrical wheel, the equation for the perturbation torque becomes,

$$M_p = \boldsymbol{\omega} \times \left[\sum \boldsymbol{\rho}_i \times (\boldsymbol{\omega}_1 \times m_i \boldsymbol{\rho}_i) \right] + \sum \boldsymbol{\rho}_i \times (\dot{\boldsymbol{\omega}}_1 \times m_i \boldsymbol{\rho}_i)$$

Since

$$\sum \boldsymbol{\rho}_i \times (\boldsymbol{\omega}_1 \times m_i \boldsymbol{\rho}_i) = \mathscr{I}_1 \cdot \boldsymbol{\omega}_1 = (C_1 \omega_1) \mathbf{k}_1 = \mathbf{h}_1$$

the equation for $\mathbf{M}_p$ can be written as

$$\mathbf{M}_p = [\dot{\mathbf{h}}_1] + \boldsymbol{\omega} \times \mathbf{h}_1$$

* Letting $\mathbf{M}_T = \sum \mathbf{r}_i \times \dot{m}_i \mathbf{u}_i$ = moment of the thrusting jet, the moment acting on the varying mass is $\mathbf{M}_0 + \mathbf{M}_T$.

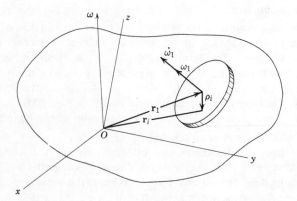

Fig. 7.11–3. Perturbation torque due to rotation of wheel.

where $\mathbf{h}_1$ is the angular momentum of the rotor wheel. Thus the perturbation torque is the result of the angular acceleration of the wheel, and the precession of the angular momentum vector of the wheel caused by the rotation $\boldsymbol{\omega}$ of the body coordinates.

PROBLEMS

1. A uniform rigid bar of length l and mass m is translating with constant velocity $\dot{\mathbf{R}}_0$ in a direction normal to its length. At the same time a mass m_0 is sliding from one end to the other with velocity $[\dot{\mathbf{r}}]$. Placing body coordinates as indicated in the sketch, verify from Eq. 7.11–3 that the moment about O is zero. Describe the motion of O.

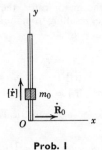

Prob. 1

2. The center of mass of a uniform rigid bar of length l is moving with constant velocity $\dot{\mathbf{R}}$ along a straight line, while the bar rotates with constant angular velocity $\boldsymbol{\omega}$. Placing body axes as shown, verify from Eq. 7.11–3 that the moment about O is zero.

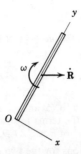

Prob. 2

3. For a system of particles which is not changing in mass, write the equation for the moment about its center of mass when relative motion betweeen particles is allowed.

4. Consider a constant-mass system such as a satellite in orbit, and assume body axes x, y, z through its center of mass and coinciding with the principal axes. There is a motor on the pitch axis x, a distance x_0 from O with its axis of rotation parallel to the y axis. Let the rotor moment of inertia be $I_{y'}$ and that of the entire satellite including the motor to be A, A, C about x, y, z respectively. If the motor is started with angular acceleration $\dot{\omega}_1$, define the terms in Eq. 7.11–11 which apply to the problem, and write the components of the moment equation.

5. Show that the angular momentum of a group of particles about an arbitrary origin is equal to

$$\mathbf{h}_0 = -\dot{\mathbf{R}}_0 \times m\bar{\mathbf{r}} + \int \mathbf{r} \times (\boldsymbol{\omega} \times \mathbf{r})\, dm + \int \mathbf{r} \times [\dot{\mathbf{r}}]\, dm$$

where $\mathscr{I} \cdot \boldsymbol{\omega} = \int \mathbf{r} \times (\boldsymbol{\omega} \times \mathbf{r})\, dm$ and $[\dot{\mathbf{r}}]$ is the velocity relative to the rotating body axes.

6. Show that the terms of the equation

$$\frac{d}{dt}(\mathscr{I} \cdot \boldsymbol{\omega}) = \mathscr{I} \cdot \dot{\boldsymbol{\omega}} + \boldsymbol{\omega} \times \mathscr{I} \cdot \boldsymbol{\omega} + [\dot{\mathscr{I}}] \cdot \boldsymbol{\omega}$$

can be identified as

$$\mathscr{I} \cdot \dot{\boldsymbol{\omega}} = \int \mathbf{r} \times \dot{\boldsymbol{\omega}} \times \mathbf{r}\, dm$$

$$\boldsymbol{\omega} \times \mathscr{I} \cdot \boldsymbol{\omega} = \int \mathbf{r} \times [\boldsymbol{\omega} \times (\boldsymbol{\omega} \times \mathbf{r})]\, dm$$

$$[\dot{\mathscr{I}}] \cdot \boldsymbol{\omega} = \int \mathbf{r} \times (\boldsymbol{\omega} \times [\dot{\mathbf{r}}])\, dm + \int [\dot{\mathbf{r}}] \times (\boldsymbol{\omega} \times \mathbf{r})\, dm + \int \mathbf{r} \times (\boldsymbol{\omega} \times \mathbf{r})\, d\dot{m}$$

7. Show that

$$\frac{d}{dt}\int \mathbf{r} \times [\dot{\mathbf{r}}]\, dm = \int \mathbf{r} \times [\ddot{\mathbf{r}}]\, dm + \int \mathbf{r} \times (\boldsymbol{\omega} \times [\dot{\mathbf{r}}])\, dm + \int (\boldsymbol{\omega} \times \mathbf{r})$$

$$\times [\dot{\mathbf{r}}]\, dm + \int \mathbf{r} \times [\dot{\mathbf{r}}]\, d\dot{m}$$

Combining the results of Prob. 5 and 6, two of the terms add to give $2\int \mathbf{r} \times (\boldsymbol{\omega} \times [\dot{\mathbf{r}}])\, dm$, while the terms $\int (\boldsymbol{\omega} \times \mathbf{r}) \times [\dot{\mathbf{r}}]\, dm$ cancel each other. Now clarify the interpretation of Eqs. 7.11–9, 7.11–10, 7.11–11, and 7.11–12.

8. Write the component equations for the two terms

$$-\sum [\dot{\mathbf{r}}_i] \times m_i(\boldsymbol{\omega} \times \mathbf{r}_i) + \sum \mathbf{r}_i \times m_i(\boldsymbol{\omega} \times [\dot{\mathbf{r}}_i])$$

of Eq. 7.11–12.

9. Write out the component terms of $[\dot{\mathscr{I}}] \cdot \boldsymbol{\omega}$ and show that they represent the time derivative of $\mathscr{I} \cdot \boldsymbol{\omega}$ relative to the body coordinates. Identify the parts due to relative motion of the particles and those due to mass variation.

10. Show that $\mathscr{I} = \sum m_i(\mathbf{r}_i \cdot \mathbf{r}_i \mathscr{E} - \mathbf{r}_i\mathbf{r}_i)$, where $\mathscr{E} = \mathbf{ii} + \mathbf{jj} + \mathbf{kk}$ is a unit diad. Show also that

$$[\dot{\mathscr{I}}] = \sum \dot{m}_i(\mathbf{r}_i \cdot \mathbf{r}_i \mathscr{E} - \mathbf{r}_i\mathbf{r}_i) + \sum m_i(\dot{\mathbf{r}}_i \cdot \mathbf{r}_i \mathscr{E} + \mathbf{r}_i \cdot \dot{\mathbf{r}}_i \mathscr{E} - \dot{\mathbf{r}}_i\mathbf{r}_i - \mathbf{r}_i\dot{\mathbf{r}}_i)$$

11. Derive Eqs. 7.9–7 and 7.9–8 as a special case of the general equation, Eq. 7.11–12. State the restrictions imposed on Eq. 7.11–12 in arriving at the above equations.

12. A symmetrical spinning satellite in orbit has moments of inertia A, A, C about the x, y, z axes (including m_0 at position $\zeta = 0$), as shown in the sketch.

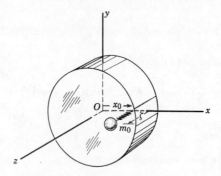

Prob. 12

If the mass m_0 is restricted to move in the z direction and has a restoring spring stiffness of k with viscous damping c, show that the differential equations of motion are:

Moment:

$$\mathbf{M}_0 = \mathbf{i}\Big\{(A + m_0\zeta^2)\dot\omega_x + (C - A - m_0\zeta^2)\omega_y\omega_z - m_0x_0\zeta\dot\omega_z - m_0x_0\zeta\omega_x\omega_y$$

$$+ 2m_0\zeta\dot\zeta\omega_x - \frac{m_0{}^2}{m}[2\dot\zeta\omega_x + \zeta(\dot\omega_x - \omega_y\omega_z)]\zeta\Big\}$$

$$+ \mathbf{j}\Big\{(A + m_0\zeta^2)\dot\omega_y - (C - A - m_0\zeta^2)\omega_x\omega_z + m_0x_0\zeta(\omega_x{}^2 - \omega_z{}^2) - m_0x_0\ddot\zeta$$

$$+ 2m_0\zeta\dot\zeta\omega_y - \frac{m_0{}^2}{m}[2\dot\zeta\omega_y - \zeta(\dot\omega_y - \omega_x\omega_z)]\zeta\Big\}$$

$$+ \mathbf{k}(C\dot\omega_z - m_0x_0\zeta\dot\omega_x + m_0x_0\zeta\omega_y\omega_z - 2m_0x_0\zeta\dot\omega_x)$$

Force on mass m_0 in z direction:

$$F_z = m_0[\ddot\zeta - x_0\dot\omega_y + x_0\omega_x\omega_z - \zeta(\omega_x{}^2 + \omega_y{}^2)] + \frac{m_0{}^2}{m}[\zeta(\omega_x{}^2 + \omega_y{}^2) - \ddot\zeta]$$

$$+ c\dot\zeta + k\zeta = 0$$

Acceleration of origin:

$$\ddot{\mathbf{R}}_0 = -\frac{m_0}{m}[2\dot\zeta\omega_y + \zeta(\dot\omega_y + \omega_x\omega_z)]\mathbf{i} + \frac{m_0}{m}[2\dot\zeta\omega_x + \zeta(\dot\omega_x - \omega_y\omega_z)]\mathbf{j}$$

$$+ \frac{m_0}{m}[\zeta(\omega_x{}^2 + \omega_y{}^2) - \ddot\zeta]\mathbf{k}$$

REFERENCES

1. H. J. Cohen, "The Effect of Jet Damping on the Motion of the Able-1 Third Stage," Space Technology Laboratories Interoffice Correspondence (April 15, 1960).
2. Ellis, J. W., and C. W. McArthur, "Applicability of Euler's Dynamical Equations to Rocket Motion," *ARS Jour.*, **29**, No. 11 (Nov. 1959), 863–864.
3. Jarmolow, K., "Dynamics of a Spinning Rocket with Varying Inertia and Applied Moment," *J. Appl. Phys.*, **28**, No. 3 (1957), 308–313.
4. Leon, H., "Angle of Attack Convergence of a Spinning Missile Descending Through the Atmosphere," *J. Aero/Space Sci.*, **25**, No. 8 (Aug. 1958), 480–484.
5. Leon, H., "Spin Dynamics of Rockets and Space Vehicles in Vacuum," *TR 59–0000–00787*, Space Technology Laboratories (Sept. 16, 1959).
6. Pilkington, W. C., "Vehicle Motion as Inferred from Radio-Signal-Strength Records," *External Publication No. 551*, Jet Propulsion Laboratory, Pasadena, Calif. (Sept. 5, 1958).
7. Reiter, G., *Space Technology Laboratories Memo* (June 17, 1959).
8. Thomson, W. T., and G. S. Reiter, "Attitude Drift of Space Vehicles," *J. Astronaut. Sci.*, **VII**, No. 2 (1960), 29–34.
9. Roberson, R. E., "Torques on a Satellite Vehicle from Internal Moving Parts," *J. Appl. Mech.*, **25**, No. 2 (June 1958), 196–200.

Performance
and Optimization

CHAPTER 8

In Chap. 4 it was shown that the problem of placing a satellite into an orbit is a matter of achieving the required velocity at a specified position in space. For earth-bound orbits the required velocity is in the neighborhood of 25,000 ft/sec, whereas for the lunar mission a velocity of approximately 35,000 ft/sec is necessary. In this chapter we discuss the basic theory of rockets and examine the problems of optimization to meet a specific performance. Missile flexibility as it affects the desired performance will be discussed in Chap. 9.

8.1 Performance of Single-Stage Rockets

A rocket is a variable mass vehicle which acquires thrust by the ejection of high-speed particles. The force equation for the rocket can be written in the general form

$$\frac{d\mathbf{v}}{dt} = \frac{\mathbf{T}}{m} + \frac{\mathbf{F}_a}{m} - \dot{\mathbf{g}} \tag{1}$$

where $\mathbf{T}$ is the thrust of the jet and $\mathbf{F}_a$ is the aerodynamic force. Since $\mathbf{F}_a/m$ varies inversely as the characteristic length of the rocket, this term is small in comparison to $\mathbf{T}/m$ for large rockets.

Certain parameters of importance can be brought out by studying the behavior of a rocket in vertical flight, neglecting aerodynamic forces, and

240

assuming the gravity field to be a constant. Referring to Fig. 8.1–1, we start with the equation,

$$\frac{dv}{dt} = \frac{T}{m} - g \tag{8.1–2}$$

Since the rocket is losing mass, dm/dt is negative, and the thrust becomes,

$$T = -u\frac{dm}{dt} \tag{8.1–3}*$$

where the small term due to the difference in pressure has been omitted.

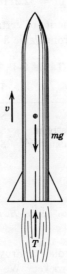

Fig. 8.1–1.   Rocket in vertical flight.

Substituting Eq. 8.1–3 into 8.1–2,

$$dv = -u\frac{dm}{m} - g\,dt \tag{8.1–4}$$

and integrating, the velocity equation becomes

$$v - v_0 = u \ln \frac{m_0}{m} - gt \tag{8.1–5}$$

where m is the mass at any time t. By substituting the burnout time t_{b0} and the burnout mass m_{b0}, the maximum attainable velocity in vertical flight is

$$v_{b0} - v_0 = u \ln \frac{m_0}{m_{b0}} - gt_{b0} \tag{8.1–6}$$

* $\mathbf{u} = -u$

For chemical propellants, the ejection speed u relative to the rocket nozzle depends on the heat energy per pound which must be high, and on the molecular weight which must be small. Its performance is rated by the *specific impulse I*, defined as the thrust of a pound of propellant multiplied by the number of seconds required to burn it. Its relationship to u is found from the equation,

$$I = \int_0^t T \, dt = \int_0^t u \frac{dm}{dt} \, dt = \int_0^{1/g} u \, dm = \frac{u}{g}$$

or

$$u = gI = 32.2I \text{ ft/sec} \tag{8.1-7}$$

Some indication as to the merits of certain fuels and their propellant combinations are obtainable from Table 8–1.

Table 8–1[a]

Chemical Propellants	Type	Specific Impulse I, sec
Ammonium nitrate rubber	Solid	170–210
Potassium perchlorite thickol or asphalt	Solid	170–210
Boron metal components and oxidant	Solid	200–250
Liquid oxygen alcohol		250–270
Liquid oxygen fluorine JP4		270–330
Fluorine hydrogen		300–385

[a] "Astronautics and its Applications," *Space Handbook*, U.S. Govt. Printing Office, Wash. D.C. (1959).

It is convenient here to introduce a thrust parameter which establishes the initial acceleration of the rocket. We define thrust ratio $\mathscr{R}$ as the thrust of the rocket divided by the initial weight,

$$\text{Thrust ratio } \mathscr{R} = \frac{T}{m_0 g} = \frac{a_0}{g} + 1 \tag{8.1-8}$$

where a_0 is the initial acceleration in vertical flight. The time duration of the powered flight is then

$$t_{b0} = gI \left(\frac{m_0 - m_{b0}}{T} \right) = \frac{I}{\mathscr{R}} \left(1 - \frac{m_{b0}}{m_0} \right) \tag{8.1-9}$$

Equation 8.1–4 can now be written as,

$$v_{b0} - v_0 = gI \left[\ln \frac{m_0}{m_{b0}} - \frac{1}{\mathscr{R}} \left(1 - \frac{m_{b0}}{m_0} \right) \right] \tag{8.1-10}$$

which indicates that the maximum attainable velocity depends on the mass fraction m_{b0}/m_0, on the specific impulse I of the fuel, and on the thrust ratio $\mathscr{R}$.

With $v_0 = 0$, it is possible to plot v_{b0}/gI versus m_{b0}/m_0, with $\mathscr{R}$ as a parameter. It is instructive, however, to plot v_{b0} versus m_{b0}/m_0 for given values of I and $\mathscr{R}$ as in Fig. 8.1–2, since such a plot indicates the inadequacy

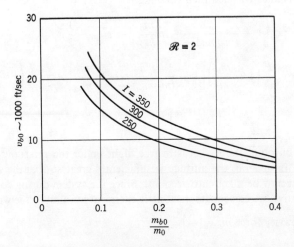

Fig. 8.1–2. Burnout velocity as function of mass ratio, specific impulse I, and thrust ratio $\mathscr{R}$.

of a single-stage rocket for placing a satellite into orbit. As in most designs, a compromise must be established between, $\mathscr{R}$, m_{b0}/m_0, and I (i.e., a large thrust ratio requires a heavier structure, and exotic fuels of high I tend to give larger values of m_{b0}/m_0. In any event, it is difficult to achieve a number less than 0.1 for m_{b0}/m_0 and a specific impulse greater than 350 for chemical propellants, which indicates the necessity of multistage rockets for satellite orbits and space missions.

To determine the distance traveled during the powered flight, the velocity equation, Eq. 8.1–5, must be integrated. Equation 8.1–5 can be integrated if the variation in g is assumed to be negligible and $m(t)$ known. A realistic assumption is that of constant rate of fuel consumption leading to constant thrust. We can then let

$$dt = \frac{dm}{\dot{m}} \qquad \text{and} \qquad -\dot{m} = \frac{m_0 - m_{b0}}{t_{b0}} = \text{constant}$$

so that

$$\int_0^{t_{b0}} (\ln m)\, dt = \frac{1}{\dot{m}} \int_{m_0}^{m_{b0}} (\ln m)\, dm = \frac{m}{\dot{m}} (\ln m - 1) \Big|_{m_0}^{m_{b0}}$$

$$= \frac{m_0 - m_{b0}}{\dot{m}} - \frac{m_0}{\dot{m}} \ln m_0 + \frac{m_{b0}}{\dot{m}} \ln m_{b0}$$

The distance traveled then becomes,

$$h_{b0} = u t_{b0} \left[1 - \frac{1}{(m_0/m_{b0}) - 1} \ln \frac{m_0}{m_{b0}} \right] + v_0 t_{b0} - \frac{1}{2} g_0 t_{b0}{}^2$$

$$= g_0 \frac{I^2}{\mathscr{R}} \left(1 - \frac{m_{b0}}{m_0} \right) \left[1 - \frac{1}{(m_0/m_{b0}) - 1} \ln \frac{m_0}{m_{b0}} \right] + v_0 \frac{I}{\mathscr{R}} \left(1 - \frac{m_{b0}}{m_0} \right)$$

$$- \frac{1}{2} g_0 \frac{I^2}{\mathscr{R}^2} \left(1 - \frac{m_{b0}}{m_0} \right)^2 \tag{8.1-11}$$

After burnout, the rocket is in free flight under the retarding force of gravity. In general, the altitude is sufficiently great so that the variation in g must now be taken into account. Since the system during coasting is conservative, we can equate the kinetic energy at burnout to the work done by the gravity force $m_{b0} g_0 \left(\dfrac{R}{r} \right)^2$.

$$g_0 \int_{r_{b0}}^{r_{b0} + h_c} \left(\frac{R}{r} \right)^2 dr = \frac{v_{b0}{}^2}{2}$$

Thus the equation for the coasting distance becomes,

$$h_c = \frac{v_{b0}{}^2}{2g} \frac{(R + h_{b0})^2}{R^2 - (v_{b0}{}^2/2g)(R + h_{b0})} \tag{8.1-12}$$

where $r_{b0} = R + h_{b0}$ has been substituted. The total height $h_{b0} + h_c$ reached by a single stage rocket is then the sum of Eqs. 8.1–11 and 8.1–12.

Equations 8.1–10 and 8.1–11 indicate that the performance of a single-stage rocket depends on the specific impulse I, the thrust ratio $\mathscr{R}$, and the mass ratio $\mu = \dfrac{m_0}{m_{b0}}$. The effect of varying these quantities on the burnout velocity or height can be found by considering Eqs. 8.1–10 and 8.1–11 to be in the form,

$$v_{b0} = f_1(I, \mathscr{R}, \mu)$$
$$h_{b0} = f_2(I, \mathscr{R}, \mu) \tag{8.1-13}$$

and differentiating. Thus the change in the burnout velocity is determined from the equation

$$dv_{bo} = \frac{\partial f_1}{\partial I} \, dI + \frac{\partial f_1}{\partial \mathscr{R}} \, d\mathscr{R} + \frac{\partial f_1}{\partial \mu} \, d\mu \qquad (8.1\text{--}14)$$

For optimum burnout velocity, $dv_{bo} = 0$, which defines the constraints imposed on the three quantities.

PROBLEMS

1. For a given mass ratio μ and specific impulse I, how does the burnout velocity of a single-stage rocket vary with the thrust ratio $\mathscr{R}$. Assume vertical flight.

2. Plot $v_{bo}/g_0 I$ versus $\mu = m_0/m_{bo}$, with $v_0 = 0$ and $\mathscr{R}$ as parameter. Use $\mathscr{R} = 1, 2, 5$.

3. For a given specific impulse and thrust ratio, plot h_{bo} versus $\mu = m_0/m_{bo}$. Use $\mathscr{R} = 2$ and $I = 150, 300$ and 400 sec.

4. Determine the burnout speed of a rocket launched vertically, using a fuel of specific impulse 250 sec and a mass fraction of 0.22 with $\mathscr{R} = 3$.

5. For $I = 300$ sec and $\mathscr{R} = 2$, determine the maximum height attained by a single-stage rocket of mass ratio $\mu = 5$.

6. Repeat Prob. 5 for $\mu = 3$ and $\mu = 10$, and plot h_{max} versus μ.

7. Determine the partial derivatives $\partial f_1/\partial I$, $\partial f_1/\partial \mathscr{R}$, and $\partial f_1/\partial \mu$ of Eq. 8.1–14. How much would the burnout velocity of Prob. 5 be changed by changing I to 250 sec; by changing μ to 6.0.

8. Determine the partial derivatives $\partial f_2/\partial I$, $\partial f_2/\partial \mathscr{R}$, and $\partial f_2/\partial \mu$ from Eq. 8.1–11 and discuss the effect of changing I, $\mathscr{R}$, or μ.

9. If the burnout velocity of a rocket fired vertically is 8500 ft/sec at a height of h miles, how high will it rise when constant gravitational acceleration is assumed?

10. Assuming a burnout velocity of v_{bo} at $r = R + h$, and an inverse square attractive force, determine the maximum height reached by a rocket. What is the maximum height for data of Prob. 9 under this assumption?

11. If the specific impulse of a rocket engine is doubled by doubling the burning time, keeping the thrust per pound of fuel constant, how does this affect the burnout velocity?

12. If the mass ratio of a rocket is doubled, keeping all other variables constant, how does this affect the burnout height for vertical flight?

13. A rocket fired vertically from rest has an initial weight of 10,000 lb and a burnout weight of 2000 lb. The flux velocity of the fuel is a constant and equal to 7500 ft/sec, and the total burning time is 55 sec. Determine the velocity and acceleration just before burnout, and calculate the height to which it will rise by equating the kinetic energy to the work done under (a) varying gravity: (b) constant gravity.

14. The efficiency of a rocket engine can be defined as the ratio of the useful power Tv (T = thrust) to the useful power plus the kinetic energy $\frac{1}{2}m'(u - v)^2$ lost to the surroundings. Show that the rocket efficiency is given by the equation,

$$\eta = \frac{2(u/v)}{1 + (u/v)^2}$$

and find the value of u/v corresponding to its maximum.

15. Repeat Prob. 14 for a jet engine.

8.2 Optimization of Multistage Rockets[2]

A simple calculation with achievable mass ratio, thrust ratio, and specific impulse indicates that satellite velocities cannot be attained by the use of a single-stage rocket. We are thus led to the multistage rocket for space missions.

In a multistage rocket the burnout velocity of the first stage becomes the initial velocity v_0 of the second stage, and, by casting off the empty first stage, the full burnout velocity of the second stage is available as an additional velocity to the burnout velocity of the first stage. The maximum velocity of a multistage rocket can then be computed as the sum of the single-stage velocities, as given by Eq. 8.1–6.

We will ignore the gravity loss in velocity due to the burning time, in which case the maximum velocity available to the multistage rocket of N stages becomes,

$$v_m = \sum_{i=1}^{N} u_i \ln \mu_i \qquad (8.2\text{--}1)$$

where $\mu_i = (m_0/m_{b0})_i$ is the mass ratio of the ith stage. Assuming that this velocity is specified, we have a choice as to how the mass ratios should be assigned to the various stages. The problem is that of minimizing the over-all mass ratio m_{01}/P, where m_{01} is the takeoff mass and P the final payload mass.

To determine the mass ratio μ_i which will lead to a minimum over-all mass ratio m_{01}/P for a specified maximum velocity v_m, it is necessary to express m_{01}/P in terms of all the μ_i. Since at each burnout the empty structure of the stage is to be discarded, the initial mass of the new stage is equal to the initial mass of the previous stage minus the fuel burned and the empty structure thrown off. For example, the initial mass of the second stage is $m_{02} = m_{01} - m_{p1} - m_{s1}$, where m_{p1} is the propellant mass of stage 1 and m_{s1} the empty structural mass of stage 1. Thus by writing m_{01}/P in the form

$$\frac{m_{01}}{P} = \frac{m_{01}}{m_{01} - m_{p1} - m_{s1}} \frac{m_{02}}{m_{02} - m_{p2} - m_{s2}} \cdots \frac{m_{0N}}{P} \qquad (8.2\text{--}2)$$

it is possible to express the over-all mass ratio in terms of all the mass ratios

$$\mu_i = \frac{m_{0i}}{m_{0i} - m_{pi}} \tag{8.2-3}$$

and an additional structural factor β_i defined as

$$\beta_i = \frac{m_{si}}{m_{pi} + m_{si}} \tag{8.2-4}$$

Examining one of the factors, we can write

$$\frac{m_{0i}}{m_{0i} - m_{pi} - m_{si}} = \frac{m_{0i}}{m_{0i} - m_{pi}} \frac{m_{pi}}{m_{pi} + m_{si}} \frac{(m_{0i} - m_{pi})(m_{pi} + m_{si})}{m_{pi}(m_{0i} - m_{pi} - m_{si})}$$

$$= \frac{\mu_i(1 - \beta_i)}{1 - \mu_i \beta_i} \tag{8.2-5}$$

Equation 8.2-2 can then be written as

$$\frac{m_{01}}{P} = \frac{\mu_1(1 - \beta_1)}{1 - \mu_1 \beta_1} \frac{\mu_2(1 - \beta_2)}{1 - \mu_2 \beta_2} \cdots \frac{\mu_N(1 - \beta_N)}{1 - \mu_N \beta_N} \tag{8.2-6}$$

If m_{01}/P is to be a minimum, $\ln(m_{01}/P)$, will also be a minimum, so that we can write,

$$\ln \frac{m_{01}}{P} = \sum_{i=1}^{N} \ln \frac{\mu_i(1 - \beta_i)}{(1 - \mu_i \beta_i)} = \sum_{i=1}^{N} [\ln \mu_i + \ln(1 - \beta_i) - \ln(1 - \mu_i \beta_i)] \tag{8.2-7}$$

The above equation by itself does not contain the constraint imposed by the specified velocity as given by Eq. 8.2-1. This constraint can be imposed on the optimization process by the Lagrange multiplier method, which requires the constraint equation to be multiplied by a constant λ and added to the above equation (adding zero), as follows:

$$\ln \frac{m_{01}}{P} = \sum_{i=1}^{N} \{\ln \mu_i + \ln(1 - \beta_i) - \ln(1 - \mu_i \beta_i) + \lambda[u_i \ln \mu_i - v_m]\} \tag{8.2-8}$$

Differentiation of this equation with respect to μ_i will lead to the optimum values of μ_i.

Carrying out the differentiation, we obtain N equations of the form

$$\frac{1}{\mu_i} + \frac{\beta_i}{1 - \mu_i \beta_i} + \lambda \frac{u_i}{\mu_i} = 0$$

leading to the result

$$\mu_i = \frac{1 + \lambda u_i}{\lambda u_i \beta_i} \tag{8.2-9}$$

Substituting this value of μ_i into Eq. 8.2–1, the constant λ can be found from the equation

$$v_m = \sum_{i=1}^{N} u_i \ln \frac{1 + \lambda u_i}{\lambda u_i \beta_i} \tag{8.2–10}$$

where u_i, β_i, and v_m are assumed to be known. With λ evaluated, the mass ratio μ_i of each stage is found from Eq. 8.2–9.

Example 8.2–1

Consider a special case where the specific impulse is the same for all stages. The u_i are then equal in all stages and Eq. 8.2–10 becomes

$$\frac{v_m}{u} = N \ln \left(\frac{1 + \lambda u}{\lambda u} \right) - \sum_{i=1}^{N} \ln \beta_i$$

$$\frac{1 + \lambda u}{\lambda u} = \exp \left[\frac{1}{N} \left(\frac{v_m}{u} + \sum_{i=1}^{N} \ln \beta_i \right) \right]$$

Since from Eq. 8.2–9

$$\mu_i \beta_i = \frac{1 + \lambda u}{\lambda u}$$

the mass ratio of stage i is

$$\mu_i = \frac{1}{\beta_i} \exp \left[\frac{1}{N} \left(\frac{v_m}{u} + \sum_{i=1}^{N} \ln \beta_i \right) \right]$$

PROBLEMS

1. Show that if u_i and β_i are the same for each stage, the optimum mass ratio is

$$\mu_i = e^{v_m/Nu}$$

2. A two-stage rocket is to attain a maximum speed of 26,000 ft/sec with $I_1 = I_2 = 300$ sec and $\beta_1 = \beta_2$. Determine the mass ratio of each stage.

Ans. $\mu = 3.85$.

3. In Prob. 2, determine the propellant mass per stage in terms of the initial mass of the stage. Also determine the structural factor β_i, assuming $m_{si} = 0.15 m_{0i}$, and show that the optimum over-all mass ratio is equal to $m_{01}/P = 82.4$.

4. In designing a two-stage rocket for a maximum speed of 26,000 ft/sec, assume that $I_1 = I_2 = 250$ sec, and $\beta_1 = 0.18$, $\beta_2 = 0.15$. Show that it is capable of boosting a payload of $0.00172 m_{01}$.

8.3 Flight Trajectory Optimization[3]

We will now consider a more difficult problem of establishing an optimum flight path to place a satellite into orbit. The rocket is assumed to be rigid, and we will neglect the aerodynamic forces. Generally, the length of the powered flight path is short compared to the earth's radius, and we are justified in replacing the central gravitational field by a constant-plane parallel-force field.

The geometry of the problem, illustrated in Fig. 8.3–1, shows the thrust attitude angle ϕ measured from the horizontal, the gravitational force along the negative y direction, and the velocity vector tangent to the flight

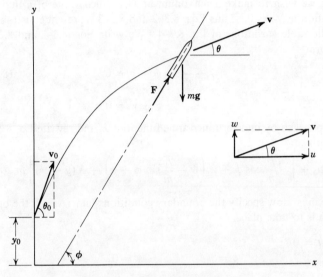

Fig. 8.3–1. Powered flight trajectory.

trajectory. Letting u and w be the x and y components of the velocity $\mathbf{v}$, the differential equation of motion in rectangular coordinates are,

$$\dot{u} = \frac{F}{m} \cos \phi \tag{8.3-1}$$

$$\dot{w} = \frac{F}{m} \sin \phi - g \tag{8.3-2}$$

$$\dot{y} = w \tag{8.3-3}$$

The quantities F/m and ϕ are functions of time, and the trajectory depends on how they vary. We will assume that F/m is a known function of time, and define the problem of selecting $\phi(t)$ for maximum horizontal velocity U at a specified altitude Y. The time T corresponding to the instant $y = Y$ will differ with different $\phi(t)$ and, therefore, will not be specified. We will assume however that all of the fuel be burned prior to the time T.

The problem which we have just defined differs from the previous one of vertical flight in that we must maximize a time integral with constraints. The equation which we are concerned with is the integral of Eq. 8.3–1,

$$U = u_0 + \int_0^T \frac{F}{m} \cos \phi \, dt \tag{8.3-4}$$

which we wish to make a maximum at a specified value of y under the condition $w_{t=T} = 0$. Thus Eqs. 8.3–2 and 8.3–3 represent constraints on the allowable variations of Eq. 8.3–4. We write Eqs. 8.3–2 and 8.3–3 in the form

$$\dot{w} - \frac{F}{m} \sin \phi + g = 0 \tag{8.3-2a}$$

$$\dot{y} - w = 0 \tag{8.3-3a}$$

multiply each by undetermined time functions λ, and rewrite Eq. 8.3–4 in the form

$$U = u_0 + \int_0^T \left[\frac{F}{m} \cos \phi + \lambda_1 \left(\dot{w} - \frac{F}{m} \sin \phi + g \right) + \lambda_2 (\dot{y} - w) \right] dt \tag{8.3-5}$$

We must now specify the boundary conditions under which the optimization is to take place.

At $t = 0$

$$y = y_0 \qquad\qquad x = 0$$

$$u_0 = v_0 \cos \theta_0 \qquad w_0 = v_0 \sin \theta_0$$

At $t = T$

$$y = Y \qquad\qquad w = 0$$

The initial velocity v_0 at initial altitude y_0 will be considered fixed, but the initial angle θ_0 will be left undetermined, thereby allowing the first term of Eq. 8.3–5 to contribute a term at the lower limit $t = 0$.

Before applying Eq. C-9 of Appendix C to Eq. 8.3–5, we will dispense with the first term $u_0 = v_0 \cos \theta_0$ by noting that its contribution to δU is $-v_0 \sin \theta_0 \, \delta \theta_0$. We are then left with the integral of Eq. 8.3–5, where in place of the variable z, we have ϕ, w, y, F/m, and T. The equation with which we are concerned is, then,

$$\delta U = -v_0 \sin \theta_0 \, \delta \theta + \delta \int_0^T f\left(\phi, w, \dot{w}, \dot{y}, \frac{F}{m} \right) dt \tag{8.3-6}$$

where the variation of the integral is to be determined from Eq. C-9.

For the first problem, F/m is assumed to be a known function of time which is not varied, and we are to find a $\phi(t)$ which will result in a maximum horizontal velocity U at a given height Y. The partial derivatives indicated in Eq. C-9 are determined from Eq. 8.3–5 to be

$$\frac{\partial f}{\partial \phi} = -\frac{F}{m}(\sin \phi + \lambda_1 \cos \phi) \qquad \frac{\partial f}{\partial \dot{\phi}} = 0$$

$$\frac{\partial f}{\partial w} = -\lambda_2 \qquad \frac{\partial f}{\partial \dot{w}} = \lambda_1$$

$$\frac{\partial f}{\partial y} = 0 \qquad \frac{\partial f}{\partial \dot{y}} = \lambda_2$$

The total variation from Eq. C-9 is then

$$\delta U = -v_0 \sin \theta_0 \, \delta\theta_0 + \lambda_1 \, \delta w \Big|_0^T + \lambda_2 \, \delta y \Big|_0^T$$

$$- \int_0^T \left[\frac{F}{m}(\sin \phi + \lambda_1 \cos \phi)\, \delta\phi + \frac{d\lambda_2}{dt}\, \delta y + \left(\frac{d\lambda_1}{dt} + \lambda_2\right)\delta w \right] dt = 0$$

$$(8.3–7)$$

Since y at $t = 0$ and $t = T$ are fixed as y_0 and Y, the variation at the end points must be zero, which eliminates the third term in Eq. 8.3–7. For the second term of Eq. 8.3–7, the variation δw at $t = 0$ is (v_0 is specified as fixed),

$$\delta w \Big|_0 = \delta(v_0 \sin \theta_0) = v_0 \cos \theta_0 \, \delta\theta_0$$

At the terminal time $t = T$, w is obtained from the integral of Eq. 8.3–2a to be

$$w = \int_0^T \frac{F}{m} \sin \phi \, dt - gT + v_0 \sin \theta_0$$

and its variation at $t = T$ due to variations $\delta\phi$ and $\delta\theta_0$ is

$$\delta w \Big|^T = \int_0^T \frac{F}{m} \cos \phi \, \delta\phi \, dt + v_0 \, \delta\theta_0 \cos \theta_0$$

We note that w for the optimum curve becomes zero at $t = T$, whereas for the varied curve w becomes zero at a different time $T + \delta T$. Since $\frac{F}{m}$ is not varied and is zero for $t > T$, the above expression for $\delta w \Big|^T$ must equal $g \, \delta T$.

$$\delta U = -v_0(\sin \theta_0 + \lambda_{1,t=0} \cos \theta_0)\, \delta\theta_0 + g\lambda_{1,t=T}\, \delta T$$

$$- \int_0^T \left[\frac{F}{m}(\sin \phi + \lambda_1 \cos \phi)\, \delta\phi + \frac{d\lambda_2}{dt}\, \delta y + \left(\frac{d\lambda_1}{dt} + \lambda_2\right)\delta w \right] dt = 0$$

$$(8.3–8)$$

Since the variations $\delta\theta_0$, δT, $\delta\phi$, etc., are arbitrary, δU can be zero only if all their coefficients are zero, or

$$\tan\theta_0 = -\lambda_{1,t=0} \qquad \frac{d\lambda_2}{dt} = 0$$

$$\lambda_{1,t=T} = 0$$

$$\tan\phi = -\lambda_1 \qquad \frac{d\lambda_1}{dt} + \lambda_2 = 0$$

From the last two equations we, obtain

$$\lambda_2 = -C_2 \qquad \text{(a constant)}$$

$$\frac{d\lambda_1}{dt} - C_2 = 0$$

$$\lambda_1 = C_1 + C_2 t$$

Substituting for λ_1 at $t = 0$, and $t = T$ from the first two equations, we find

$$C_1 = -\tan\theta_0$$

$$C_2 = \frac{1}{T}\tan\theta_0$$

so that

$$\lambda_1 = -\left(1 - \frac{t}{T}\right)\tan\theta_0$$

The equation for the thrust attitude is then obtained from the third equation to be

$$\tan\phi = \left(1 - \frac{t}{T}\right)\tan\theta_0 \tag{8.3–9}$$

and ϕ varies from θ_0 to zero according to Eq. 8.3–9.

8.4 Optimum Program for Propellant Utilization

In the problem just discussed, it was found that, for an arbitrarily defined time variation of F/m, the optimum program for the thrust attitude, where the initial angle θ_0 was also allowed a variation, was found to be

$$\tan\phi = \left(1 - \frac{t}{T}\right)\tan\theta_0 \tag{8.4–1}$$

If the initial angle θ_0 is a given quantity which is not allowed a variation, the optimum thrust attitude will still be a linear function of time but of the form (see Prob. 8.4–1),

$$\tan\phi = \tan\phi_0 - C_2 t \tag{8.4–2}$$

In either case, the above thrust attitude programs result in a maximum horizontal velocity U for the specified program of F/m. The maximum U attained will, however, differ with different programs of F/m, or the manner of propellant utilization, and we wish now to establish the optimum program for propellant utilization to obtain the largest of the maximum U attained under optimum thrust attitude variation.

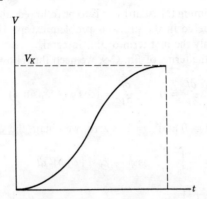

Fig. 8.4–1. Possible variation of V.

We recognize here that the theoretically attainable velocity V_k in the absence of all forces other than thrust, is essentially a function of the specific impulse I of the fuel and the mass ratio of the rocket (see Sec. 8.1)

$$V_k = gI \ln \frac{m_0}{m_{b0}}$$

where it is assumed that the specific impulse is independent of the manner in which the fuel is utilized. It is then convenient to make the following substitution

$$\frac{dV}{dt} = \frac{F}{m}$$

$$V = \int_0^t \frac{F}{m} \, dt \tag{8.4–3}$$

where V is the rocket thrust velocity with the maximum attainable value equal to V_k. It is evident that V is a function only of time, and depends on the manner in which the fuel is consumed. Thus for any program of propellant utilization, any curve between $V = 0$ and V_k with positive slope (including zero), as shown in Fig. 8.4–1, is a possible curve. Since F and m are both positive quantities, the slope dV/dt is bounded between 0 (for $F = 0$) to infinity (for instantaneous burning).

We will assume the initial value of v_0 and θ_0 to be fixed and, that the optimum thrust attitude variation indicated by Eq. 8.4–2 is to be followed. The integral to be maximized is from Eq. 8.3–5,

$$U = \int_0^T \left[(\cos \phi - \lambda_1 \sin \phi) \frac{dV}{dt} + \lambda_1(\dot{w} + g) + \lambda_2(\dot{y} - w) \right] dt \quad (8.4–4)$$

and since the optimum thrust attitude is to be followed, all of the variations have been considered in the previous problem except δV, which requires us to consider only the first term of this integral.

To determine the terms of Eq. C-9, we need the following:

$$\frac{\partial f}{\partial V} = 0 \qquad \frac{\partial f}{\partial \dot{V}} = (\cos \phi - \lambda_1 \sin \phi)$$

and since δV at $t = 0$ and $t = T$ is zero, we obtain the variation

$$\delta U = -\int_0^T \frac{d}{dt} (\cos \phi - \lambda_1 \sin \phi) \, \delta V \, dt$$

$$= \int_0^T \left[(\sin \phi + \lambda_1 \cos \phi)\dot{\phi} + \sin \phi \frac{d\lambda_1}{dt} \right] \delta V \, dt \quad (8.4–5)$$

From Eqs. 8.4–5 and 8.4–2 we have

$$\lambda_1 = -\tan \phi = -\tan \phi_0 + C_2 t$$

$$\frac{d\lambda_1}{dt} = -\frac{\dot{\phi}}{\cos^2 \phi} = C_2$$

from which

$$\delta U = C_2 \int_0^T (\sin \phi) \, \delta V \, dt$$

$$= C_2 \int_0^T \frac{\tan \phi_0 - C_2 t}{\sqrt{1 + (\tan \phi_0 - C_2 t)^2}} \, \delta V \, dt$$

$$= C_2 \int_0^T \Phi \, \delta V \, dt = 0 \quad (8.4–6)$$

It is evident here that the integrand

$$\Phi = \frac{\tan \phi_0 - C_2 t}{\sqrt{1 + (\tan \phi_0 - C_2 t)^2}} \quad (8.4–7)$$

can be positive, negative, or vary from plus to minus as t varies from 0 to T, so that, for the integral δU to be zero, the function $V(t)$ must be a discontinuous function of time (i.e., the fuel must be burned instantaneously). We will consider three possible cases:

1. If Φ is always positive, all of the propellant must be burned instantaneously at the beginning, and the optimum program for V is the heavy discontinuous curve of Fig. 8.4–2. For $V(t)$ to be optimum, δU

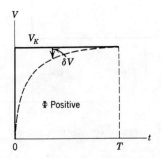

Fig. 8.4–2. Instantaneous burning at $t = 0$.

must be negative for any variation of V from the optimum. For instance, any variation δV from the optimum curve of Fig. 8.4–2, such as the dotted curve, must be negative, which means that δU would be negative. A positive variation δV is not possible from the optimum curve shown, since V_k is the maximum available rocket thrust velocity fixed by the specific

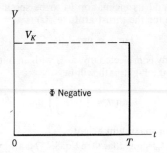

Fig. 8.4–3. Instantaneous burning at $t = T$.

impulse and the mass ratio. These arguments, therefore, establish the optimum program for the propellant utilization corresponding to positive Φ to be that of instantaneous burning at $t = 0$.

2. If Φ is negative, the optimum curve for $V(t)$ must be as shown in Fig. 8.4–3, which implies instantaneous burning at $t = T$. Any variation δV from this curve would be positive, which would result in a negative δU or a smaller value of U.

3. If Φ changes sign from plus to minus, the optimum curve for V must appear as in Fig. 8.4–4, and part of the fuel must be burned instantaneously

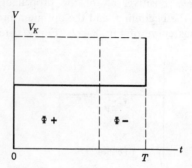

Fig. 8.4–4. Instantaneous burning at $t = 0$ and $t = T$.

at $t = 0$, and the remainder at $t = T$. Again, any varied curve which has zero or positive slope is allowable, and if a positive δU cannot be produced, the discontinuous curve of Fig. 8.4–4 must stand as the optimum curve.

PROBLEMS

1. For a specified program of fuel utilization, F/m is a known function of time which allows no variation in the quantity F/m. Assume now that both the initial velocity v_0 and its inclination θ_0 to be specified, and show that the optimum program for the thrust attitude $\phi(t)$ is

$$\tan \phi = \tan \phi_0 - C_2 t$$

2. Show that, if a body is projected upwards with an initial velocity at an angle θ_0, the tangent to the flight path will be

$$\tan \theta = \left(1 - \frac{t}{T}\right) \tan \theta_0$$

   where $T =$ time for maximum height.

3. For the case $\phi_0 = \theta_0$, show that Φ in Eq. 8.4–7 is always positive, and that for optimum utilization of the propellant all of the fuel must be burned instantaneously at the beginning.

4. Assuming that v_0 is zero, and all of the fuel is burned at the initial instant, determine the time T at which maximum horizontal velocity is attained at height Y.

5. Determine the time required to deliver a satellite to a height $Y = 300$ miles with a horizontal velocity of $V = 22,000$ ft/sec.

6. In Prob. 4, show that the initial thrust attitude should be $\sin \phi_0 = (1/V_k) \times \sqrt{2gY}$, and that V_k must be greater than the velocity necessary for projecting a body vertically upwards to a height Y.

7. In Prob. 5, determine the initial angle ϕ_0 and the total rocket velocity V_k.

8. For the case where the variation of Φ is from positive to negative, discuss how you would find V_1 or the portion of the fuel to be burned at the initial instant.

9. Determine the optimum fuel utilization program to achieve maximum height for a rocket shot vertically.

10. Determine the optimum fuel utilization program for maximum height of a two-stage rocket of equal stages.

11. Discuss the problem of obtaining the greatest altitude for a specified horizontal velocity.

12. Discuss the problem of minimum propellant consumption for a specified horizontal velocity and altitude.

8.5 Gravity Turn[1]

In the previous section it was shown that the optimum thrust attitude for placing a satellite into orbit is given by the equation,

$$\tan \phi = \left(1 - \frac{t}{T}\right) \tan \theta_0 \qquad (8.5\text{-}1')$$

Likewise, the optimum thrust attitude for maximum range (ballistic missile) can be shown to be $\phi = $ constant. These conditions may be satisfactory for a rocket traveling in vacuum but, owing to the large angle of attack $(\phi - \theta)$ which results from such trajectories, they are not feasible through the atmosphere. Thus for flight through the atmosphere, a trajectory known as *gravity turn* is generally used. In a gravity turn the thrust vector is kept parallel to the velocity vector at all times, starting with some nonvertical initial velocity v_0. Thus the gravity turn is also one of zero angle of attack or zero lift.

It is convenient here to measure the angle made by the velocity vector from the vertical, as shown in Fig. 8.5-1. Again assuming zero aerodynamic drag and constant gravity field, we write equations for the forces in the tangential and normal directions to the trajectory.

$$\frac{1}{g}\frac{dv}{dt} = \frac{F}{mg} - \cos \psi \qquad (8.5\text{-}1)$$

$$\frac{v}{g}\frac{d\psi}{dt} = \sin \psi \qquad (8.5\text{-}2)$$

These equations are nonlinear and no analytical solution is known when F/mg varies with time. A reasonable assumption for F/mg is $F = $ constant, and $m = m_0 - m't$.

When F/mg is a constant, these nonlinear equations can be solved analytically. For F/mg to be constant, the thrust F must decrease with time to conform to the decreasing mass. If F/mg is a varying function of time, we can assume F/mg to be constant over short intervals of time, and we can carry out a step-by-step numerical integration, using the analytical solution over each interval. It is evident, then, that the analytical solution

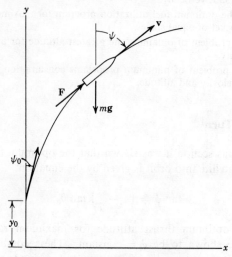

Fig. 8.5–1. Gravity turn trajectory.

for constant F/mg is of practical interest, and we consider its development as follows:

Let $F/mg = n$ over a short increment of the flight path. We introduce

$$z = \tan \frac{1}{2} \psi = \sqrt{\frac{1 - \cos \psi}{1 + \cos \psi}} = \frac{\sin \psi}{1 + \cos \psi} \qquad (8.5\text{–}3)$$

Then

$$z^2 = \frac{1 - \cos \psi}{1 + \cos \psi}$$

and

$$\frac{1 - z^2}{1 + z^2} = \cos \psi$$

Differentiating Eq. 8.5–3,

$$\frac{dz}{dt} = \frac{1}{2 \cos^2 (\psi/2)} \frac{d\psi}{dt}$$

Therefore,

$$\frac{d\psi}{dt} = (1 + \cos \psi) \frac{dz}{dt}$$

By substitution into Eqs. 8.5–1 and 8.5–2, they become,

$$\frac{1}{g}\frac{dv}{dt} = n - \frac{1 - z^2}{1 + z^2} \tag{8.5-4}$$

$$\frac{v}{g}\frac{dz}{dt} = z \tag{8.5-5}$$

Eliminating dt between the above equations,

$$\frac{dv}{v} = n\frac{dz}{z} - \frac{1 - z^2}{1 + z^2}\frac{dz}{z} \tag{8.5-6}$$

Integrating, we have (see Peirce* nos. 53 and 55),

$$\ln v = \ln z^n + \ln \frac{1 + z^2}{z} + \ln C'$$

or

$$v = Cz^{n-1}(1 + z^2) \tag{8.5-7}$$

The constant C can be evaluated from initial conditions to be,

$$C = \frac{v_0}{z_0^{n-1}(1 + z_0^2)} \tag{8.5-8}$$

Substituting Eq. 8.5–7 into 8.5–5 and integrating,

$$t = \frac{C}{g}\int_{z_0}^{z} z^{n-2}(1 + z^2)\, dz$$

$$= \frac{C}{g} z^{n-1}\left(\frac{1}{n-1} + \frac{z^2}{n+1}\right)\Bigg|_{z_0}^{z} \tag{8.5-9}$$

Equation 8.5–9 gives t as a function of $z = \tan \frac{1}{2}\psi$ for any initial condition C. Thus, conversely, ψ is known in terms of t and C. Equation 8.5–7 gives v as a function of z and C, so that v is also known in terms of t and C. Thus Eqs. 8.5–7, 8.5–8, and 8.5–9 represent the solution for the gravity turn trajectory when the thrust-to-weight ratio F/mg is a constant n.

To apply these equations for a varying F/mg, we start with the initial conditions expressed by Eqs. 8.5–3 and 8.5–8, and $F/mg = n$ at $t = 0$. Choosing a value of ψ slightly greater than ψ_0, determine v from Eq. 8.5–7 and Δt from Eq. 8.5–9. The increment in the displacement is, then,

$$\Delta x = \tfrac{1}{2}(v_0 \sin \psi_0 + v \sin \psi)\, \Delta t$$
$$\Delta y = \tfrac{1}{2}(v_0 \cos \psi_0 + v \cos \psi)\, \Delta t$$

The procedure can now be repeated with the values at the new point as initial condition.

---

* Short Table of Integrals (3rd rev. ed), Ginn & Co., Boston (1929).

PROBLEMS

1. By reversing the force F, show that the motion of a body hurled into a resisting medium can be solved by the techniques of this section.
2. If F/mg is constant throughout flight, what is the effect of changing the initial angle ψ_0 on; (a) the final velocity; (b) the time of flight.
3. For constant value of F/mg, the velocity and its inclination ψ were observed at a given time. If the initial angle is doubled, how would the above quantities differ at the same time.
4. From Eqs. 8.5–1 and 8.5–2 derive the ballistic equation of a rocket

$$\frac{1}{v}\frac{dv}{d\psi} = \left(\frac{F - D}{mg}\right)\frac{1}{\sin \psi} - \frac{1}{\tan \psi}$$

where D is the aerodynamic drag.

Hint: The normal and tangential accelerations can be written as

$$v\dot{\psi} = v^2/R$$

$$\dot{v} = \frac{dv}{d\psi}\frac{d\psi}{ds}\frac{ds}{dt} = \frac{v}{R}\frac{dv}{d\psi}$$

where R is the radius of curvature of the trajectory, and $ds = R\,d\psi$.

REFERENCES

1. Culler, G. J., and B. D. Fried, "Universal Gravity Turn Trajectories," *J. Appl. Physics*, **28**, No. 6 (June 1957), 672–676
2. Hall, H. H., and E. D. Zambelli, "On the Optimization of Multistage Rockets," *Jet Propulsion*, **28** (July 1958), 463–465.
3. Okhotsimskii, D. E., and T. M. Eneev, "Some Variation Problems Connected with the Launching of Artificial Satellites of the Earth," *J. British Interplanetary Soc.*, **16** (5), (Jan.–Feb. 1958), 263–294.
4. Bellman, R. and Dreyfus, S., "An Application of Dynamic Programming of the Determination of Optimal Satellite Trajectories" *J. British Interplanetary Society*, Vol. 16 No. 3–4 (1959), p. 78–83.
5. Leitmann, G., "Optimization Techniques with Applications to Aerospace Systems" Academic Press, New York, 1962.

Generalized Theories
of Mechanics

CHAPTER 9

9.1 Introduction

Experience indicates that our learning process consists of first assimilating simple bits of information and, second, of comprehending the relationships between the various bits of information. As a result there begins to emerge an over-all pattern of behavior predictable from a theory. Simple theories are necessary for the beginner, in spite of the fact that they are limited in scope and incapable of extension beyond the bounds for which they are intended.

Beyond this stage must be a more general theory which encompasses and unites all special theories into a harmonious understanding. Such a generalized theory of mechanics was developed by Hamilton and Lagrange. It encompasses all of classical mechanics, and an understanding of this important work is an essential part of advanced dynamics.

Preliminary to the discussion of the generalized theories, it is necessary to have clearly in mind the basic concepts of coordinates and their classification.

9.2 System with Constraints

The degrees of freedom of a body correspond to the minimum number of independent coordinates required to define its position. For a particle free to move in space, three coordinates are necessary to define its position. They may be rectangular coordinates x, y, z, spherical coordinates r, θ, ϕ, or some other system of coordinates, but three are necessary, and each coordinate may be varied independently. We say then that the free particle has three degrees of freedom.

If, next, the particle is constrained to move on a specified surface, only two coordinates are necessary to define its position, and we say that it has two degrees of freedom. For instance, the latitude and longitude completely define a position of a particle on the earth's surface. If the particle is further constrained to move along a specified line on the surface, one coordinate—such as the distance along this line—will define its position, and such a particle will have one degree of freedom. Here we have placed two constraints on the particle, one to restrict it to a surface, and another to confine the motion along some line on the surface. In each case the three degrees of freedom of a free particle have been reduced by the number of constraints imposed on the particle.

The constraints of a system can be expressed analytically in terms of its geometry. For example, a particle a distance l from the fixed end of a string is constrained to move on a spherical surface, the equation of which is,

$$x^2 + y^2 + z^2 = l^2$$

In the general case, the equation of constraint restricting a particle to any surface is,

$$f(x, y, z) = 0 \tag{9.2-1}$$

When the particle is constrained to move along a curve in space, the curve can be considered to be the intersection of two surfaces, so that the two constraint equations to be satisfied are,

$$f_1(x, y, z) = 0$$
$$f_2(x, y, z) = 0 \tag{9.2-2}$$

An elementary example of two constraints is illustrated by the simple pendulum whose circular path in the vertical plane is the intersection of a sphere and a vertical plane through its center. The constraint equation, $x^2 + y^2 = l^2$, is actually the result of two equations,

Sphere $\qquad\qquad\qquad x^2 + y^2 + z^2 = l^2$

Plane $\qquad\qquad\qquad\qquad z = 0$

with the coordinates oriented as shown in Fig. 9.2-1.

Sometimes a constraint equation will also depend on time. For instance, if the support point of the simple pendulum is given a motion $x_0(t)$, the constraint equation would have to be written as

$$[x - x_0(t)]^2 + y^2 = l^2$$

It is evident, then, that for a time-dependent constraint, the equation may be written in the form,

$$C(x, y, z, t) = 0 \qquad (9.2\text{-}3)$$

Constraints may also be imposed between particles; e.g., two particles whose distance between them is always constant. The first particle has

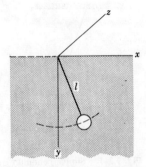

Fig. 9.2–1. Intersection of the x, y plane and a sphere of radius l defines the path of the simple pendulum mass.

three degrees of freedom. The second particle has two degrees of freedom. Thus we have $3 + 2 = 5$ degrees of freedom for the two particles bound to each other a given distance apart. Actually two free particles would have six degrees of freedom; however a constraint has been introduced, specifying the distance between them, so that the degrees of freedom have been reduced by one.

The position of a rigid body is known if we know the position of three noncollinear points on it. The three points, if free, would have nine degrees of freedom, but since there are three rigid constraints between them, a rigid body has six degrees of freedom. Another way of arriving at the same result is to choose the first point arbitrarily, in which case it will have three degrees of freedom. The second point must move in a sphere about the first, so it has two degrees of freedom. The third point must now move in a circular line about 1 and 2 as axis, which adds another degree of freedom, making a total of six degrees of freedom.

The general rule for N particles can be stated as follows. If N particles have c constraints restricting their freedom, the number of degrees of freedom n will be,

$$n = 3N - c \qquad (9.2\text{-}4)$$

9.3 Generalized Coordinates

A simple pendulum is defined as a point mass on the end of a weightless, inextensible string, which is made to move in a vertical plane, as shown in Fig. 9.3–1. The point mass must move on a circular line in the plane, and

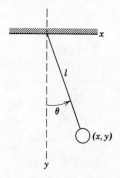

Fig. 9.3–1. Simple pendulum.

has one degree of freedom. The position of the mass may be specified by rectangular coordinates x, y, which are not independent but subject to the constraint equation,

$$x^2 + y^2 = l^2 \qquad (9.3\text{-}1)$$

It is simpler, however, to specify its position by the angle θ, which is independent and free of any constraint equation. Such independent coordinates are called *generalized coordinates*, and the number of such coordinates corresponds to the degrees of freedom of the system.

Consider next the double pendulum of Fig. 9.3–2, which is a system of two degrees of freedom. If the position of each mass is to be defined in terms of rectangular coordinates, four coordinates, x_1, y_1, x_2, y_2, would be necessary. There are however, two constraint equations between the coordinates,

$$x_1^2 + y_1^2 = l_1^2$$
$$(x_2 - x_1)^2 + (y_2 - y_1)^2 = l_2^2 \qquad (9.3\text{-}2)$$

and the number of coordinates minus the number of constraints again agree with the degrees of freedom of the system. Any two of the four rectangular coordinates can be considered independent, but the remaining two must be related by the above constraint equations.

The double pendulum can also be defined by two angles, θ_1 and θ_2. Each θ can be varied independently and, therefore, no constraints exist between them. θ_1 and θ_2 are thus generalized coordinates for the double pendulum.

For a system of n degrees of freedom, there are n generalized coordinates, $q_1, q_2, q_3, q_4, \ldots, q_n$. They are independent coordinates free from any

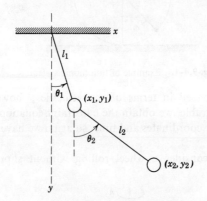

Fig. 9.3–2. Double pendulum.

constraints. They are not necessarily lengths or angles, but can be any independent set of quantities which describe completely the motion of the system.

It is always possible to relate the rectangular coordinates as some function of the generalized coordinates. For instance, in the case of the double pendulum, the rectangular coordinates expressed in terms of the generalized coordinates $q_1 = \theta_1$ and $q_2 = \theta_2$ are

$$
\begin{aligned}
x_1 &= l_1 \sin \theta_1 \\
y_1 &= l_1 \cos \theta_1 \\
x_2 &= l_1 \sin \theta_1 + l_2 \sin \theta_2 \\
y_2 &= l_1 \cos \theta_1 + l_2 \cos \theta_2
\end{aligned}
\tag{9.3-3}
$$

In the more general case, the relationships between the various position coordinates x, y, z and the generalized coordinates can be expressed by the functional equation,

$$
x_i = f_i(q_1, q_2, q_3, \ldots, q_n, t)
\tag{9.3-4}
$$

where n is the number of degrees of freedom of the system.

9.4 Holonomic and Nonholonomic System

When the constraints are expressible as functions of the coordinates or coordinates and time, the system is said to be holonomic. Sometimes the

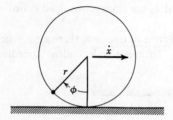

Fig. 9.4–1. Example of holomonic system.

constraints are expressed in terms of the velocities; however, if such expressions are integrable, we obtain the constraint equations as function of the coordinates or coordinates and time, so again we have a holonomic system.

As an example, consider a wheel rolling without slipping along a

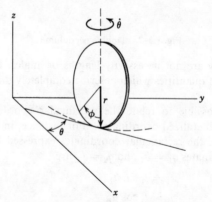

Fig. 9.4–2. Example of nonholonomic system.

specified straight line, as shown in Fig. 9.4–1. The velocity of the center is $\dot{x} = r\dot{\phi}$, which can be integrated to $x = r\phi + c$. The system is hence holonomic.

If the same wheel rolls without slipping on a plane, and is allowed to pivot about a vertical axis through the point of contact, as shown in Fig. 9.4–2, the relationship between the velocities will be found to be non-integrable, and hence the system must be classified as nonholonomic. We have, for instance, the equations for the velocity normal and parallel to the path as,

$$\dot{x} \sin \theta - \dot{y} \cos \theta = 0$$
$$\dot{x} \cos \theta + \dot{y} \sin \theta = r\dot{\phi}$$

and it is not possible to integrate these expressions to obtain relationships between the coordinates. It is possible, for instance, to roll the wheel to another point with a different value of x, but with the other coordinates y, θ, and ϕ unchanged. It is evident, then, that x cannot be functionally related to the remaining coordinates. Each of the other coordinates can be singled out in the same manner, so that there can be no unique relationship existing between them. We have only a relationship between the infinitesimals which can be written as

$$\sin \theta \, dx - \cos \theta \, dy = 0$$
$$\cos \theta \, dx + \sin \theta \, dy - r \, d\phi = 0$$

In a holonomic system, the constraint equations are in the form,

$$C(r_1, r_2, r_3, \ldots, t) = 0 \qquad (9.4\text{--}1)$$

Moreover, the differential of the constraint equation is exact and expressible as,

$$dC = \frac{\partial C}{\partial r_1} dr_1 + \frac{\partial C}{\partial r_2} dr_2 + \cdots \frac{\partial C}{\partial t} dt = 0 \qquad (9.4\text{--}2)$$

It is also possible in the holonomic system to reduce the number of the dependent variables r_i by the number of constraint equations to the number of the degrees of freedom of the system. These are then expressible in terms of the n generalized coordinates, equal to the number of degrees of freedom of the system, and independent of each other. Thus, provided we are able to find the n generalized coordinates of the system, the problem reduces to that of solving n independent equations in q_i, without concern of the constraints of the system.

In the nonholonomic system, the constraints are not expressible in terms of the coordinates or coordinates and time, as in Eq. 9.4–1. The constraint equations are available only as relationships between the infinitesimals,

$$a_1 \, dr_1 + a_2 \, dr_2 + \cdots + a_m \, dt = 0 \qquad (9.4\text{--}3)$$

which are nonintegrable. Thus, if we transform the dependent variables r_i into the equation

$$r_i = r_i(q_1, q_2, \ldots, t)$$

we would find that not all the q are independent. Special procedures to be used under these conditions are discussed in Sec. 9.9.

9.5 Principle of Virtual Work

A virtual displacement δx, $\delta\theta$, δq, etc., is an infinitesimal change in the coordinate, which may be conceived in any manner irrespective of the time t. It may or may not coincide with the actual displacement, dx, $d\theta$, dq.

In the case of constrained motion, the virtual displacement must be compatible with the constraints. For instance, if a particle is constrained to move on a surface, the virtual displacement must be confined to the surface. If the constraint equation for this case is

$$f(x, y, z, t) = 0 \qquad (9.5\text{--}1)$$

the virtual displacement must satisfy the equation,

$$\frac{\partial f}{\partial x}\,\delta x + \frac{\partial f}{\partial y}\,\delta y + \frac{\partial f}{\partial z}\,\delta z = 0 \qquad (9.5\text{--}2)$$

For a nonholonomic constraint, the restriction is

$$a_1\,\delta x + a_2\,\delta y + a_3\,\delta z = 0 \qquad (9.5\text{--}3)$$

Since virtual displacements are made irrespective of time, the above expression must be independent of time t.

Consider a particle acted upon by several forces. If the particle is in equilibrium, the resultant $\mathbf{R}$ of the forces must vanish, and the work done by the forces in a virtual displacement $\delta\mathbf{r}$ is zero.

$$\mathbf{R} \cdot \delta\mathbf{r} = 0 \qquad (9.5\text{--}4)$$

If the particle is constrained, the force $\mathbf{R}$ may be separated into an applied force $\mathbf{F}$ and a constraint force $\mathbf{f}$. For equilibrium,

$$\mathbf{R} = \mathbf{F} + \mathbf{f} = 0 \qquad (9.5\text{--}5)$$

and the applied force is balanced by the constraint force. The virtual work is then,

$$\mathbf{F} \cdot \delta\mathbf{r} + \mathbf{f} \cdot \delta\mathbf{r} = 0 \qquad (9.5\text{--}6)$$

But now the virtual displacement $\delta\mathbf{r}$ must be consistent with the constraint, which requires that $\mathbf{f} \cdot \delta\mathbf{r} = 0$. For instance, a particle made to move along a smooth wire would have a constraint force of the wire acting normal to the wire and hence to the virtual displacement. Thus the constraint force cannot contribute to the work, and we are left with the result,

$$\mathbf{F} \cdot \delta\mathbf{r} = 0 \qquad (9.5\text{--}7)$$

which states that, if the particle is in equilibrium, the work done by the applied forces due to a virtual displacement is zero.

For a system of particles in equilibrium, the sum of the forces acting on each particle must vanish. The virtual work of the system is the sum of the virtual work done on each particle, which must also be zero.

$$\sum_i \mathbf{R}_i \cdot \delta\mathbf{r}_i = 0 \qquad (9.5\text{--}8)$$

The force $\mathbf{R}_i$ can again be separated into the applied force $\mathbf{F}_i$ and the force of constraint $\mathbf{f}_i$, and since $\sum_i \mathbf{f}_i \cdot \delta\mathbf{r}_i = 0$, we obtain the virtual work for the system of particles to be,

$$\sum_i \mathbf{F}_i \cdot \delta\mathbf{r}_i = 0 \qquad (9.5\text{--}9)$$

where $\mathbf{F}_i$ and $\delta\mathbf{r}_i$ are the applied force and the virtual displacement associated with particle i. Thus the principle of virtual work as presented by Jean Bernoulli (1717) can be stated as follows: If a system of forces are in equilibrium, the work done by the applied forces in a virtual displacement compatible with the constraints is zero.

For a rigid body or a system of interconnected rigid bodies, internal forces, which always appear in equal and opposite pairs, must do no work. Thus, with the principle of virtual work, we can ignore all internal forces and reaction forces of constraints, and equate the virtual work of the applied forces to zero.

9.6 D'Alembert's Principle

The principle of virtual work, established for the case of static equilibrium, can be extended to dynamics by a reasoning advanced by D'Alembert (1743). We will let $\mathbf{p}$ be the momentum of a particle in the system, and separate the forces acting on it into an applied force $\mathbf{F}$ and a constraint force $\mathbf{f}$. The equation of motion of the particle can then be written as

$$\mathbf{F} + \mathbf{f} - \dot{\mathbf{p}} = 0 \qquad (9.6\text{--}1)$$

which states that the forces are in equilibrium with the kinetic reaction $-\dot{\mathbf{p}}$. The quantity $-\dot{\mathbf{p}}$ is sometimes referred to as the "reverse effective force" because the force effective in producing the motion is equal to $\dot{\mathbf{p}}$, and, if such a force is applied in the reverse sense, the motion could be nullified to produce a state of static equilibrium. As before, the virtual work of the constraint force is zero since $\mathbf{f}$ and $\delta\mathbf{r}$ are mutually perpendicular. The virtual work of the forces acting on the particle is, then,

$$(\mathbf{F} - \dot{\mathbf{p}}) \cdot \delta\mathbf{r} = 0 \qquad (9.6\text{--}2)$$

and for the system of N particles, we sum to obtain the result

$$\sum_{i=1}^{N} (\mathbf{F} - \dot{\mathbf{p}})_i \cdot \delta\mathbf{r}_i = 0 \qquad (9.6\text{--}3)$$

PROBLEMS

1. A constraint is scleronomic if time-independent, and it is rheonomic if time-dependent. If a particle of mass m is free to slide on a smooth hoop of radius r, which is rotated with constant speed Ω about a verticle diameter, discuss the degrees of freedom and the type of constraint, holonomic, nonholonomic, scleronomic, or rheonomic.

2. The vertical diameter of a wheel rolling on a rough horizontal floor is rotated at a constant rate to form a circular path for the contact point. Is the constraint holonomic or nonholonomic?

3. A sphere of radius a, rolling on a rough, horizontal plane is a nonholonomic system. Attach body axes x, y, z to the center of the sphere, defining their position by the Euler angles θ, ψ, φ. Show that the angular velocity of the sphere with respect to fixed axes X, Y, Z is $\boldsymbol{\omega} = \omega_X \mathbf{I} + \omega_Y \mathbf{J} + \omega_z \mathbf{K}$ where,

$$\omega_X = \dot\theta \cos\psi + \dot\varphi \sin\theta \sin\psi$$

$$\omega_Y = \dot\theta \sin\psi - \dot\varphi \sin\theta \cos\psi$$

$$\omega_Z = \dot\psi + \dot\varphi \cos\theta$$

Show also that the constraint equation for no slipping is

$$\mathbf{V}_0 + \boldsymbol{\omega} \times \mathbf{r} = 0$$

where $\mathbf{V}_0$ is the velocity of the center of the sphere, and $\mathbf{r} = -a\mathbf{K}$.

4. In Prob. 3, determine the variation of the constraint equation along the X and Y axes.

5. Virtual work for a system of N particles can be expressed by the equation $\sum_{i=1}^{N} (\mathbf{F}_i - m_i\ddot{\mathbf{r}}_i) \cdot \delta\mathbf{r}_i = 0$. Discuss the interpretation of this equation for a system of free particles versus a system of constrained particles.

9.7 Hamilton's Principle

The principle of virtual work together with D'Alembert's principle was viewed by W. R. Hamilton (1805–1865) as a basis for his variational approach, leading to one of the most general statements of mechanics known as the Hamilton's Principle. It reduces the formulation of problems in dynamics to that of the variation of a scalar integral, irrespective of coordinates and for conservative and nonconservative systems.

We start with N discrete mass particles, coupled by either holonomic or nonholonomic forces of constraints, and write the virtual work equation,

$$\sum_{i=1}^{N} (m_i\ddot{\mathbf{r}}_i - \mathbf{F}_i) \cdot \delta\mathbf{r}_i = 0 \qquad (9.7\text{-}1)$$

We recognize in this equation that $\sum_{i=1}^{N} F_i \cdot \delta r_i$ is the work done by the external forces in a virtual displacement, which could include nonconservative forces.

$$\sum_{i=1}^{N} \mathbf{F}_i \cdot \delta \mathbf{r}_i = \delta W \qquad (9.7\text{--}2)$$

Relating to the first term of Eq. 9.7–1, the following differential relationships exist:

$$\frac{d}{dt}(\dot{\mathbf{r}}_i \cdot \delta \mathbf{r}_i) = \dot{\mathbf{r}}_i \cdot \frac{d}{dt}\delta \mathbf{r}_i + \ddot{\mathbf{r}}_i \cdot \delta \mathbf{r}_i$$

$$= \dot{\mathbf{r}}_i \cdot \delta \dot{\mathbf{r}}_i + \ddot{\mathbf{r}}_i \cdot \delta \mathbf{r}_i$$

$$= \delta(\tfrac{1}{2}\dot{\mathbf{r}}_i{}^2) + \ddot{\mathbf{r}}_i \cdot \delta \mathbf{r}_i \qquad (9.7\text{--}3)$$

The first term of Eq. 9.7–1 can then be written as

$$\sum_{i=1}^{N} m_i \ddot{\mathbf{r}}_i \cdot \delta \mathbf{r}_i = \sum_{i=1}^{N} \frac{d}{dt}(m_i \dot{\mathbf{r}}_i \cdot \delta \mathbf{r}_i) - \sum_{i=1}^{N} \delta\left(\frac{1}{2} m_i \dot{r}_i{}^2\right)$$

$$= \sum_{i=1}^{N} \frac{d}{dt}(m_i \dot{\mathbf{r}}_i \cdot \delta \mathbf{r}_i) - \delta T \qquad (9.7\text{--}4)$$

and Eq. 9.7–1 becomes

$$\sum_{i=1}^{N} \frac{d}{dt}(m_i \dot{\mathbf{r}}_i \cdot \delta \mathbf{r}_i) = \delta T + \delta W \qquad (9.7\text{--}5)$$

Consider times $t = 0$ and $t = t_1$ at which $\delta r_i = 0$, and integrate Eq. 9.7–5:

$$\int_0^{t_1} \sum_{i=1}^{N} \frac{d}{dt}(m_i \dot{\mathbf{r}}_i \cdot \delta \mathbf{r}_i)\, dt = \int_0^{t_1}(\delta T + \delta W)\, dt \qquad (9.7\text{--}6)$$

The left side of this equation is equal to the integrand evaluated at the upper and lower limits:

$$\sum_{i=1}^{N} m_i \dot{\mathbf{r}}_i \cdot \delta \mathbf{r}_i \bigg|_0^{t_1} = \int_0^{t_1}(\delta T + \delta W)\, dt \qquad (9.7\text{--}7)$$

But since δr_i equals zero at $t = 0$ and t_1, the left side of Eq. 9.7–7 is zero, and we arrive at the final result

$$\int_0^{t_1}(\delta T + \delta W)\, dt = \delta \int_0^{t_1} T\, dt + \int_0^{t_1} \delta W\, dt = 0 \qquad (9.7\text{--}8)$$

We will review now the variational principle leading to Eq. 9.7–8. The motion of the system defined by the time variation of the N $\mathbf{r}_i$ is defined as the dynamical path. At any time t between $t = 0$ and t_1, the N-valued $\mathbf{r}_i$ are given a virtual displacement δr, thereby varying the dynamical path under the restriction $\delta t = 0$. In the varied configuration, T and W undergo variations δT and δW due to the variation in the coordinates and their velocities. Of all the possible variations, the dynamical path corresponds to the one which leads to a stationary value of the integral in Eq. 9.7–8 which is Hamilton's principle.

When the system is conservative the work can be expressed in terms of the potential energy in which case Eq. 9.7–8 reduces to

$$\delta \int_0^{t_1} (T + W)\, dt = 0 \qquad (9.7\text{–}9)$$

Hamilton's principle states that, if the configuration of the system at two instants $t = 0$ and t_1 is known, the motion of the system is given by the stationarity of the scalar integral. Hamilton's principle does not provide the solution to the dynamical problem, but formulates the equations of motion in a general manner irrespective of the coordinate system. It embodies both the Lagrange equation and the theorem of conservation of energy.

9.8 Lagrange's Equation (Holonomic System)

For a holonomic system, the dependent-variables $\mathbf{r}_i$ can be expressed entirely in terms of the n-generalized coordinates q_k and time t, corresponding to the n degrees of freedom.

$$\mathbf{r}_i = \mathbf{r}_i(q_1, q_2, \ldots, q_n, t) \qquad (9.8\text{–}1)$$

The q_k in the above equation are independent and no constraint equation exists between them.

Differentiating Eq. 9.8–1, the velocity can be written as

$$\dot{\mathbf{r}}_i = \frac{\partial \mathbf{r}_i}{\partial q_1} \dot{q}_1 + \frac{\partial \mathbf{r}_i}{\partial q_2} \dot{q}_2 + \cdots + \frac{\partial \mathbf{r}_i}{\partial q_n} \dot{q}_n + \frac{\partial \mathbf{r}_i}{\partial t} \qquad (9.8\text{–}2)$$

By squaring and summing over all the particles of the system, the kinetic energy becomes

$$T = \frac{1}{2} \sum_{i=1}^{N} m_i \dot{\mathbf{r}}_i{}^2 = \frac{1}{2} \sum_{i=1}^{N} m_i \left[\sum_k^n \sum_l^n \frac{\partial \mathbf{r}_i}{\partial q_k} \frac{\partial \mathbf{r}_i}{\partial q_l} \dot{q}_k \dot{q}_l + 2 \frac{\partial \mathbf{r}_i}{\partial t} \sum_k^n \frac{\partial \mathbf{r}_i}{\partial q_k} \dot{q}_k + \left(\frac{\partial \mathbf{r}_i}{\partial t} \right)^2 \right]$$

$$(9.8\text{–}3)$$

It is evident then that T is a function of q_k, $\dot{q}_k$, and t, and we may write

$$T = T(q_1, q_2, \ldots, \dot{q}_1, \dot{q}_2, \ldots, t) \qquad (9.8\text{–}4)$$

We will now consider the variation δT in Hamilton's equation, holding t fixed

$$\delta T = \sum_{i=1}^{n} \frac{\partial T}{\partial q_i} \delta q_i + \sum_{i=1}^{n} \frac{\partial T}{\partial \dot{q}_i} \delta \dot{q}_i \tag{9.8-5}$$

and

$$\int_0^{t_1} \delta T \, dt = \sum_{i=1}^{n} \int_0^{t_1} \frac{\partial T}{\partial q_i} \delta q_i \, dt + \sum_{i=1}^{n} \int_0^{t_1} \frac{\partial T}{\partial \dot{q}_i} \delta \dot{q}_i \, dt \tag{9.8-6}$$

Writing the last integral in the form,

$$\int_0^{t_1} \frac{\partial T}{\partial \dot{q}_i} \delta \dot{q}_i \, dt = \int_0^{t_1} \frac{\partial T}{\partial \dot{q}_i} \frac{d}{dt} \delta q_i \, dt$$

we integrate by parts, letting

$$u = \frac{\partial T}{\partial \dot{q}_i} \qquad\qquad dv = \frac{d}{dt} \delta q_i \, dt$$

Then

$$du = \frac{d}{dt} \frac{\partial T}{\partial \dot{q}_i} \, dt \qquad\qquad v = \delta q_i$$

and the integral becomes

$$\int_0^{t_1} \frac{\partial T}{\partial \dot{q}_i} \frac{d}{dt} \delta q_i \, dt = \frac{\partial T}{\partial \dot{q}_i} \delta q_i \Big|_0^{t_1} - \int_0^{t_1} \delta q_i \frac{d}{dt} \frac{\partial T}{\partial \dot{q}_i} \, dt$$

Since $\delta q_i = 0$ at $t = 0$ and t_1, the first term on the right is zero and Eq. 9.8-6 becomes,

$$\delta \int_0^{t_1} T \, dt = -\sum_{i=1}^{n} \int_0^{t_1} \left(\frac{d}{dt} \frac{\partial T}{\partial \dot{q}_i} - \frac{\partial T}{\partial q_i} \right) \delta q_i \, dt \tag{9.8-7}$$

Consider next the variation δW due to the m forces acting on the system

$$\delta W = \sum_{j=1}^{m} \mathbf{F}_j \cdot \delta \mathbf{r}_j \tag{9.8-8}$$

The virtual displacement δr_j is

$$\delta \mathbf{r}_j = \frac{\partial \mathbf{r}_j}{\partial q_1} \delta q_1 + \frac{\partial \mathbf{r}_j}{\partial q_2} \delta q_2 + \cdots + \frac{\partial \mathbf{r}_j}{\partial q_n} \delta q_n \tag{9.8-9}$$

so that δW becomes

$$\delta W = \sum_{j=1}^{m} \left(\mathbf{F}_j \cdot \frac{\partial \mathbf{r}_j}{\partial q_1} \delta q_1 + \mathbf{F}_j \cdot \frac{\partial \mathbf{r}_j}{\partial q_2} \delta q_2 + \cdots \mathbf{F}_j \cdot \frac{\partial \mathbf{r}_j}{\partial q_n} \delta q_n \right) \tag{9.8-10}$$

We can now define the generalized force Q_i associated with q_i to be

$$Q_i = \sum_{j=1}^{m} \mathbf{F}_j \cdot \frac{\partial \mathbf{r}_j}{\partial q_i} \qquad (9.8\text{--}11)$$

which enables Eq. 9.8–10 to be written as

$$\delta W = Q_1 \, \delta q_1 + Q_2 \, \delta q_2 + \cdots Q_n \, \delta q_n = \sum_{i=1}^{n} Q_i \, \delta q_i \qquad (9.8\text{--}12)$$

We now substitute Eqs. 9.8–7 and 9.8–12 into Hamilton's equation,

$$\delta \int_0^{t_1} (T + W) \, dt = - \sum_{i=1}^{n} \int_0^{t_1} \left(\frac{d}{dt} \frac{\partial T}{\partial \dot{q}_i} - \frac{\partial T}{\partial q_i} - Q_i \right) \delta q_i \, dt = 0$$

$$(9.8\text{--}13)$$

and since all the δq_i are independent and arbitrary, we can let all the δq_i equal zero except for δq_k which will be specified as not equal to zero. Then in order to satisfy the above equation, the coefficient of δq_k must be zero, and we arrive at Lagrange's equation for the holonomic system.

$$\frac{d}{dt} \frac{\partial T}{\partial \dot{q}_k} - \frac{\partial T}{\partial q_k} = Q_k \qquad (9.8\text{--}14)$$

So far we have not stated whether the forces $\mathbf{F}_j$ are conservative or nonconservative. For a conservative system, the work is expressible in terms of the potential energy U

and $$W = -U(q_i) \qquad (9.8\text{--}15)$$

$$\delta W = -\sum \frac{\partial U}{\partial q_i} \delta q_i \qquad (9.8\text{--}16)$$

It is evident then that Lagrange's equation for the conservative system is

$$\frac{d}{dt} \frac{\partial T}{\partial \dot{q}_k} - \frac{\partial T}{\partial q_k} + \frac{\partial U}{\partial q_k} = 0 \qquad (9.8\text{--}17)$$

It is now convenient to define the Lagrangian as

$$L = T - U \qquad (9.8\text{--}18)$$

and since $\partial U / \partial \dot{q}_k = 0$, Eq. 9.8–17 can be written in terms of the Lagrangian as

$$\frac{d}{dt} \frac{\partial L}{\partial \dot{q}_k} - \frac{\partial L}{\partial q_k} = 0 \qquad (9.8\text{--}19)$$

When both conservative and nonconservative forces act on a system, we can separate the virtual work into terms like Eqs. 9.8–12 and 9.8–16, and write Lagrange's equation as

$$\frac{d}{dt}\frac{\partial L}{\partial \dot{q}_k} - \frac{\partial L}{\partial q_k} = Q_k \tag{9.8–20}$$

Example 9.8–1

The pendulum analogy is used in the simplified analysis of many dynamical problems, including the sloshing of liquid fuel in missiles. As an application of Lagrange's equation, we will consider here the spherical pendulum of Fig. 9.8–1.

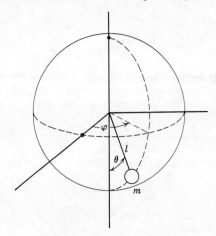

Fig. 9.8–1. Spherical pendulum.

The position of the mass, at a distance l from the center, can be specified by the generalized coordinates θ and φ, which can be varied independently.

The kinetic and potential energies are,

$$T = \tfrac{1}{2}m[(l\dot{\theta})^2 + (l\dot{\varphi}\sin\theta)^2] \tag{a}$$

$$U = mgl(1 - \cos\theta) \tag{b}$$

and the Lagrangian becomes,

$$L = \tfrac{1}{2}m[(l\dot{\theta})^2 + (l\dot{\varphi}\sin\theta)^2] - mgl(1 - \cos\theta) \tag{c}$$

Substituting into Lagrange's equation, the equations of motion are,

$$ml^2\left(\ddot{\theta} - \dot{\varphi}^2\sin\theta\cos\theta + \frac{g}{l}\sin\theta\right) = 0 \tag{d}$$

$$ml^2(\ddot{\varphi}\sin^2\theta + 2\dot{\varphi}\dot{\theta}\sin\theta\cos\theta) = 0 \tag{e}$$

The solution of these equations can be obtained in the following manner. Equation e can be written as

$$\frac{d}{dt}\dot{\varphi}\sin^2\theta = 0$$

$$\dot{\varphi}\sin^2\theta = C_1 \tag{f}$$

If we examine the Lagrangian, we would note that L is independent of φ, so that

$$\frac{\partial L}{\partial \varphi} = 0$$

We should then expect Lagrange's equation for the coordinate to reduce to

$$\frac{d}{dt}\frac{\partial L}{\partial \dot{\varphi}} = 0$$

and

$$\frac{\partial L}{\partial \dot{\varphi}} = \text{constant}$$

Equation f then is the direct consequence of $\partial L / \partial \varphi = 0$.

We can now make a general statement as follows: If the Lagrangian is not a function of the generalized coordinate q_k, then $\partial L / \partial q_k = 0$, and Lagrange's equation for q_k becomes

$$\frac{d}{dt}\frac{\partial L}{\partial \dot{q}_k} = 0 \tag{g}$$

Its integral is then immediately available as

$$\frac{\partial L}{\partial \dot{q}_k} = \text{constant} = p_k \tag{h}$$

where p_k is the generalized momentum for coordinate q_k. Such coordinates are called cyclic coordinates.

Returning to the solution of the two equations of motion, we substitute $\dot{\varphi}$ from Eq. f into Eq. d,

$$\ddot{\theta} - C_1{}^2\frac{\cos\theta}{\sin^3\theta} + \frac{g}{l}\sin\theta = 0 \tag{i}$$

We solve this equation in the usual way by multiplying by $2\dot{\theta}$ and integrating.

$$2\dot{\theta}\ddot{\theta} = 2\dot{\theta}C_1{}^2\frac{\cos\theta}{\sin^3\theta} - \frac{2g}{l}\dot{\theta}\sin\theta$$

$$\int d(\dot{\theta}^2) = 2C_1{}^2\int\frac{\cos\theta\, d\theta}{\sin^3\theta} - \frac{2g}{l}\int\sin\theta\, d\theta$$

The solution for θ is then,

$$\dot{\theta}^2 = -\frac{C_1{}^2}{\sin^2\theta} + \frac{2g}{l}\cos\theta + C_2 \tag{j}$$

We have so far identified the constant C_1 as the generalized momentum p_φ. We will show now that the constant C_2 is associated with the total energy E of the conservative system, which is

$$E = T + U = 2T - L \tag{k}$$

We have from Eqs. a, b, and c,

$$ml^2\left[\dot\theta^2 + \dot\varphi^2\sin^2\theta - \frac{1}{2}\dot\theta^2 - \frac{1}{2}\dot\varphi^2\sin^2\theta + \frac{g}{l}(1 - \cos\theta)\right] = E$$

Eliminating $\dot\varphi$ from Eq. f, we arrive at the result

$$\dot\theta^2 = -\frac{p_\varphi^2}{\sin^2\theta} + \frac{2g}{l}\cos\theta + 2\left(\frac{E}{ml^2} - \frac{g}{l}\right) \tag{l}$$

By comparison with Eq. j, we find that

$$C_2 = 2\left(\frac{E}{ml^2} - \frac{g}{l}\right)$$

Example 9.8–2

A spinning satellite with moments of inertia A, A, C, with $C < A$, has whip antennas which are free to vibrate in the z direction, as shown in Fig. 9.8–2. Set

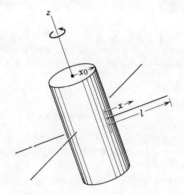

Fig. 9.8–2. Energy dissipation by whip antennas of a satellite.

up the vibration equation, using generalized coordinates associated with the normal modes of the antenna beam, and outline a procedure to establish the attitude drift of the spin axis.

We will assume the beam to have structural damping, which can be accounted for by a complex stiffness $EI(1 + i\alpha)$, where α is the structural damping factor. Letting m be the mass per unit length of the antenna, and w the elastic deflection in the z direction, the differential equation of motion is

$$EI(1 + i\alpha)\frac{\partial^4 w}{\partial x^4} + m\frac{\partial^2 w}{\partial t^2} = ma_z \tag{a}$$

The acceleration a_z is that of a point along the undeformed antenna, which from Eq. 7.6–14a is

$$a_z = 2\frac{C}{A}(x_0 + x)\omega_0^2\sin\theta\cos\theta\sin\dot\varphi t \tag{b}$$

We will express the deflection in terms of generalized coordinates $q_n(t)$ and the normal modes $\varphi_n(x)$ of the antenna,

$$w(x, t) = \sum_{n=1}^{\infty} q_n(t)\,\varphi_n(x) \tag{c}$$

The normal modes are vibration shapes associated with the undamped harmonic oscillations at the natural frequencies Ω_n, which obey the equation

$$EI\frac{d^4\varphi_n}{dx^4} - m\Omega_n{}^2\varphi_n = 0 \tag{d}$$

We now substitute Eq. c into a and replace $EI(d^4\varphi_n/dx^4)$ by $m\Omega_n{}^2\varphi_n$ of Eq. d to obtain

$$\sum_{n=1}^{\infty} m\varphi_n\ddot{q}_n + \sum_{n=1}^{\infty} m\Omega_n{}^2(1 + i\alpha)\varphi_n q_n = ma_z \tag{e}$$

Multiplying Eq. e by $\varphi_k\,dx$ and integrating over $x = 0$ to l, and noting the orthogonality relationship of the normal modes,

$$\int_0^l \varphi_n\varphi_k m\,dx = \begin{cases} 0 & \text{for } n \neq k \\ M & \text{for } n = k \end{cases} \tag{f}$$

The result is

$$\ddot{q}_k + (1 + i\alpha)\Omega_n{}^2 q_k = \frac{1}{M}\int_0^l ma_z\varphi_k\,dx$$

$$= F\sin\dot{\varphi}t \tag{g}$$

where

$$F = \frac{2}{M}\left(\frac{C}{A}\right)\omega_0{}^2\sin\theta\cos\theta\int_0^l (x_0 + x)\varphi_k m\,dx$$

The steady-state oscillation of the antenna is then established as

$$q_k = \frac{F\sin(\dot{\varphi}t + \epsilon_k)}{\Omega_k{}^2\sqrt{[1 - (\dot{\varphi}/\Omega_k)^2]^2 + \alpha^2}} \tag{h}$$

To determine the energy dissipated per cycle, we start with the strain energy

$$U = \frac{1}{2}EI\int_0^l \left(\frac{d^2w}{dx^2}\right)^2 dx = \frac{EI}{2}\left(q_1{}^2\int_0^l \varphi_1''{}^2\,dx + q_2{}^2\int_0^l \varphi_2''{}^2\,dx + 2q_1q_2\int_0^l \varphi_1''\varphi_2''\,dx + \cdots\right) \tag{i}$$

Using only the first mode and noting that (see Sec. 7.6)

$$\frac{\dot{\varphi}}{\Omega_1} = \left(1 - \frac{C}{A}\right)\frac{\omega_0}{\Omega_1}\cos\theta = R_1\cos\theta \tag{j}$$

the equation for the energy dissipated per cycle is

$$\frac{\gamma EI}{\pi M^2}\left(1 - \frac{C}{A}\right)\left(\frac{C}{A}\right)^2\left(\frac{\omega_0}{\Omega_1}\right)^5 \frac{\sin^2\theta\cos^3\theta}{(1 - R_1{}^2\cos^2\theta)^2 + \alpha^2}\left(\int_0^l (x_0 + x)\varphi_1 m\,dx\right)^2\int_0^l \varphi''^2\,dx \tag{k}$$

Equating this to $\dot{T}$ of Eq. 7.6–6, the rate of drift of the attitude angle can be expressed as

$$\dot{\theta} = \frac{K \sin \theta \cos^2 \theta}{(1 - R_1{}^2 \cos^2 \theta)^2 + \alpha^2} \qquad (l)$$

where the many constants of the problem have been lumped into K.

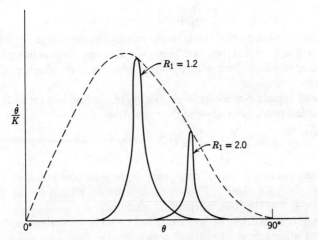

Fig. 9.8–3. Resonance in attitude drift rate due to whip antennas.

This result should be compared to that of Example 7.6–1. Since θ generally starts from some small angle θ_0 and increases for $C/A < 1$, there is a possibility of resonance if R_1 is greater than unity. The peak values, however, are expressible as

$$\dot{\theta} = \frac{K}{\alpha^2} \sin \theta \cos^2 \theta \qquad (m)$$

which is similar in form to Eq. f of Ex. 7.6–1, although the K are different. A plot of the drift rate is shown in Fig. 9.8–3.

9.9 Nonholonomic Systems

The development of the previous section for the holonomic system, Eqs. 9.8–1 to 9.8–13, apply equally well to nonholonomic systems. The difference here is that for the nonholonomic system the q_i of Eq. 9.8–1 are not all independent. However, the requirement for the independence of the q_i for the holonomic system was not imposed in the development of the previous section until the step between Eqs. 9.8–13 and 9.8–14 was required.

For the nonholonomic system, the q_i are restricted by the constraint equations of the form

$$a_{11} \, dq_1 + a_{21} \, dq_2 + \cdots a_{n1} \, dq_n + a_{01} \, dt = 0$$

$$\vdots$$

$$\text{(9.9-1)}$$

$$a_{1m} \, dq_1 + a_{2m} \, dq_2 + \cdots a_{nm} \, dq_n + a_{0m} \, dt = 0$$

which are nonintegrable. Holonomic constraints may also be present, but we will assume that they have been used to reduce the q_i to independent quantities, of which there will be $n - m$, where m is the number of nonholonomic constraints.

We will assume for convenience that there are just two ($m = 2$) nonholonomic constraints and write their variation,

$$a_{11} \, \delta q_1 + a_{21} \, \delta q_2 + \cdots a_{n1} \, \delta q_n = 0$$
$$a_{12} \, \delta q_1 + a_{22} \, \delta q_2 + \cdots a_{n2} \, \delta q_n = 0 \tag{9.9-2}$$

Since the variation is one of configuration, holding time constant, t does not enter into Eq. 9.9-2. Two of the q_i are now related by Eq. 9.9-2, leaving $n - 2$ of the q_i as independent quantities.

We will now multiply each of Eq. 9.9-2 by an undetermined multiplier λ and integrate between $t = 0$ and t_1 as follows:

$$\sum_{i=1}^{n} \int_0^{t_1} \lambda_1 a_{i1} \, \delta q_i \, dt = 0$$
$$\sum_{i=1}^{n} \int_0^{t_1} \lambda_2 a_{i2} \, \delta q_i \, dt = 0 \tag{9.9-3}$$

(Note that the λ could be a function of time as well as constants.) Including these terms in Eq. 9.8-13, we can write,

$$\delta \int_0^{t_1} (T + W) \, dt = -\sum_{i=1}^{n} \int_0^{t_1} \left(\frac{d}{dt} \frac{\partial T}{\partial \dot{q}_i} - \frac{\partial T}{\partial q_i} - Q_i - \lambda_1 a_{i1} - \lambda_2 a_{i2} \right) \delta q_i \, dt$$
$$= 0 \tag{9.9-4}$$

Since $n - 2$ of the δq_i are independent, we will separate the above integrals to

$$\sum_{i=1}^{n-2} \int_0^{t_1} \left(\frac{d}{dt} \frac{\partial T}{\partial \dot{q}_i} - \frac{\partial T}{\partial q_i} - Q_i - \lambda_1 a_{i1} - \lambda_2 a_{i2} \right) \delta q_i \, dt$$
$$+ \sum_{i=n-1}^{n} \int_0^{t_1} \left(\frac{d}{dt} \frac{\partial T}{\partial \dot{q}_i} - \frac{\partial T}{\partial q_i} - Q_i - \lambda_1 a_{i1} - \lambda_2 a_{i2} \right) \delta q_i \, dt = 0$$

$$\text{(9.9-5)}$$

where the two δq_i of the last integral are fixed by the constraint equations and, therefore, are not arbitrary. We can however, make the integrand of the last integral zero by a proper choice of λ_1 and λ_2. Each of the integrand of the first integral can be shown to be zero by assuming one of the arbitrary δq_i to be nonzero and the remaining δq_i to be all zero, repeating this procedure for each of the $(n - 2)$ δq_i. We then arrive at n equations of the form,

$$\frac{d}{dt}\frac{\partial T}{\partial \dot{q}_k} - \frac{\partial T}{\partial q_k} = Q_k + \lambda_1 a_{k1} + \lambda_2 a_{k2} \qquad (9.9\text{-}6)$$

Since we have two additional unknowns λ_1 and λ_2 besides the n q_i, we need two other equations, which are furnished by the constraint equations, Eq. 9.9-1, written in the form,

$$\begin{aligned} a_{11}\dot{q}_1 + a_{21}\dot{q}_2 + \cdots a_{n1}\dot{q}_n + a_{01} = 0 \\ a_{12}\dot{q}_1 + a_{22}\dot{q}_2 + \cdots a_{n2}\dot{q}_n + a_{02} = 0 \end{aligned} \qquad (9.9\text{-}7)$$

These $n + 2$ equations are then sufficient for the solution of the problem.

In the development of this section, we have illustrated the method of Lagrange's undetermined multipliers, λ, which is not restricted to the nonholonomic system, and can be applied equally well to holonomic systems. Occasionally it is not convenient to reduce the variables of the holonomic system to independent quantities by the use of the constraint equations, in which case the Lagrange multiplier method can be used. In such a case, the Lagrange multiplier method will also provide a solution for the constraint forces, which are sometimes required.

Example 9.9-1

A thin disk of radius r rolls down an inclined plane of small angle α with the horizontal. If the plane of the disk is always normal to the inclined plane, and capable of rotation about the normal, determine the x, y motion of the disk.

Referring to Fig. 9.9-1, the coordinates of the problem are ψ, x, and ϕ. The equations for the kinetic and potential energies are,

$$T = \tfrac{1}{2}m[(r\dot{\phi})^2 + \tfrac{1}{4}(r\dot{\psi})^2 + \tfrac{1}{2}(r\dot{\phi})^2] \qquad (a)$$

$$U = -mgx \sin \alpha \qquad (b)$$

and the constraint equation is

$$r\, d\phi - \frac{dx}{\cos \psi} = 0 \qquad (c)$$

The Lagrange equations are then,

$$\frac{d}{dt}\left[m\frac{r^2}{4}\dot{\psi}\right] = 0 \qquad (d)$$

$$-mg \sin \alpha + \frac{\lambda}{\cos \psi} = 0 \qquad (e)$$

$$\frac{d}{dt}[m(\tfrac{3}{2}r^2)\dot{\phi}] - r\lambda = 0 \qquad (f)$$

From (d), $\dot{\psi}$ is a constant n, and its integral is $\psi = \psi_0 + nt$. Substituting for ψ in (e), the equation for λ becomes

$$\lambda = mg \sin \alpha \cos (\psi_0 + nt) \qquad (g)$$

With λ substituted into (f), its integral is

$$\frac{3}{2} mr^2(\phi - \phi_0) - mgr \sin \alpha \left[\frac{1}{n} \sin (\psi_0 + nt) - \frac{1}{n} \sin \psi_0 \right] = 0 \qquad (h)$$

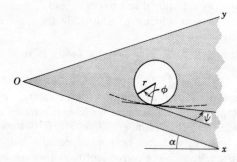

Fig. 9.9–1. Coordinates y, x, and ϕ are related by nonholonomic constraint.

The x and y displacements can be found from the integrals of

$$dx = r\dot{\phi} \cos \psi \, dt \qquad (i)$$

$$dy = r\dot{\phi} \sin \psi \, dt \qquad (j)$$

9.10 Lagrange's Equation for Impulsive Forces

During impact of one body on another, a very large force acts for a very short time. Such forces are said to be impulsive. As the time of contact diminishes to zero, the force tends to infinity; however, like the delta function, the time integral of the impulsive force is finite.

When an impulsive force acts on a body, the velocity of the body undergoes an instantaneous change over an infinitesimal change in displacement. Thus, generalized velocities $\dot{q}_i$ will change instantaneously, whereas generalized coordinates q_i will not. The system is nonconservative since energy is generally dissipated during impact.

In applying Lagrange's equation to impulsive force systems, we approach it by a limiting process. Starting with the equation,

$$\frac{d}{dt} \frac{\partial T}{\partial \dot{q}_i} - \frac{\partial T}{\partial q_i} = Q_i \qquad (9.10\text{--}1)$$

we multiply by dt and integrate over the impact time.

$$\int_0^\epsilon d\,\frac{\partial T}{\partial \dot q_i} - \int_0^\epsilon \frac{\partial T}{\partial q_i}\,dt = \int_0^\epsilon Q_i\,dt \qquad (9.10\text{--}2)$$

The second integral contain terms associated with the generalized coordinate q_i which do not change during the impact. Thus, in the limiting case when $\epsilon \to 0$, the second integral vanishes, and we obtain the relationship,

$$\Delta \frac{\partial T}{\partial \dot q_i} = \lim_{\epsilon \to 0} \int_0^\epsilon Q_i\,dt = \hat Q_i \qquad (9.10\text{--}3)$$

This equation states that the change in the generalized momentum is equal to the generalized impulse.

Example 9.10–1
Four equal bars, each of mass m and length $2a$, lie on a smooth, horizontal floor, hinged together in the form of a rhombus, as shown in the sketch. If an

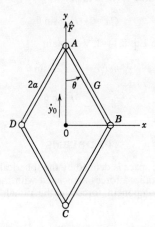

Fig. 9.10–1. Impulsively loaded structure.

impulsive force $\hat F$ lb-sec. is applied at A in the direction CA, determine the initial angular velocity of the bars.

Place coordinates x, y with origin O at the center of mass.

The x and y coordinates of the center G of the bar are,

$$x_G = a \sin \theta \qquad y_G = a \cos \theta$$

and the velocity of the center of mass of the bars becomes

$$\mathbf{v}_G = (a\dot\theta \cos \theta)\mathbf{i} + (\dot y_0 \pm a\dot\theta \sin \theta)\mathbf{j}$$

where the minus sign applies to AB and AD and the plus sign to CB and CD.

The kinetic energy of the bars is

$$T = \frac{1}{2}\, 4m\, v_G{}^2 + \frac{1}{2}\left(4m\, \frac{a^2}{3}\right)\dot\theta^2$$
$$= 2m(\dot{y}_0{}^2 + \tfrac{4}{3}a^2\dot\theta^2)$$

and the change in the generalized momentum becomes

$$\Delta\, \frac{\partial T}{\partial \dot\theta} = \tfrac{16}{3}ma^2\dot\theta$$

The generalized force $\hat{Q}$ is found from the virtual work of the impulse. Due to virtual displacements δy_0 and $\delta\theta$, the point A undergoes a displacement

$$\delta(y_0 + 2a \cos\theta) = \delta y_0 - 2a \sin\theta\, \delta\theta$$

and the virtual work of $\hat{F}$ is

$$\delta W = \hat{F}(\delta y_0 - 2a \sin\theta\, \delta\theta)$$

The generalized force due to $\delta\theta$ is then

$$\hat{Q}_\theta = -\hat{F}2a \sin\theta$$

and, by substituting into Eq. 8.14–3, we obtain

$$\tfrac{16}{3}ma^2\dot\theta = -\hat{F}2a \sin\theta$$

and

$$\dot\theta = -\frac{3\hat{F}\sin\theta}{8ma}$$

PROBLEMS

1. A particle moving in space is defined by the spherical coordinates r, θ, and φ. Determine the generalized forces associated with the spherical coordinates, and establish the component forces in the radial, meridian, and latitude directions.

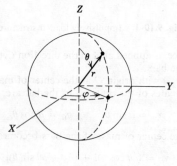

Prob. I

2. A satellite moves in a plane orbit under the influence of an inverse square attraction. Derive the orbit equation from Lagrange's formulation. Is there a cyclic coordinate for the system and, if so, what does it imply?

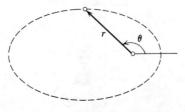

Prob. 2

3. A spherical pendulum of length l is set up on the earth's surface at latitude λ, with the z axis in the vertical direction and x axis pointing north. Show that the Lagrangian is

$$L = \frac{m}{2}\left[\dot{x}^2 + \dot{y}^2 + 2(x\dot{y} - \dot{x}y)\Omega \sin \lambda - \frac{g}{l}(x^2 + y^2) - 2\dot{y}R\Omega \cos \lambda + 2l\dot{y}\Omega \cos \lambda\right]$$

where Ω is the rotation speed of the earth and R = radius of earth.

4. Determine the Lagrangian for the symmetric top spinning about a fixed pivot on the floor. Establish the cyclic (ignorable) coordinates and write directly the resulting integrals.

5. For a system of N particles, the kinetic energy can be written in the form

$$T - \frac{1}{2}\sum_{k=1}^{n}\sum_{l=1}^{n}A_{kl}\dot{q}_k\dot{q}_l + \sum_{k=1}^{n}B_k\dot{q}_k + C$$

where

$$A_{kl} = \sum_{i=1}^{N}m_i\frac{\partial r_i}{\partial q_k}\frac{\partial r_i}{\partial q_l} \qquad B_k = \sum_{i=1}^{N}m_i\frac{\partial r_i}{\partial t}\frac{\partial r_i}{\partial q_k} \qquad C = \frac{1}{2}\sum_{i=1}^{N}m_i\left(\frac{\partial r_i}{\partial t}\right)^2$$

Prove that for a scleronomic conservative system,

$$2T = \sum_{k=1}^{n}\dot{q}_k\frac{\partial L}{\partial \dot{q}_k}$$

6. For a conservative system of N particles, prove that the conservation of energy holds only if the time t does not appear explicitly in the Lagrangian, in which case,

$$\frac{dE}{dt} = \frac{d}{dt}\left(\sum_{k=1}^{n}\dot{q}_k\frac{\partial L}{\partial \dot{q}_k} - L\right) = 0$$

7. Using the Lagrangian approach, deduce Eulers dynamical equations for an arbitrary rigid body subjected to moments about the body axes.
   Hint: $T = \frac{1}{2}(A\omega_x^2 + B\omega_y^2 + C\omega_z^2)$. Express the angular velocites in terms of Eulers angles. The generalized force can be determined by giving each of the Euler angles virtual displacement with the other two equal to zero.

8. A mass is suspended from a weightless spring of stiffness k lb/in. Write Lagrange's equations and discuss the motion for small oscillations in the vertical plane.

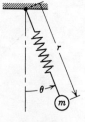

Prob. 8

9. Using Lagrange's method, set up the equations for the motion of a bar suspended by a string and oscillating in a plane.

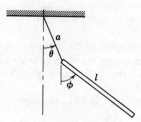

Prob. 9

10. A uniform bar of mass m and length l is suspended from one end by a spring of stiffness k lb/in. The bar can swing freely only in one vertical plane and the spring x is constrained to move only in the vertical direction. Set up the Lagrange equations of motion.

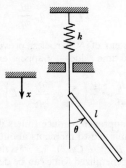

Prob. 10

11. A centrifugal pendulum of mass m and length r is attached to a flywheel of moment of inertia I. Show that the kinetic energy of the system is,

$$T = mr^2\left[\frac{1}{2}\dot{\varphi}^2 + \left(A + \frac{R}{r}\cos\varphi\right)\dot{\theta}^2 + \left(1 + \frac{R}{r}\cos\varphi\right)\dot{\theta}\dot{\varphi}\right]$$

where

$$A = \frac{1}{2}\left(\frac{I}{mr^2} + \frac{R^2}{r^2} + 1\right)$$

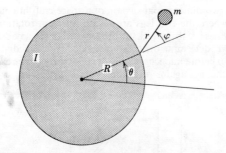

Prob. 11

12. A particle of mass m slides without friction on a hoop of radius r, which is rotated with constant speed Ω about a vertical diameter. A single coordinate θ is sufficient to describe the position of m; however, we have a moving constraint. Compare the total energy of the system with that determined from $T + U = \dot{\theta}(\partial L/\partial\dot{\theta}) - L$ and discuss this discrepancy.

Prob. 12

13. A uniform bar of length $2l$ and mass m is dropped from a height h onto a horizontal floor. The bar descends without rotation and at an angle θ_0 with the floor. If the coefficient of restitution between the floor and bar is e, determine the velocity of the center of the bar and its angular velocity immediately after impact.

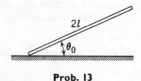

Prob. 13

14. In Prob. 13, determine the time elapsed after the first impact when the opposite side of the bar strikes the floor. What is the angle θ when it strikes?

15. A simplified two-dimensional version of a space craft in landing is shown in the sketch, where the two legs are restrained from rotation by a torsional spring of stiffness K lb-in./rad. If the legs strike the smooth inelastic plane with velocity v, determine the rotational velocity $\dot\theta$ immediately after impact. Assume $\dot\theta = 0$ before impact and the torsional spring to be exerting a moment C_0, holding the legs against the stop.

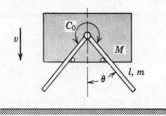

Prob. 15

16. For the system shown, show that, if one incorrectly takes y and θ as independent generalized coordinates q_1 and q_2, then the resulting two Lagrange equations are also incorrect. Demonstrate the use of the Lagrangian multiplier by establishing the correct equation, using y and θ, and a constraint equation.

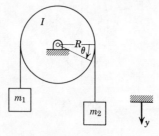

Prob. 16

17. Two uniform bars of mass m and length l are hinged as shown, and lie on a smooth horizontal plane. If an impulsive force strikes the bar normally at

one end, determine the angular velocities of the two bars and the velocity of the hinge.

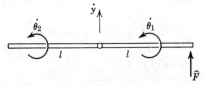

Prob. 17

18. Four hinged bars, each of length l and mass m, fall in translation and strike a horizontal inelastic ground. Taking y and θ as generalized coordinates, find the motion immediately after impact.

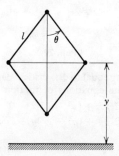

Prob. 18

19. A rigid pendulum of length l and mass m can swing around the horizontal axis AA which is mounted in a frame that can rotate freely around the vertical axis BB. The moment of inertia of the frame and horizontal axle about the vertical axis is I. At the instant $t = 0$, the pendulum in the vertical position is given an initial velocity v_0, and the frame is given an initial angular velocity ω_0. (a) Set up the equations of motion, noting that the momentum and energy are conserved; (b) simplify the results of (a) by assuming small angles, $\sin \theta = \theta$, and $\cos \theta = 1 - \frac{1}{2}\theta^2$; (c) discuss the motion of the simplified system for $\omega_0^2 < g/l$ and $\omega_0^2 > g/l$.

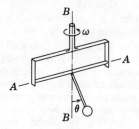

Prob. 19

20. A spherical pendulum of length l and mass m is suspended from end 0 of a horizontal arm of length e that rotates with constant angular velocity ω about a fixed vertical axis C, as shown in the sketch. Assuming small oscillatory

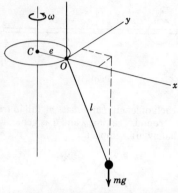

Prob. 20

amplitudes and a damping equal to $\zeta \times$ critical damping in a plane pendulum oscillation, show that the equations of motion of the pendulum bob are

$$\ddot{x} + 2\zeta p\dot{x} + (p^2 - \omega^2)x - 2\omega\dot{y} = e\omega^2$$
$$\ddot{y} + 2\zeta p\dot{y} + (p^2 - \omega^2)y + 2\omega\dot{x} = 0$$

where $p^2 = g/l$.

21. In Prob. 20, let $z = x - x_0 + iy$ and $x_0 = e\omega^2/(p^2 - \omega^2)$, where x_0 denotes the position of static equilibrium (not necessarily stable), and show that the two equations can be reduced to a single homogeneous equation in z with a general solution

$$z = Ae^{(-\zeta-i)(p+\omega)t} + Be^{(-\zeta+i)(p-\omega)t} \quad \zeta << 1$$

where terms of order ζ^2 have been neglected. Discuss the criteria for stability of small oscillations.

22. A one-wheel trailer shown in the sketch is towed with velocity v. If the trailer hitch has a lateral stiffness k, and the radius of gyration of the trailer about the center of mass G is ρ, write the two equations of motion and the constraint equation for no lateral sliding of the trailer wheel. Determine the stability of the trailer for small θ (i.e., roots of the characteristic equation must not have positive real parts).

23. Carry out the solution for x and y in Example 9.9–1. Show that, if the initial conditions are zero, the curve traced out will be a cycloid.

24. The spin axis of a space station with $C = kA$ (for a thin disk $k = 2$) is initially pointing to the north star. It is desired to change the direction of the spin axis by $\theta°$ in a specified direction without inducing a precession about the new direction. This can be done by two properly timed impulsive blasts from a single rocket engine on the rim and pointing parallel to the spin axis. If the initial angular momentum of the station is $h = C\omega_0$, determine the timing of the first and second blasts in relation to the desired direction change and their magnitude.

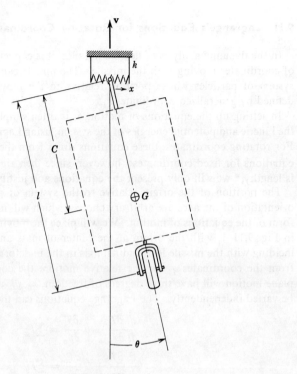

Prob. 22

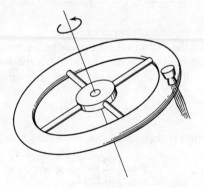

Prob. 24

9.11 Lagrange's Equations for Rotating Coordinates

In the dynamic analysis of flexible missiles, it is convenient to use a set of coordinates moving with the missile. The missile can be considered a system of particles whose position relative to the moving axes can be defined by generalized coordinates q_i.

In setting up the equations of motion, it is often simpler to start from the kinetic and potential energies of the system, using Lagrange's equation. For rotating coordinates, these equations differ from the usual Lagrange equations for fixed coordinates; however, since their rigorous derivation is lengthy,* we will only present the equations and justify each term.

The position of the origin relative to the system of particles and the orientation of the axes are arbitrary choices which will influence the final form of the equations of motion. We will place the moving axes as shown in Fig. 9.11–1, with the origin at the center of mass and the x axis coinciding with the missile longitudinal axis in the undeformed state. Aside from the coordinates q_i for the relative motion, the coordinate axes in plane motion will have three degrees of freedom, x_0, y_0, and θ, which can be varied independently. The Lagrange equations can then be written as,

$$\frac{d}{dt}\frac{\partial T}{\partial \dot{x}_0} - \dot{\theta}\,\frac{\partial T}{\partial \dot{y}_0} = \sum F_x \qquad (9.11\text{–}1)$$

$$\frac{d}{dt}\frac{\partial T}{\partial \dot{y}_0} + \dot{\theta}\,\frac{\partial T}{\partial \dot{x}_0} = \sum F_y \qquad (9.11\text{–}2)$$

$$\frac{d}{dt}\frac{\partial T}{\partial \dot{\theta}} + \dot{x}_0\,\frac{\partial T}{\partial \dot{y}_0} - \dot{y}_0\,\frac{\partial T}{\partial \dot{x}_0} = \sum M_0 \qquad (9.11\text{–}3)$$

The equation for q_i remain unaltered in form.

In accounting for the terms in these equations, we recognize $\partial T/\partial \dot{x}_0$ and $\partial T/\partial \dot{y}_0$ as the generalized momenta, and $\partial T/\partial \dot{\theta}$ as the generalized angular momentum. We will then represent the linear momentum by

$$\mathbf{p} = \frac{\partial T}{\partial \dot{x}_0}\mathbf{i} + \frac{\partial T}{\partial \dot{y}_0}\mathbf{j} \qquad (9.11\text{–}4)$$

As the force equation is the rate of change of the linear momentum,

$$\mathbf{F} = \left[\frac{d\mathbf{p}}{dt}\right] + \boldsymbol{\omega} \times \mathbf{p} \qquad (9.11\text{–}5)$$

the terms of Eq. 9.11–1 and 9.11–2 are immediately accounted for.

* Thomson, W. T., "Lagrange's Equations for Moving Coordinates," Space Technology Laboratories Report No. *EM 9–15 TR–59–000–00768*, Los Angeles (July 1959).

The terms of Eq. 9.11-3 can be identified from Eq. 7.11-3, which can be written as

$$\mathbf{M}_0 = \dot{\mathbf{h}}_0 + \dot{\mathbf{R}}_0 \times \sum m_i \mathbf{r}_i \qquad (9.11\text{-}6)$$

The term $\partial T/\partial \dot{\theta}$ is the angular momentum $\mathbf{h}_0$, and the remaining two terms are equal to $\dot{\mathbf{R}}_0 \times \sum m_i \mathbf{r}_i$, where $\dot{\mathbf{R}}_0 = \dot{x}_0 \mathbf{i} + \dot{y}_0 \mathbf{j}$.

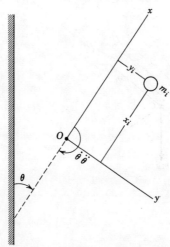

Fig. 9.11-1. Rotating coordinate system.

9.12 Missile Dynamic Analysis

Missiles are, in general, flexible structures, weight being of primary concern. Vibrational problems are thus likely to plague their performance, and it is in general necessary to make a dynamical analysis of a vehicle in flight. Such an analysis consists of formulating the equations of motion in all their details to account for the interaction of the bending of the flexible missile with the rigid body motion, the reaction of the swiveling engine, the sloshing of the propellant, the excitation of the aerodynamic forces and gusts, and the coupling of the servosystem controlling the stability of its attitude. The equations of motion must then be linearized and programmed for machine computation, i.e., the missile is theoretically flown on the high-speed computer.

It is evident that a detailed formulation of all the factors pertinent to the dynamical analysis is beyond the scope of this text. It is possible, however, to discuss a greatly simplified problem which will serve to illustrate the dynamical techniques employed in the general analysis.

We will consider here the problem of a flexible missile of constant mass, where the propellant is treated as a solid to avoid the complications of sloshing. The engine will be considered to be gimballed, and the motion

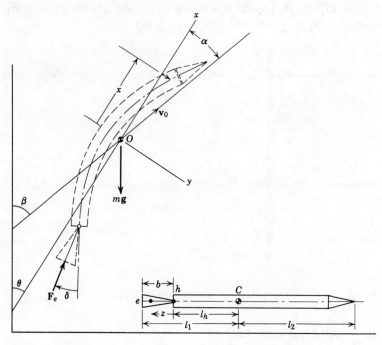

Fig. 9.12–1. Geometry of flexible missile and coordinates.

will be restricted to the vertical plane of the trajectory, as shown in Fig. 9.12–1. The coupling of the servosystem, actuating the gimballed engine attitude, will also be neglected.

Coordinate system

The choice of the coordinate system is an important one which depends on the vibrational data available. The missile can be treated as a free-free beam, and its normal modes, computed with engine locked on, will be assumed to be available from a previous analysis. Such modes are orthogonal and possess the property of zero linear and angular momenta about the missile vibrational axis (axis coinciding with the undeformed missile with engine locked on, from which vibrational displacements are measured) passing through its center of mass. In actual analysis, the shear deformation and rotatory inertia terms are accounted for in the normal mode analysis; however, for simplicity of discussion we will omit these terms.

The x, y coordinate axes will be chosen with the origin at the center of mass, the x axis coinciding with the missile vibrational axis. It is evident, then, that the missile longitudinal axis will undergo an additional displacement due to the rotation of the engine. The nature of this additional

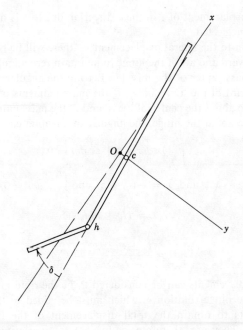

Fig. 9.12–2. Conservation of linear and angular momenta.

displacement is readily understood by considering a rigid missile whose engine section is given a rotational velocity $\dot{\delta}$ by an internal hinge moment, as shown in Fig. 9.12–2. If m and m_e are the total mass and the engine mass respectively, the maintenance of zero linear momentum (since there is no external force) can be expressed by the equation,

$$\int_{-l_1}^{l_2} \dot{y}\, dm - \dot{\delta} \int_o^b z\, dm = 0 \qquad (9.12\text{–}1)$$

Letting c be a point on the missile longitudinal axis coinciding with the center of mass of the missile in the undeformed state, and e be the center of mass of the engine, or section aft of the hinge, the above equation becomes,

$$\dot{y}_c = \frac{m_e z_e}{m} \dot{\delta} \qquad (9.12\text{--}2)$$

It is evident then that point c is displaced laterally by the amount

$$y_c = \frac{m_e z_e}{m} \delta = C_1 \delta \qquad (9.12\text{--}3)$$

The small displacement of c in the x direction due to δ is of second order and will be neglected.

In addition to the lateral displacement y_c, there will be rotations of the sections forward and aft of the hinge to maintain zero angular momentum about the mass center of the missile. Letting the angular velocity of the missile forward of the hinge be $\dot{\theta}_f$, I_0 the mass moments of inertia of the entire missile about the center of mass, and I_h the mass moment of inertia of the engine about the hinge, the angular momentum equation is

$$\int_{-l_1}^{l_2} \dot{y}x \, dm - \dot{\delta} \int_{0}^{b} zx \, dm = 0 \qquad (9.12\text{--}4)$$

Since $\dot{y} = \dot{y}_c - x\dot{\theta}_f$ and $x = -(l_h + z)$ and $\int_{-l_1}^{l_2} x \, dm = 0$, the equation reduces to

$$I_0 \dot{\theta}_f = (I_h + m_e l_h z_e)\dot{\delta}$$

from which

$$\theta_f = \frac{I_h + m_e l_h z_e}{I_0} \delta = C_2 \delta \qquad (9.12\text{--}5)$$

These displacements can be considered to be the rigid missile displacements due to hinge rotation δ, which must be added to the vibrational displacements to obtain the total displacement of the flexible missile center line. They are, therefore, equivalent to translating and rotating the missile vibrational axis by y_c and θ_f, and the total lateral displacement y of the flexible center line from the moving coordinate axis x is, at point x

$$y = (C_1 - C_2 x)\delta + \sum_i q_i(t) \, \phi_i(x) \qquad (9.12\text{--}6)$$

where the vibrational displacement is represented by the sum of the normal modes $\phi_i(x)$ multiplied by the generalized coordinate $q_i(t)$ associated with the mode.

With the displacement relative to the coordinate system established, we next examine the motion of the coordinate axes themselves. The x axis, which coincides with the missile vibrational axis or the missile longitudinal axis with $\delta = 0$, makes an angle θ with the vertical, as shown in Fig. 9.12–1. The rate of rotation of the coordinate axes is then $\omega = \dot{\theta}$.

The origin of the coordinate axes, coinciding with the center of mass at all times, has a velocity $\mathbf{v}_0$ tangent to the trajectory. The angle α between $\mathbf{v}_0$ and the x axis is the angle of attack of the missile's longitudinal axis (the local angle of attack will differ from α by dy/dx). Due to the changing direction of the trajectory tangent, the acceleration of the origin will be $\dot{v}_0$ parallel to $\mathbf{v}_0$, and $v_0\dot{\beta}$ perpendicular to $\mathbf{v}_0$, where β is the angle made by $\mathbf{v}_0$ and the vertical.

Equations of motion

With the coordinate system defined in the foregoing, the equations of motion can be formulated by determining the kinetic and potential energies of the missile to be substituted into Lagrange's equation, taking note of the fact that the coordinates are rotating (see Sec. 9.11).

The velocity of a point $\mathbf{r} = x\mathbf{i} + y\mathbf{j}$ in the rotating coordinate system is

$$\mathbf{v} = \mathbf{v}_0 + [\mathbf{v}] + \boldsymbol{\omega} \times \mathbf{r} \tag{9.12-7}$$

For any point x, y forward of the hinge h, the x and y components are

$$v_x = \dot{x}_0 - y\dot{\theta}$$
$$v_y = \dot{y}_0 + \dot{y} + x\dot{\theta}$$

(The relative velocity $\dot{x}$ of a point on the missile is of second order compared to $\dot{y}$ and is, therefore, neglected.) The velocity of a point aft of the hinge can be found from the above equations by replacing y by $y - z\delta$.

In writing the kinetic energy equation of the missile with the swiveling engine, we need the squares of the velocity as follows:

Forward of the hinge

$$v^2 = (\dot{x}_0 - y\dot{\theta})^2 + (\dot{y}_0 + \dot{y} + x\dot{\theta})^2 \tag{9.12-8}$$

Aft of the hinge

$$\begin{aligned} v^2 &= [(\dot{x}_0 - y\dot{\theta}) + z\dot{\theta}\delta]^2 + [(\dot{y}_0 + \dot{y} + x\dot{\theta}) - z\dot{\delta}]^2 \\ &= (\dot{x}_0 - y\dot{\theta})^2 + (\dot{y}_0 + \dot{y} + x\dot{\theta})^2 \\ &\quad + 2(\dot{x}_0 - y\dot{\theta})z\dot{\theta}\delta + (z\dot{\theta}\delta)^2 - 2(\dot{y}_0 + \dot{y} + x\dot{\theta})z\dot{\delta} + (z\dot{\delta})^2 \end{aligned} \tag{9.12-9}$$

Thus part of v^2 aft of the hinge has the same form as that forward of the hinge, which enables T to be written in the form $T = T_0 + T_\delta$, where the quantity δ appears only in T_δ.

$$T_0 = \frac{1}{2} \int_{-l_1}^{l_2} [(\dot{x}_0 - y\dot{\theta})^2 + (\dot{y}_0 + \dot{y} + x\dot{\theta})^2]m(x)\,dx \tag{9.12-10}$$

$$T_\delta = \frac{1}{2} \int_0^b [2(\dot{x}_0 - y\dot{\theta})z\dot{\theta}\delta + z^2\delta^2\dot{\theta}^2 - 2(\dot{y}_0 + \dot{y} + x\dot{\theta})z\dot{\delta} + z^2\dot{\delta}^2]m(z)\,dz \tag{9.12-11}$$

In these equations the generalized coordinates are x_0, y_0, θ, δ, and q_i, and it must be remembered that y is related to δ and q_i through Eq. 9.12–6.

The equations of motion relating to the generalized coordinates x_0, y_0, and θ can be determined from Lagrange's equations, Eqs. 9.11–1, 9.11–2, and 9.11–3, however, since the linear and angular momenta relative to the coordinate system are zero, the sums of the external forces and moments must equal the rate of change of the linear and angular momenta of the rigid missile, which are related to the linear and angular accelerations of the x, y coordinate system with origin at the center of mass. These equations can therefore be written as

$$ma_{0x} = -mg \cos \theta + F_e + D^* \tag{9.12–12}$$

$$ma_{0y} = mg \sin \theta + F_e(\delta + y_h') - L^* \tag{9.12–13}$$

$$I\ddot{\theta} = -F_e(\delta + y_h')l_h - F_e y_h + M_0^* \tag{9.12–14}$$

where F_e is the engine thrust, D^*, L^*, and M_0^* are the drag and lift components of the aerodynamic force and its moment about 0, $y_h' = (dy/dx)_h$, and a_{0x} and a_{0y} represent the acceleration of the center of mass 0, which is

$$a_{0x} = \ddot{x}_0 - \dot{y}_0\dot{\theta} = \dot{v}_0 \cos \alpha - v_0\dot{\beta} \sin \alpha \tag{9.12–15}$$

$$a_{0y} = \ddot{y}_0 + \dot{x}_0\dot{\theta} = \dot{v}_0 \sin \alpha + v_0\dot{\beta} \cos \alpha \tag{9.12–16}$$

To demonstrate that these equations can also be determined from Lagrange's equations for rotating coordinates, the first of the above equation, for the sum of the forces in the x direction, will be derived from Eq. 9.11–1. Differentiating the kinetic energy, we have,

$$\frac{\partial T}{\partial \dot{x}_0} = \int_{-l_1}^{l_2} (\dot{x}_0 - y\dot{\theta})m(x)\,dx + \dot{\theta}\delta \int_0^b z\,m(z)\,dz \tag{9.12–17}$$

$$\frac{\partial T}{\partial \dot{y}_0} = \int_{-l_1}^{l_2} (\dot{y}_0 + \dot{y} + x\dot{\theta})\,m(x)\,dx - \dot{\delta} \int_0^b z\,m(z)\,dz \tag{9.12–18}$$

Substituting into equation

$$\frac{d}{dt}\frac{\partial T}{\partial \dot{x}_0} - \dot{\theta}\frac{\partial T}{\partial \dot{y}_0} = \sum F_x \tag{9.12–19}$$

we obtain

$$\int_{-l_1}^{l_2} [(\ddot{x}_0 - y\ddot{\theta} - \dot{y}\dot{\theta}) - \dot{\theta}(\dot{y}_0 + \dot{y} + x\dot{\theta})]m(x)\,dx$$
$$+ (2\dot{\theta}\dot{\delta} + \ddot{\theta}\delta)\int_0^b z\,m(z)\,dz = \sum F_x \tag{9.12–20}$$

This equation can be rearranged as follows:

$$m(\ddot{x}_0 - \dot{\theta}\dot{y}_0) - 2\dot{\theta}\left[\int_{-l_1}^{l_2} \dot{y}\, m(x)\, dx - \delta\int_0^b z\, m(z)\, dz\right]$$

$$- \ddot{\theta}\left[\int_{-l_1}^{l_2} y\, m(x)\, dx - \delta\int_0^b z\, m(z)\, dz\right]$$

$$- \dot{\theta}^2\int_{-l_1}^{l_2} x\, m(x)\, dx = \sum F_x \qquad (9.12\text{--}21)$$

Since the second and third terms on the left side of this equation are the linear momentum relative to the coordinates, or the condition for the x axis passing through the center of mass (see Eq. 9.12–4), they are zero. The term $\dot{\theta}^2$ is also a negligibly small term. We have, therefore, demonstrated the use of the Lagrange's equation for rotating coordinates.

For the beam equation relating to the generalized coordinate q_i and the engine rotation equation relating to δ, the usual form of Lagrange's equation applies. We need, however, the equation for the potential or strain energy due to bending, which is,

$$U = \frac{1}{2}\int_{-l_1}^{l_2} EI_A\left(\frac{\partial^2 y}{\partial x^2}\right)^2 dx \qquad (9.12\text{--}22$$

From Eq. 9.12–6 the curvature is

$$\frac{\partial^2 y}{\partial x^2} = \sum_i q_i\varphi_i{}'' \qquad (9.12\text{--}23)$$

where the primes stand for differentiation with respect to x. Substituting into U and making use of the orthogonality relation,

$$\int_{-l_1}^{l_2} EI_A\varphi_i{}''\varphi_j{}''\, dx = \begin{cases} 0 & \text{for } j \neq i \\ m\omega_i{}^2 & \text{for } j = i \end{cases} \qquad (9.12\text{--}24)$$

The equation for the strain energy becomes,

$$U = \frac{1}{2}m\sum_i \omega_i{}^2 q_i{}^2 \qquad (9.12\text{--}25)$$

The various partial derivatives needed for Lagrange's equation are:

$$\frac{\partial T}{\partial \dot{q}_i} = \int_{-l_1}^{l_2} (\dot{y}_0 + \dot{y} + x\dot{\theta})\,\frac{\partial \dot{y}}{\partial \dot{q}_i}\, m(x)\, dx - \delta\int_0^b \frac{\partial \dot{y}}{\partial \dot{q}_i} z\, m(z)\, dz$$

$$\frac{\partial T}{\partial q_i} = -\dot{\theta}\int_{-l_1}^{l_2} (\dot{x}_0 - y\dot{\theta})\frac{\partial y}{\partial q_i}\, m(x)\, dx - \dot{\theta}^2\delta\int_0^b \frac{\partial y}{\partial q_i} z\, m(z)\, dz$$

$$\frac{\partial U}{\partial q_i} = m\omega_i{}^2 q_i$$

We note here that $\dot{\theta}^2$ is a very small quantity, and terms multiplied by it can be neglected. Substituting for y from Eq. 9.12–6, taking into account the orthogonality condition, and noting that the linear and angular momenta of the vibration modes are zero,

$$\int_{-l_1}^{l_2} \varphi_i \, m(x) \, dx = \int_{-l_1}^{l_2} x\varphi_i \, m(x) \, dx = 0 \qquad (9.12–26)$$

we arrive at the result,

$$m(\ddot{q}_i + \omega_i^2 q_i) - \ddot{\delta} \int_0^b \varphi_i z \, dm$$
$$= -\int_{-l_1}^{l_2} L^* \varphi_i \, dx + F_e(\delta + y_h')\varphi_i(-l_h) \qquad (9.12–27)$$

where the right side of the equation is established from the generalized force

$$Q_{qi} = \int \bar{F} \frac{\delta y}{\delta q_i} \, dx = \int \bar{F} \frac{\partial y}{\partial q_i} \, dx \qquad (9.12–28)$$

In a similar manner, the engine rotation equation is determined from

$$\frac{d}{dt} \frac{\partial T}{\partial \dot{\delta}} - \frac{\partial T}{\partial \delta} + \frac{\partial U}{\partial \delta} = Q_\delta \qquad (9.12–29)$$

which gives

$$[I_h - C_1 m_e z_e - C_2(I_h + m_e l_h z_e)]\ddot{\delta}$$
$$+ m_e z_e(\ddot{x}_0 - \dot{\theta}\dot{y}_0 + g \cos \theta)(\delta + y_h') + (I_h + m_e l_h z_e)\ddot{\theta}$$
$$- m_e z_e(\ddot{y}_0 + \dot{\theta}\dot{x}_0) - \sum_i \ddot{q}_i \left[\int_0^b \varphi_i z \, m(z) \, dz \right]$$
$$= M_A(\delta_A - \delta) - M_D \dot{\delta} - M_\delta \delta \qquad (9.12–30)$$

where the generalized force $Q_\delta = M_A(\delta_A - \delta) - M_D \dot{\delta} - M_\delta \delta$ is associated with M_δ, the spring moment per unit angle of the hinge when locked, M_D with the damping moment of the hinge, M_A with the engine actuator moment, and δ_A with the engine actuator position called by the autopilot. The autopilot actuating the engine attitude δ operates from signals generated by the rate gyros in the missile in such a way that the missile motion at the position of the rate gyro, which depends on all of the generalized coordinates, is coupled to the previous equation through the engine attitude δ.

PROBLEMS

1. Assume that the mass of a portion of the missile is represented by a lumped mass m_j mounted on a spring of stiffness k against lateral motion ζ_j from the center line at x_j. Derive the equation for the additional kinetic energy of the missile due to ζ_j.

2. Derive Eq. 9.12–13 by the use of Lagrange's equation,

$$\frac{d}{dt}\frac{\partial T}{\partial \dot{y}_0} + \theta\frac{\partial T}{\partial \dot{x}_0} = \sum F_y$$

3. Derive Eq. 9.12–14 by the use of Lagrange's equation,

$$\frac{d}{dt}\frac{\partial T}{\partial \theta} + \dot{x}_0\frac{\partial T}{\partial \dot{y}_0} - \dot{y}_0\frac{\partial T}{\partial \dot{x}_0} = \sum M_0$$

4. Prove Eqs. 9.12–24 and 9.12–25.

5. Derive Eq. 9.12–30 for the engine rotation.

6. An elastic uniform bar of mass m and length l is supported from its upper end by a smooth pin. A constant force P is suddenly applied normal to the bar at its mid-length. Using generalized coordinates q_i, and three arbitrary modes,

$$\varphi_1 = \frac{x}{l} \qquad \varphi_2 = \sin\frac{\pi x}{l} \qquad \varphi_3 = \sin\frac{2\pi x}{l}$$

determine the equation of motion, the natural frequencies, and the mode shapes.

7. It is proposed to determine the natural frequencies of the two span beams of unequal length by assuming the deflection to be expressible by the equation,

$$y = \sum_i q_i \varphi_i(x)$$

where

$$\varphi_i(x) = \sqrt{2}\sin\frac{i\pi x}{l}$$

$$\omega_i = (i_n)^2\sqrt{\frac{EI}{ml^4}}$$

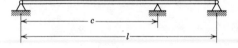

Prob. 7

Show that the constraint equation $\sum_i q_i \varphi_i(c) = 0$, must be imposed and that the equation for the natural frequencies is,

$$\frac{\varphi_1^2(c)}{\omega_1^2 - \omega^2} + \frac{\varphi_2^2(c)}{16\omega_1^2 - \omega^2} + \frac{\varphi_3^2(c)}{81\omega_1^2 - \omega^2} = 0$$

General References

1. Davis, L., D. Follin, and L. Blitzer, Exterior Ballistics of Rockets, D. Van Nostrand, Princeton, N. J. (1958).
2. Deimel, R. F., *Mechanics of the Gyroscope*, The Macmillan Book Co., New York (1929). Reprinted by Dover Publications.
3. Ehricke, H. A., *Space Flight*, Vol. 1, D. Van Nostrand, New York (1960).
4. Goldstein, H., *Classical Mechanics*, Addison-Wesley Publishing Co., Reading, Mass. (1951).
5. Gray, A., *A Treatise on Gyrostatics and Rotational Motion*, The Macmillan Book Co., New York (1918). Reprinted by Dover Publications.
6. Lanczos, C., *The Variational Principle of Mechanics*, The University of Toronto Press, Toronto (1949).
7. McCuskey, S. W., *Introduction to Advanced Dynamics*, Addison-Wesley Publishing Co., Reading, Mass. (1958).
8. Moulton, F. R., *Introduction to Celestial Mechanics*, 2nd. ed., The Macmillan Book Co., New York (1914).
9. Osgood, W. F., *Mechanics*, The Macmillan Book Co., New York (1948).
10. Rosser, J. B., R. R. Newton, and G. L. Gross, *Mathematical Theory of Rocket Flight*, McGraw-Hill Book Co., New York (1947).
11. Routh, E. J., *Advanced Dynamics of a System of Rigid Bodies*, 6th ed., The Macmillan Book Co., London (1905). Reprinted by Dover Publications.
12. Scarborough, J. B., *The Gyroscope*, Interscience Publishers, New York (1958).
13. Seifert, H. S., *et al.*, *Space Technology*, John Wiley and Sons, New York (1959).
14. Smart, E. H., *Advanced Dynamics*, The Macmillan Book Co., New York (1951).
15. Sommerfeld, A., *Mechanics*, Academic Press, New York (1952).
16. Synge, J. L., and B. A. Griffith, *Principles of Mechanics*, 3rd ed., McGraw-Hill Book Co., New York (1959).
17. Timoshenko, S., and D. H. Young, *Advanced Dynamics*, McGraw-Hill Book Co., New York (1948).
18. Webster, A. G., *The Dynamics of Particles*, 2nd ed. B. G. Teubner, Leipzig, (1912). Reprinted by Dover Publications.

Matrices

APPENDIX A

A system of linear equations

$$y_1 = a_{11}x_1 + a_{12}x_2 + a_{13}x_3$$
$$y_2 = a_{21}x_1 + a_{22}x_2 + a_{23}x_3 \qquad \text{(A-1)}$$
$$y_3 = a_{31}x_1 + a_{32}x_2 + a_{33}x_3$$

can be arranged into the matrix notation

$$\begin{bmatrix} y_1 \\ y_2 \\ y_3 \end{bmatrix} = \begin{bmatrix} a_{11} & a_{12} & a_{13} \\ a_{21} & a_{22} & a_{23} \\ a_{31} & a_{32} & a_{33} \end{bmatrix} \begin{bmatrix} x_1 \\ x_2 \\ x_3 \end{bmatrix} \qquad \text{(A-2)}$$

where the rule for the matrix multiplication is evident from the original equations. For Eq. A–2 to equal Eq. A–1, the terms of each row must be multiplied by the terms of the column x_1, x_2, x_3. We can then view the matrix equation, Eq. A–2, as a convenient notation which may be further abbreviated to

$$\{y\} = [a]\{x\} \qquad \text{(A-3)}$$

We will next consider another set of linear equations like that of Eq. A–1, relating x to z, and write its matrix form as

$$\{x\} = [b]\{z\} \qquad \text{(A-4)}$$

where $[b]$ is a square matrix like that of Eq. A–2. If we wish to relate y to z, Eq. A–4 can be substituted into Eq. A–3 as follows:

$$\{y\} = [a][b]\{z\}$$
$$= [c]\{z\} \qquad \text{(A-5)}$$

The elements of $[c]$ can then be shown to be available from the equation

$$c_{ij} = \sum_k a_{ik}b_{kj} \tag{A-6}$$

i.e., the third element of the second row is

$$c_{23} = a_{21}b_{13} + a_{22}b_{23} + a_{23}b_{33}$$

There are many theorems relating to the manipulation of matrix equations; however for the purposes of linear transformation of co-ordinates, as treated in this text, the simple algebraic concepts discussed above are sufficient.

REFERENCES

1. Frazer, R. A., W. J. Duncan, and A. R. Collar, *Elementary Matrices*, Cambridge University Press, New York (1938).
2. Pipes, L. A., *Applied Mathematics for Engineers and Physicists*, 2nd ed., McGraw-Hill Book Co., New York (1958), Chap. 4.

Dyadics

APPENDIX B

We occasionally encounter a quantity which has nine components in a three-dimensional space. In elasticity we encounter nine components of stress at a point, whereas in dynamics we find nine components of inertia.

For our purposes we can define a dyadic as a nine-component quantity which can be formed by multiplying two vectors, ignoring the dot- or cross-product rule. Thus the product of two vectors **a** and **b** is,

$$\begin{aligned} \mathbf{ab} = {}& \mathbf{ii}a_x b_x + \mathbf{ij}a_x b_y + \mathbf{ik}a_x b_z \\ & + \mathbf{ji}a_y b_x + \mathbf{jj}a_y b_y + \mathbf{jk}a_y b_z \\ & + \mathbf{ki}a_z b_x + \mathbf{kj}a_z b_y + \mathbf{kk}a_z b_z \end{aligned} \tag{B-1}$$

Although the above dyadic was formed by the multiplication of the two vectors **a** and **b**, the elements of the dyadic (called dyads) need not be related to the two vectors. Furthermore, it is convenient to arrange such terms in matrix form, so that a dyadic is in general expressible as,

$$\mathscr{C} = \begin{bmatrix} \mathbf{ii}c_{xx} & \mathbf{ij}c_{xy} & \mathbf{ik}c_{xz} \\ \mathbf{ji}c_{yx} & \mathbf{jj}c_{yy} & \mathbf{jk}c_{yz} \\ \mathbf{ki}c_{zx} & \mathbf{kj}c_{zy} & \mathbf{kk}c_{zz} \end{bmatrix} \tag{B-2}$$

As an example of a dyadic not related to any vector, we have the inertia dyadic,

$$\mathscr{I} = \begin{bmatrix} \mathbf{ii}I_{xx} & -\mathbf{ij}I_{xy} & -\mathbf{ik}I_{xz} \\ -\mathbf{ji}I_{yx} & \mathbf{jj}I_{yy} & -\mathbf{jk}I_{yz} \\ -\mathbf{ki}I_{zx} & -\mathbf{kj}I_{zy} & \mathbf{kk}I_{zz} \end{bmatrix} \tag{B-3}$$

307

To illustrate the general rule for the dot or cross product of a dyadic with a vector, we assume that the dyadic is formed by the product of two vectors as

$$\mathscr{C} = \mathbf{ab} \tag{B-4}$$

Its dot and cross product with a vector $\mathbf{r}$ is then dependent on the order of the product, and its interpretation is made clear by the following examples:

$$\mathscr{C} \cdot \mathbf{r} = (\mathbf{ab}) \cdot \mathbf{r} = \mathbf{a}(\mathbf{b} \cdot \mathbf{r}) = \text{a vector in the direction of } \mathbf{a} \tag{B-5}$$

$$\mathbf{r} \cdot \mathscr{C} = \mathbf{r} \cdot (\mathbf{ab}) = (\mathbf{r} \cdot \mathbf{a})\mathbf{b} = \text{a vector in the direction of } \mathbf{b} \tag{B-6}$$

$$\mathscr{C} \times \mathbf{r} = (\mathbf{ab}) \times \mathbf{r} = \mathbf{a}(\mathbf{b} \times \mathbf{r}) = \text{another dyadic} \tag{B-7}$$

$$\mathbf{r} \times \mathscr{C} = \mathbf{r} \times (\mathbf{ab}) = (\mathbf{r} \times \mathbf{a})\mathbf{b} = \text{another dyadic} \tag{B-8}$$

If we form the dot product of the inertia dyadic with the angular velocity vector $\boldsymbol{\omega} = \omega_x \mathbf{i} + \omega_y \mathbf{j} + \omega_z \mathbf{k}$, the result will be the angular momentum vector.

$$\begin{aligned}
\mathbf{h} = \mathscr{I} \cdot \boldsymbol{\omega} = &\mathbf{i}(I_{xx}\omega_x - I_{xy}\omega_y - I_{xz}\omega_z) \\
&+ \mathbf{j}(-I_{xy}\omega_x + I_{yy}\omega_y - I_{yz}\omega_z) \\
&+ \mathbf{k}(-I_{xz}\omega_x - I_{yz}\omega_y + I_{zz}\omega_z)
\end{aligned} \tag{B-9}$$

Here we encounter dot products such as,

$$\mathbf{ji} \cdot \mathbf{i} = \mathbf{j}(\mathbf{i} \cdot \mathbf{i}) = \mathbf{j}$$

$$\mathbf{ji} \cdot \mathbf{j} = \mathbf{j}(\mathbf{i} \cdot \mathbf{j}) = 0, \text{ etc.}$$

which are evident from the general rule, and recognize that the subscripts of the inertia are interchangeable, i.e., $I_{xy} = I_{yx}$. If, furthermore, we dot the angular velocity vector into the angular momentum vector, the result is a scalar, which in this case is twice the kinetic energy.

$$2T = \boldsymbol{\omega} \cdot \mathscr{I} \cdot \boldsymbol{\omega} \tag{B-10}$$

In summary, the dyadic is a special form of a tensor; however, our simple definition of the dyadic and its product with a vector requires no new rules of vector algebra, which appears to be adequate for the dynamical problems encountered in this text. For further reading on the subject, see reference.

REFERENCE

1. Weatherburn, C. E., *Advanced Vector Analysis*, G. Bell & Sons, Ltd., London (1947), Chapter 5, and p. 207.

The Variational Calculus

APPENDIX C

Many problems in Dynamics are formulated in terms of maxima and minima of quantities expressed by an integral. In this section we will briefly discuss the essentials of the variational calculus which are encountered for such problems.

Consider the integral,

$$I = \int_a^b f\left(t, z, \frac{dz}{dt}\right) dt \qquad (C\ 1)$$

taken along a curve $z = \psi(t)$. The quantity z can stand for any number of variables, such as position x, y, and the thrust attitude ϕ of the missile problem. The value of the integral I will depend on the curve $z = \psi(t)$ which we wish to find for the condition of maxima or minima of the integral C–1.

Assuming that $z = \psi(t)$ along curve ab of Fig. C–1 to be the optimum curve, we draw curve 1 along ab_1 as the varied curve. The quantity z along the varied curve is represented by

$$z_1 = z + \delta z \qquad (C–2)$$

where δz is the variation of z. The variation δz differs from dz in that dz is the increment along the curve z due to an increment dt, whereas δz is the difference in the z between the two curves for any given time t. We can also define the difference in the slopes of the z curves at any time t to be

$$\delta z' = z_1' - z' \qquad (C–3)$$

If δz is assumed to be a continuous function of time, we can differentiate Eq. C–2 and obtain

$$\frac{d}{dt}\,\delta z = z_1{}' - z' \tag{C–4}$$

Comparing Eqs. C–3 and C–4, we find that

$$\frac{d}{dt}\,\delta z = \delta\frac{dz}{dt} \tag{C–5}$$

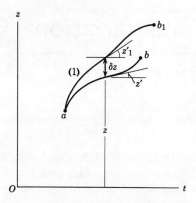

Fig. C–I.   Curve (1) is the varied curve of *ab*.

which indicates that the orders of operation of δ and d/dt are interchangeable.   Likewise, it can be shown that the interchangeability rule applies to integrals.

$$\delta\int z\,dt = \int \delta z\,dt \tag{C–6}$$

With this understanding of δ, we now express f along the varied curve by expanding it about the original curve.   By Taylor series we can write,

$$f(t, z + \delta z, z' + \delta z') = f(t, z, z') + \frac{\partial f}{\partial z}\,\delta z + \frac{\partial f}{\partial z'}\,\delta z' + \cdots \tag{C–7}$$

Considering only the first order variation, the variation of the integral I is

$$\delta I = \int_a^b \left(\frac{\partial f}{\partial z}\,\delta z + \frac{\partial f}{\partial z'}\,\delta z'\right) dt \tag{C–8}$$

The second term in this variation can be integrated by parts so that the final expression for δI becomes

$$\delta I = \frac{\partial f}{\partial z'}\,\delta z\,\Big|_a^b + \int_a^b \left(\frac{\partial f}{\partial z} - \frac{d}{dt}\frac{\partial f}{\partial z'}\right) \delta z\,dt \tag{C–9}$$

For the curve $z = \psi(t)$ to be the optimum curve resulting in a maximum or minimum of I, the variation δI must equal zero.

Most texts treat the case where the variation δz is zero at the end points a and b (i.e., b_1 coincides with b). In such cases the first term of Eq. C–9 vanishes, and we are left only with the integral. Since δz in the integral is arbitrary and not zero over the interval a to b, for the integrand to be zero, $\delta I = 0$, and we obtain Euler's equation

$$\frac{d}{dt}\frac{\partial f}{\partial z'} - \frac{\partial f}{\partial z} = 0 \tag{C–10}$$

The satisfaction of Euler's equation insures that the integral I is a maximum or minimum. For the more general case where δz is not zero at the end points, we must retain the first term and consider the entire expression $\delta I = 0$, of Eq. C–9.

Variation with Constraints

We often encounter problems of optimization where the integral I must be maximized or minimized under conditions of constraints,

$$g\left(t, z, \frac{dz}{dt}\right) = 0$$
$$h\left(t, z, \frac{dz}{dt}\right) = 0 \tag{C–11}$$

The procedure to be followed is then to multiply the constraint equations by arbitrary functions λ, and maximize or minimize the equation

$$I = \int_a^b [f(t, z, z') + \lambda_1 g(t, z, z') + \lambda_2 h(t, z, z') + \cdots] \, dt \tag{C–12}$$

which insures the satisfaction of the conditions of constraints in the process of optimization. The quantities λ are also functions of the variables of the problem and, since they are multiplied by zeros, the expression for I is unaltered. Equation C–9 is again applicable where f now stands for the entire integrand of Eq. C–12.

Index

313

A CATALOG OF SELECTED DOVER
BOOKS IN ALL FIELDS OF INTEREST

LASERS AND HOLOGRAPHY, Winston E. Kock. Sound introduction to burgeoning field, expanded (1981) for second edition. 84 illustrations. 160pp. 5⅜ × 8¼. (EUK)   24041-X Pa. $3.50

FLORAL STAINED GLASS PATTERN BOOK, Ed Sibbett, Jr. 96 exquisite floral patterns—irises, poppie, lilies, tulips, geometrics, abstracts, etc.—adaptable to innumerable stained glass projects. 64pp. 8¼ × 11.   24259-5 Pa. $3.50

THE HISTORY OF THE LEWIS AND CLARK EXPEDITION, Meriwether Lewis and William Clark. Edited by Eliott Coues. Great classic edition of Lewis and Clark's day-by-day journals. Complete 1893 edition, edited by Eliott Coues from Biddle's authorized 1814 history. 1508pp. 5⅜ × 8½.
21268-8, 21269-6, 21270-X Pa. Three-vol. set $22.50

ORLEY FARM, Anthony Trollope. Three-dimensional tale of great criminal case. Original Millais illustrations illuminate marvelous panorama of Victorian society. Plot was author's favorite. 736pp. 5⅜ × 8½.   24181-5 Pa. $10.95

THE CLAVERINGS, Anthony Trollope. Major novel, chronicling aspects of British Victorian society, personalities. 16 plates by M. Edwards; first reprint of full text. 412pp. 5⅜ × 8½.   23464-9 Pa. $6.00

EINSTEIN'S THEORY OF RELATIVITY, Max Born. Finest semi-technical account; much explanation of ideas and math not readily available elsewhere on this level. 376pp. 5⅜ × 8½.   60769-0 Pa. $5.00

COMPUTABILITY AND UNSOLVABILITY, Martin Davis. Classic graduate-level introduction th theory of computability, usually referred to as theory of recurrent functions. New preface and appendix. 288pp. 5⅜ × 8½. 61471-9 Pa. $6.50

THE GODS OF THE EGYPTIANS, E.A. Wallis Budge. Never excelled for richness, fullness: all gods, goddesses, demons, mythical figures of Ancient Egypt; their legends, rites, incarnations, etc. Over 225 illustrations, plus 6 color plates. 988pp. 6⅛ × 9¼. (EBE)   22055-9, 22056-7 Pa., Two-vol. set $20.00

THE I CHING (THE BOOK OF CHANGES), translated by James Legge. Most penetrating divination manual ever prepared. Indispensable to study of early Oriental civilizations, to modern inquiring reader. 448pp. 5⅜ × 8½.
21062-6 Pa. $6.50

THE CRAFTSMAN'S HANDBOOK, Cennino Cennini. 15th-century handbook, school of Giotto, explains applying gold, silver leaf; gesso; fresco painting, grinding pigments, etc. 142pp. 6⅛ × 9¼.   20054-X Pa. $3.50

AN ATLAS OF ANATOMY FOR ARTISTS, Fritz Schider. Finest text, working book. Full text, plus anatomical illustrations; plates by great artists showing anatomy. 593 illustrations. 192pp. 7⅛ × 10¼.   20241-0 Pa. $6.50

EASY-TO-MAKE STAINED GLASS LIGHTCATCHERS, Ed Sibbett, Jr. 67 designs for most enjoyable ornaments: fruits, birds, teddy bears, trumpet, etc. Full size templates. 64pp. 8¼ × 11.   24081-9 Pa. $3.95

TRIAD OPTICAL ILLUSIONS AND HOW TO DESIGN THEM, Harry Turner. Triad explained in 32 pages of text, with 32 pages of Escher-like patterns on coloring stock. 92 figures. 32 plates. 64pp. 8¼ × 11.   23549-1 Fa. $2.95

REASON IN ART, George Santayana. Renowned philosopher's provocative, seminal treatment of basis of art in instinct and experience. Volume Four of *The Life of Reason*. 230pp. 5⅜ × 8. 24358-3 Pa. $4.50

LANGUAGE, TRUTH AND LOGIC, Alfred J. Ayer. Famous, clear introduction to Vienna, Cambridge schools of Logical Positivism. Role of philosophy, elimination of metaphysics, nature of analysis, etc. 160pp. 5⅜ × 8½. (USCO) 20010-8 Pa. $2.75

BASIC ELECTRONICS, U.S. Bureau of Naval Personnel. Electron tubes, circuits, antennas, AM, FM, and CW transmission and receiving, etc. 560 illustrations. 567pp. 6½ × 9¼. 21076-6 Pa. $8.95

THE ART DECO STYLE, edited by Theodore Menten. Furniture, jewelry, metalwork, ceramics, fabrics, lighting fixtures, interior decors, exteriors, graphics from pure French sources. Over 400 photographs. 183pp. 8⅜ × 11¼. 22824-X Pa. $6.95

THE FOUR BOOKS OF ARCHITECTURE, Andrea Palladio. 16th-century classic covers classical architectural remains, Renaissance revivals, classical orders, etc. 1738 Ware English edition. 216 plates. 110pp. of text. 9½ × 12¾. 21308-0 Pa. $11.50

THE WIT AND HUMOR OF OSCAR WILDE, edited by Alvin Redman. More than 1000 ripostes, paradoxes, wisecracks: Work is the curse of the drinking classes, I can resist everything except temptations, etc. 258pp. 5⅜ × 8½. (USCO) 20602-5 Pa. $3.95

THE DEVIL'S DICTIONARY, Ambrose Bierce. Barbed, bitter, brilliant witticisms in the form of a dictionary. Best, most ferocious satire America has produced. 145pp. 5⅜ × 8½. 20487-1 Pa. $2.50

ERTÉ'S FASHION DESIGNS, Erté. 210 black-and-white inventions from *Harper's Bazar*, 1918-32, plus 8pp. full-color covers. Captions. 88pp. 9 × 12. 24203-X Pa. $6.50

ERTÉ GRAPHICS, Erté. Collection of striking color graphics: *Seasons, Alphabet, Numerals, Aces* and *Precious Stones*. 50 plates, including 4 on covers. 48pp. 9⅜ × 12¼. 23580-7 Pa. $6.95

PAPER FOLDING FOR BEGINNERS, William D. Murray and Francis J. Rigney. Clearest book for making origami sail boats, roosters, frogs that move legs, etc. 40 projects. More than 275 illustrations. 94pp. 5⅜ × 8½. 20713-7 Pa. $2.25

ORIGAMI FOR THE ENTHUSIAST, John Montroll. Fish, ostrich, peacock, squirrel, rhinoceros, Pegasus, 19 other intricate subjects. Instructions. Diagrams. 128pp. 9 × 12. 23799-0 Pa. $4.95

CROCHETING NOVELTY POT HOLDERS, edited by Linda Macho. 64 useful, whimsical pot holders feature kitchen themes, animals, flowers, other novelties. Surprisingly easy to crochet. Complete instructions. 48pp. 8¼ × 11. 24296-X Pa. $1.95

CROCHETING DOILIES, edited by Rita Weiss. Irish Crochet, Jewel, Star Wheel, Vanity Fair and more. Also luncheon and console sets, runners and centerpieces. 51 illustrations. 48pp. 8¼ × 11. 23424-X Pa. $2.50

DECORATIVE NAPKIN FOLDING FOR BEGINNERS, Lillian Oppenheimer and Natalie Epstein. 22 different napkin folds in the shape of a heart, clown's hat, love knot, etc. 63 drawings. 48pp. 8¼ × 11. 23797-4 Pa. $1.95

DECORATIVE LABELS FOR HOME CANNING, PRESERVING, AND OTHER HOUSEHOLD AND GIFT USES, Theodore Menten. 128 gummed, perforated labels, beautifully printed in 2 colors. 12 versions. Adhere to metal, glass, wood, ceramics. 24pp. 8¼ × 11. 23219-0 Pa. $2.95

EARLY AMERICAN STENCILS ON WALLS AND FURNITURE, Janet Waring. Thorough coverage of 19th-century folk art: techniques, artifacts, surviving specimens. 166 illustrations, 7 in color. 147pp. of text. 7⅞ × 10¾. 21906-2 Pa. $9.95

AMERICAN ANTIQUE WEATHERVANES, A.B. & W.T. Westervelt. Extensively illustrated 1883 catalog exhibiting over 550 copper weathervanes and finials. Excellent primary source by one of the principal manufacturers. 104pp. 6⅛ × 9¼. 24396-6 Pa. $3.95

ART STUDENTS' ANATOMY, Edmond J. Farris. Long favorite in art schools. Basic elements, common positions, actions. Full text, 158 illustrations. 159pp. 5⅜ × 8½. 20744-7 Pa. $3.95

BRIDGMAN'S LIFE DRAWING, George B. Bridgman. More than 500 drawings and text teach you to abstract the body into its major masses. Also specific areas of anatomy. 192pp. 6½ × 9¼. (EA) 22710-3 Pa. $4.50

COMPLETE PRELUDES AND ETUDES FOR SOLO PIANO, Frederic Chopin. All 26 Preludes, all 27 Etudes by greatest composer of piano music. Authoritative Paderewski edition. 224pp. 9 × 12. (Available in U.S. only) 24052-5 Pa. $7.50

PIANO MUSIC 1888-1905, Claude Debussy. Deux Arabesques, Suite Bergamesque, Masques, 1st series of Images, etc. 9 others, in corrected editions. 175pp. 9⅜ × 12¼. (ECE) 22771-5 Pa. $5.95

TEDDY BEAR IRON-ON TRANSFER PATTERNS, Ted Menten. 80 iron-on transfer patterns of male and female Teddys in a wide variety of activities, poses, sizes. 48pp. 8¼ × 11. 24596-9 Pa. $2.25

A PICTURE HISTORY OF THE BROOKLYN BRIDGE, M.J. Shapiro. Profusely illustrated account of greatest engineering achievement of 19th century. 167 rare photos & engravings recall construction, human drama. Extensive, detailed text. 122pp. 8¼ × 11. 24403-2 Pa. $7.95

NEW YORK IN THE THIRTIES, Berenice Abbott. Noted photographer's fascinating study shows new buildings that have become famous and old sights that have disappeared forever. 97 photographs. 97pp. 11⅜ × 10. 22967-X Pa. $7.50

MATHEMATICAL TABLES AND FORMULAS, Robert D. Carmichael and Edwin R. Smith. Logarithms, sines, tangents, trig functions, powers, roots, reciprocals, exponential and hyperbolic functions, formulas and theorems. 269pp. 5⅜ × 8½. 60111-0 Pa. $4.95

HANDBOOK OF MATHEMATICAL FUNCTIONS WITH FORMULAS, GRAPHS, AND MATHEMATICAL TABLES, edited by Milton Abramowitz and Irene A. Stegun. Vast compendium: 29 sets of tables, some to as high as 20 places. 1,046pp. 8 × 10½. 61272-4 Pa. $19.95

CATALOG OF DOVER BOOKS

TWENTY-FOUR ART NOUVEAU POSTCARDS IN FULL COLOR FROM CLASSIC POSTERS, Hayward and Blanche Cirker. Ready-to-mail postcards reproduced from rare set of poster art. Works by Toulouse-Lautrec, Parrish, Steinlen, Mucha, Cheret, others. 12pp. 8¼× 11. 24389-3 Pa. $2.95

READY-TO-USE ART NOUVEAU BOOKMARKS IN FULL COLOR, Carol Belanger Grafton. 30 elegant bookmarks featuring graceful, flowing lines, foliate motifs, sensuous women characteristic of Art Nouveau. Perforated for easy detaching. 16pp. 8¼ × 11. 24305-2 Pa. $2.95

FRUIT KEY AND TWIG KEY TO TREES AND SHRUBS, William M. Harlow. Fruit key covers 120 deciduous and evergreen species; twig key covers 160 deciduous species. Easily used. Over 300 photographs. 126pp. 5⅜ × 8½. 20511-8 Pa. $2.25

LEONARDO DRAWINGS, Leonardo da Vinci. Plants, landscapes, human face and figure, etc., plus studies for Sforza monument, *Last Supper*, more. 60 illustrations. 64pp. 8¼ × 11⅛. 23951-9 Pa. $2.75

CLASSIC BASEBALL CARDS, edited by Bert R. Sugar. 98 classic cards on heavy stock, full color, perforated for detaching. Ruth, Cobb, Durocher, DiMaggio, H. Wagner, 99 others. Rare originals cost hundreds. 16pp. 8¼ × 11. 23498-3 Pa. $3.25

TREES OF THE EASTERN AND CENTRAL UNITED STATES AND CANADA, William M. Harlow. Best one-volume guide to 140 trees. Full descriptions, woodlore, range, etc. Over 600 illustrations. Handy size. 288pp. 4½ × 6⅜. 20395-6 Pa. $3.95

JUDY GARLAND PAPER DOLLS IN FULL COLOR, Tom Tierney. 3 Judy Garland paper dolls (teenager, grown-up, and mature woman) and 30 gorgeous costumes highlighting memorable career. Captions. 32pp. 9¼ × 12¼. 24404-0 Pa. $3.50

GREAT FASHION DESIGNS OF THE BELLE EPOQUE PAPER DOLLS IN FULL COLOR, Tom Tierney. Two dolls and 30 costumes meticulously rendered. Haute couture by Worth, Lanvin, Paquin, other greats late Victorian to WWI. 32pp. 9¼ × 12¼. 24425-3 Pa. $3.50

FASHION PAPER DOLLS FROM GODEY'S LADY'S BOOK, 1840-1854, Susan Johnston. In full color: 7 female fashion dolls with 50 costumes. Little girl's, bridal, riding, bathing, wedding, evening, everyday, etc. 32pp. 9¼ × 12¼. 23511-4 Pa. $3.95

THE BOOK OF THE SACRED MAGIC OF ABRAMELIN THE MAGE, translated by S. MacGregor Mathers. Medieval manuscript of ceremonial magic. Basic document in Aleister Crowley, Golden Dawn groups. 268pp. 5⅜ × 8½. 23211-5 Pa. $5.00

PETER RABBIT POSTCARDS IN FULL COLOR: 24 Ready-to-Mail Cards, Susan Whited LaBelle. Bunnies ice-skating, coloring Easter eggs, making valentines, many other charming scenes. 24 perforated full-color postcards, each measuring 4¼ × 6, on coated stock. 12pp. 9 × 12. 24617-5 Pa. $2.95

CELTIC HAND STROKE BY STROKE, A. Baker. Complete guide creating each letter of the alphabet in distinctive Celtic manner. Covers hand position, strokes, pens, inks, paper, more. Illustrated. 48pp. 8¼ × 11. 24336-2 Pa. $2.50

THE BOOK OF WOOD CARVING, Charles Marshall Sayers. Still finest book for beginning student. Fundamentals, technique; gives 34 designs, over 34 projects for panels, bookends, mirrors, etc. 33 photos. 118pp. 7¾ × 10⅝.   23654-4 Pa. $3.95

CARVING COUNTRY CHARACTERS, Bill Higginbotham. Expert advice for beginning, advanced carvers on materials, techniques for creating 18 projects—mirthful panorama of American characters. 105 illustrations. 80pp. 8⅜ × 11.
23135-1 Pa. $2.50

300 ART NOUVEAU DESIGNS AND MOTIFS IN FULL COLOR, C.B. Grafton. 44 full-page plates display swirling lines and muted colors typical of Art Nouveau. Borders, frames, panels, cartouches, dingbats, etc. 48pp. 9⅜ × 12¼.
24354-0 Pa. $6.95

SELF-WORKING CARD TRICKS, Karl Fulves. Editor of *Pallbearer* offers 72 tricks that work automatically through nature of card deck. No sleight of hand needed. Often spectacular. 42 illustrations. 113pp. 5⅜ × 8½.   23334-0 Pa. $3.50

CUT AND ASSEMBLE A WESTERN FRONTIER TOWN, Edmund V. Gillon, Jr. Ten authentic full-color buildings on heavy cardboard stock in H-O scale. Sheriff's Office and Jail, Saloon, Wells Fargo, Opera House, others. 48pp. 9¼ × 12¼.
23736-2 Pa. $3.95

CUT AND ASSEMBLE AN EARLY NEW ENGLAND VILLAGE, Edmund V. Gillon, Jr. Printed in full color on heavy cardboard stock. 12 authentic buildings in H-O scale: Adams home in Quincy, Mass., Oliver Wight house in Sturbridge, smithy, store, church, others. 48pp. 9¼ × 12¼.   23536-X Pa. $4.95

THE TALE OF TWO BAD MICE, Beatrix Potter. Tom Thumb and Hunca Munca squeeze out of their hole and go exploring. 27 full-color Potter illustrations. 59pp. 4¼ × 5½. (Available in U.S. only)   23065-1 Pa. $1.75

CARVING FIGURE CARICATURES IN THE OZARK STYLE, Harold L. Enlow. Instructions and illustrations for ten delightful projects, plus general carving instructions. 22 drawings and 47 photographs altogether. 39pp. 8⅜ × 11.
23151-8 Pa. $2.50

A TREASURY OF FLOWER DESIGNS FOR ARTISTS, EMBROIDERERS AND CRAFTSMEN, Susan Gaber. 100 garden favorites lushly rendered by artist for artists, craftsmen, needleworkers. Many form frames, borders. 80pp. 8¼ × 11.
24096-7 Pa. $3.50

CUT & ASSEMBLE A TOY THEATER/THE NUTCRACKER BALLET, Tom Tierney. Model of a complete, full-color production of Tchaikovsky's classic. 6 backdrops, dozens of characters, familiar dance sequences. 32pp. 9⅜ × 12¼.
24194-7 Pa. $4.50

ANIMALS: 1,419 COPYRIGHT-FREE ILLUSTRATIONS OF MAMMALS, BIRDS, FISH, INSECTS, ETC., edited by Jim Harter. Clear wood engravings present, in extremely lifelike poses, over 1,000 species of animals. 284pp. 9 × 12.
23766-4 Pa. $9.95

MORE HAND SHADOWS, Henry Bursill. For those at their 'finger ends," 16 more effects—Shakespeare, a hare, a squirrel, Mr. Punch, and twelve more—each explained by a full-page illustration. Considerable period charm. 30pp. 6½ × 9¼.
21384-6 Pa. $1.95

SURREAL STICKERS AND UNREAL STAMPS, William Rowe. 224 haunting, hilarious stamps on gummed, perforated stock, with images of elephants, geisha girls, George Washington, etc. 16pp. one side. 8¼ × 11.   24371-0 Pa. $3.50

GOURMET KITCHEN LABELS, Ed Sibbett, Jr. 112 full-color labels (4 copies each of 28 designs). Fruit, bread, other culinary motifs. Gummed and perforated. 16pp. 8¼ × 11.   24087-8 Pa. $2.95

PATTERNS AND INSTRUCTIONS FOR CARVING AUTHENTIC BIRDS, H.D. Green. Detailed instructions, 27 diagrams, 85 photographs for carving 15 species of birds so life-like, they'll seem ready to fly! 8¼ × 11.   24222-6 Pa. $2.75

FLATLAND, E.A. Abbott. Science-fiction classic explores life of 2-D being in 3-D world. 16 illustrations. 103pp. 5⅜ × 8.   20001-9 Pa. $2.00

DRIED FLOWERS, Sarah Whitlock and Martha Rankin. Concise, clear, practical guide to dehydration, glycerinizing, pressing plant material, and more. Covers use of silica gel. 12 drawings. 32pp. 5⅜ × 8½.   21802-3 Pa. $1.00

EASY-TO-MAKE CANDLES, Gary V. Guy. Learn how easy it is to make all kinds of decorative candles. Step-by-step instructions. 82 illustrations. 48pp. 8¼ × 11.
23881-4 Pa. $2.50

SUPER STICKERS FOR KIDS, Carolyn Bracken. 128 gummed and perforated full-color stickers: GIRL WANTED, KEEP OUT, BORED OF EDUCATION, X-RATED, COMBAT ZONE, many others. 16pp. 8¼ × 11.   24092-4 Pa. $2.50

CUT AND COLOR PAPER MASKS, Michael Grater. Clowns, animals, funny faces...simply color them in, cut them out, and put them together, and you have 9 paper masks to play with and enjoy. 32pp. 8¼ × 11.   23171-2 Pa. $2.25

A CHRISTMAS CAROL: THE ORIGINAL MANUSCRIPT, Charles Dickens. Clear facsimile of Dickens manuscript, on facing pages with final printed text. 8 illustrations by John Leech, 4 in color on covers. 144pp. 8⅜ × 11¼.
20980-6 Pa. $5.95

CARVING SHOREBIRDS, Harry V. Shourds & Anthony Hillman. 16 full-size patterns (all double-page spreads) for 19 North American shorebirds with step-by-step instructions. 72pp. 9¼ × 12¼.   24287-0 Pa. $4.95

THE GENTLE ART OF MATHEMATICS, Dan Pedoe. Mathematical games, probability, the question of infinity, topology, how the laws of algebra work, problems of irrational numbers, and more. 42 figures. 143pp. 5⅜ × 8½. (EBE)
22949-1 Pa. $3.50

READY-TO-USE DOLLHOUSE WALLPAPER, Katzenbach & Warren, Inc. Stripe, 2 floral stripes, 2 allover florals, polka dot; all in full color. 4 sheets (350 sq. in.) of each, enough for average room. 48pp. 8¼ × 11.   23495-9 Pa. $2.95

MINIATURE IRON-ON TRANSFER PATTERNS FOR DOLLHOUSES, DOLLS, AND SMALL PROJECTS, Rita Weiss and Frank Fontana. Over 100 miniature patterns: rugs, bedspreads, quilts, chair seats, etc. In standard dollhouse size. 48pp. 8¼ × 11.   23741-9 Pa. $1.95

THE DINOSAUR COLORING BOOK, Anthony Rao. 45 renderings of dinosaurs, fossil birds, turtles, other creatures of Mesozoic Era. Scientifically accurate. Captions. 48pp. 8¼ × 11.   24022-3 Pa. $2.50

25 KITES THAT FLY, Leslie Hunt. Full, easy-to-follow instructions for kites made from inexpensive materials. Many novelties. 70 illustrations. 110pp. 5⅜ × 8½.
22550-X Pa. $2.25

PIANO TUNING, J. Cree Fischer. Clearest, best book for beginner, amateur. Simple repairs, raising dropped notes, tuning by easy method of flattened fifths. No previous skills needed. 4 illustrations. 201pp. 5⅜ × 8½. 23267-0 Pa. $3.50

EARLY AMERICAN IRON-ON TRANSFER PATTERNS, edited by Rita Weiss. 75 designs, borders, alphabets, from traditional American sources. 48pp. 8¼ × 11.
23162-3 Pa. $1.95

CROCHETING EDGINGS, edited by Rita Weiss. Over 100 of the best designs for these lovely trims for a host of household items. Complete instructions, illustrations. 48pp. 8¼ × 11. 24031-2 Pa. $2.25

FINGER PLAYS FOR NURSERY AND KINDERGARTEN, Emilie Poulsson. 18 finger plays with music (voice and piano); entertaining, instructive. Counting, nature lore, etc. Victorian classic. 53 illustrations. 80pp. 6½ × 9¼. 22588-7 Pa. $1.95

BOSTON THEN AND NOW, Peter Vanderwarker. Here in 59 side-by-side views are photographic documentations of the city's past and present. 119 photographs. Full captions. 122pp. 8¼ × 11. 24312-5 Pa. $6.95

CROCHETING BEDSPREADS, edited by Rita Weiss. 22 patterns, originally published in three instruction books 1939-41. 39 photos, 8 charts. Instructions. 48pp. 8¼ × 11. 23610-2 Pa. $2.00

HAWTHORNE ON PAINTING, Charles W. Hawthorne. Collected from notes taken by students at famous Cape Cod School; hundreds of direct, personal *apercus,* ideas, suggestions. 91pp. 5⅜ × 8½. 20653-X Pa. $2.50

THERMODYNAMICS, Enrico Fermi. A classic of modern science. Clear, organized treatment of systems, first and second laws, entropy, thermodynamic potentials, etc. Calculus required. 160pp. 5⅜ × 8½. 60361-X Pa. $4.00

TEN BOOKS ON ARCHITECTURE, Vitruvius. The most important book ever written on architecture. Early Roman aesthetics, technology, classical orders, site selection, all other aspects. Morgan translation. 331pp. 5⅜ × 8½. 20645-9 Pa. $5.50

THE CORNELL BREAD BOOK, Clive M. McCay and Jeanette B. McCay. Famed high-protein recipe incorporated into breads, rolls, buns, coffee cakes, pizza, pie crusts, more. Nearly 50 illustrations. 48pp. 8¼ × 11. 23995-0 Pa. $2.00

THE CRAFTSMAN'S HANDBOOK, Cennino Cennini. 15th-century handbook, school of Giotto, explains applying gold, silver leaf; gesso; fresco painting, grinding pigments, etc. 142pp. 6⅛ × 9¼. 20054-X Pa. $3.50

FRANK LLOYD WRIGHT'S FALLINGWATER, Donald Hoffmann. Full story of Wright's masterwork at Bear Run, Pa. 100 photographs of site, construction, and details of completed structure. 112pp. 9¼ × 10. 23671-4 Pa. $6.95

OVAL STAINED GLASS PATTERN BOOK, C. Eaton. 60 new designs framed in shape of an oval. Greater complexity, challenge with sinuous cats, birds, mandalas framed in antique shape. 64pp. 8¼ × 11. 24519-5 Pa. $3.50

READY-TO-USE BORDERS, Ted Menten. Both traditional and unusual interchangeable borders in a tremendous array of sizes, shapes, and styles. 32 plates. 64pp. 8¼ × 11. 23782-6 Pa. $3.50

THE WHOLE CRAFT OF SPINNING, Carol Kroll. Preparing fiber, drop spindle, treadle wheel, other fibers, more. Highly creative, yet simple. 43 illustrations. 48pp. 8¼ × 11. 23968-3 Pa. $2.50

HIDDEN PICTURE PUZZLE COLORING BOOK, Anna Pomaska. 31 delightful pictures to color with dozens of objects, people and animals hidden away to find. Captions. Solutions. 48pp. 8¼ × 11. 23909-8 Pa. $2.25

QUILTING WITH STRIPS AND STRINGS, H.W. Rose. Quickest, easiest way to turn left-over fabric into handsome quilt. 46 patchwork quilts; 31 full-size templates. 48pp. 8¼ × 11. 24357-5 Pa. $3.25

NATURAL DYES AND HOME DYEING, Rita J. Adrosko. Over 135 specific recipes from historical sources for cotton, wool, other fabrics. Genuine premodern handicrafts. 12 illustrations. 160pp. 5⅜ × 8½. 22688-3 Pa. $2.95

CARVING REALISTIC BIRDS, H.D. Green. Full-sized patterns, step-by-step instructions for robins, jays, cardinals, finches, etc. 97 illustrations. 80pp. 8¼ × 11. 23484-3 Pa. $3.00

GEOMETRY, RELATIVITY AND THE FOURTH DIMENSION, Rudolf Rucker. Exposition of fourth dimension, concepts of relativity as Flatland characters continue adventures. Popular, easily followed yet accurate, profound. 141 illustrations. 133pp. 5⅜ × 8½. 23400-2 Pa. $3.00

READY-TO-USE SMALL FRAMES AND BORDERS, Carol B. Grafton. Graphic message? Frame it graphically with 373 new frames and borders in many styles: Art Nouveau, Art Deco, Op Art. 64pp. 8¼ × 11. 24375-3 Pa. $3.50

CELTIC ART: THE METHODS OF CONSTRUCTION, George Bain. Simple geometric techniques for making Celtic interlacements, spirals, Kellstype initials, animals, humans, etc. Over 500 illustrations. 160pp. 9 × 12. (Available in U.S. only) 22923-8 Pa. $6.00

THE TALE OF TOM KITTEN, Beatrix Potter. Exciting text and all 27 vivid, full-color illustrations to charming tale of naughty little Tom getting into mischief again. 58pp. 4¼ × 5½. (USO) 24502-0 Pa. $1.75

WOODEN PUZZLE TOYS, Ed Sibbett, Jr. Transfer patterns and instructions for 24 easy-to-do projects: fish, butterflies, cats, acrobats, Humpty Dumpty, 19 others. 48pp. 8¼ × 11. 23713-3 Pa. $2.50

MY FAMILY TREE WORKBOOK, Rosemary A. Chorzempa. Enjoyable, easy-to-use introduction to genealogy designed specially for children. Data pages plus text. Instructive, educational, valuable. 64pp. 8¼ × 11. 24229-3 Pa. $2.50

Prices subject to change without notice.

Available at your book dealer or write for free catalog to Dept. GI, Dover Publications, Inc., 31 East 2nd St. Mineola, N.Y. 11501. Dover publishes more than 175 books each year on science, elementary and advanced mathematics, biology, music, art, literary history, social sciences and other areas.